Studies in the History of Civil Engineering

Volume 12

I0867041

Structural and Civil Engineering Design

STUDIES IN THE HISTORY OF CIVIL ENGINEERING

General Editor: Joyce Brown

with the assistance of M.M. Chrimes, A.W. Skempton,
N.A.F. Smith and R.J.M. Sutherland

Twelve volumes are to appear from the Autumn of 1997:

1. **The Engineering of Medieval Cathedrals**
   Edited by LYNN T. COURTENAY, University of Wisconsin

2. **Masonry Bridges, Viaducts and Aquaducts**
   Edited by TED RUDDOCK, University of Edinburgh

3. **Land Drainage and Irrigation**
   Edited by SALVATORE CIRIACONO, University of Padova

4. **Dams**
   Edited by DONALD C. JACKSON, Lafayette College

5. **Water-Supply and Public Health Engineering**
   Edited by DENIS SMITH, formerly University of East London

6. **Port and Harbour Engineering**
   Edited by ADRIAN JARVIS, Merseyside Maritime Museum and Liverpool University

7. **The Civil Engineering of Canals and Railways before 1850**
   Edited by MIKE CHRIMES, Institution of Civil Engineers

8. **The Development of Timber as a Structural Material**
   Edited by DAVID T. YEOMANS, Liverpool University

9. **Structural Iron, 1750–1850**
   Edited by R.J.M. SUTHERLAND, Consultant, Harris and Sutherland, London

10. **Structural Iron and Steel, 1850–1900**
    Edited by ROBERT THORNE, Editor, *Construction History Journal*

11. **Early Reinforced Concrete**
    Edited by FRANK NEWBY, formerly F.J. Samuely and Partners, London

12. **Structural and Civil Engineering Design**
    Edited by WILLIAM ADDIS, University of Reading

Studies in the History of Civil Engineering
General Editor: Joyce Brown

Volume 12

Structural and Civil Engineering Design

edited by
William Addis

Routledge
Taylor & Francis Group
LONDON AND NEW YORK

First published 1999 by Ashgate Publishing

Published 2016 by Routledge
2 Park Square, Milton Park, Abingdon, Oxfordshire OX14 4RN
711 Third Avenue, New York, NY 10017, USA

First issued in paperback 2016

Routledge is an imprint of the Taylor & Francis Group, an informa business

British Library CIP data
Structural and Civil Engineering Design.
(Studies in the History of Civil Engineering; 12).
1. Civil engineering–History. 2. Structural engineering–History. 3. Engineering design–History.
I. Addis, William, 1949–
624

US Library of Congress CIP data
Structural and Civil Engineering Design/edited by William Addis.
p. cm. – (Studies in the History of Civil Engineering; v. 12).
Includes index.
1. Structural design. 2. Structural design–History.
3. Engineering design–History.
I. Addis, William, 1949– II. Series.
TA658. S83 1999 98–49124
624. 1'771–dc21 CIP

ISBN 13: 978-1-138-27865-3 (vol 12, pbk)
ISBN 13: 978-0-86078-761-7 (vol 12, hbk)

ISBN 13: 978-0-754-60654-3 (set)

STUDIES IN THE HISTORY OF CIVIL ENGINEERING – VOL. 12

Contents

Acknowledgements

The chapters in this volume are taken from the sources listed below, for which the editors and publishers wish to thank their authors, original publishers or other copyright holders for permission to use their material as follows:

Chapter 1: Sergio Luis Sanabria, 'The mechanization of design in the 16th century: the structural formulae of Rodrigo Gil de Hontañón', *Journal of the Society of Architectural Historians*, 41, no. 4 (1982), pp. 281–293. Copyright © 1982 by the Society of Architectural Historians, Chicago, IL.

Chapter 2: David T. Yeomans, 'Designing the beam: from rules of thumb to calculations', *Journal of the Institute of Wood Science*, 11, no.1, issue 61 (1987), pp. 43–49. Copyright © 1987 by the Institute of Wood Science, High Wycombe, Bucks.

Chapter 3: J. Kerisel, 'History of retaining wall design', Proceedings of the conference *Retaining Structures* organized by the Institution of Civil Engineers and held at Robinson College, Cambridge, 20–23 July 1992, ed. C.R.I. Clayton (London, 1993), pp. 1–16. Copyright © 1993 by J. Kerisel and the Institution of Civil Engineers, London.

Chapter 4: Denis Smith, 'The use of models in nineteenth-century British suspension bridge design', *History of Technology*, 2 (1977), pp. 169–214. Copyright © 1977 by Denis Smith and Mansell Publishing Limited, London. Reproduced by permission of Mansell, a Cassell imprint, Wellington House, 125 Strand, London, WC2R 0BB, England.

Chapter 5: R.F.D. Porter Goff, 'Brunel and the design of the Clifton Suspension Bridge', *Minutes of the Proceedings of the Institution of Civil Engineers*, 56, pt. 1 (1974), pp. 303–321. Copyright © 1974 by the Institution of Civil Engineers, London.

Chapter 6: Ralph B. Peck, 'History of building foundations in Chicago', *University of Illinois, Engineering Experiment Station, Bulletin Series*, no. 373 (1948), pp. 9–62. Copyright © 1948 by the University of Illinois, Chicago.

Chapter 7: Otto G. von Simson, 'The Gothic cathedral: design and meaning', *Journal of the Society of Architectural Historians*, 11, no. 3 (1952), pp. 6–16. Copyright © 1952 by the Society of Architectural Historians, Chicago, IL.

Chapter 8: Rowland J. Mainstone, 'Stability concepts from Renaissance to today', *Stable-Unstable*, Centre for Conservation of Historic Buildings (1988), pp. 65–78. Copyright © 1988 by Leuven University Press.

Chapter 9: Stanley B. Hamilton, 'The place of Sir Christopher Wren in the history of structural engineering', *Transactions of the Newcomen Society*, 14 (1933–34), pp. 27–38. Copyright © 1933–34 by the Newcomen Society, London.

Chapter 10: A.W. Skempton, 'The origin of iron beams', *Actes du VIIIe Congrès International d'Histoire des Sciences* (Florence, 3–9 September 1956), pp. 1029–1039. Copyright © 1956 by A.W. Skempton.

Chapter 11: L.G. Booth, 'Thomas Tredgold (1788–1829): some aspects of his work (part 2: carpentry)', *Transactions of the Newcomen Society*, 51 (1979–80), pp. 65–71. Copyright © 1979–80 by the Newcomen Society, London.

Chapter 12: R.J.M. Sutherland, 'Thomas Tredgold (1788–1829): some aspects of his work (part 3: cast iron)', *Transactions of the Newcomen Society*, 51 (1979–80), pp. 71–82. Copyright © 1979–80 by the Newcomen Society, London.

Chapter 13: R.J.M. Sutherland, 'The birth of stress: a historical review', *The Art and Practice of Structural Design* (1984), pp. 5–15. Copyright © 1984 by the Institution of Structural Engineers, London.

Chapter 14: A. Rupert Hall, 'The genesis of engineering: theory and practice', *Society of Engineers Journal*, 60 (1969), 17–31. Copyright © 1969 by A. Rupert Hall.

Chapter 15: A.J.S. Pippard, 'The functions of engineering research in the university', *Proceedings of the Institution of Civil Engineers*, 33, no. 4 (1950), pp. 265–284. Copyright © 1950 by the Institution of Civil Engineers, London.

Chapter 16: Stanley Smith, 'The design of structural ironwork 1850–1890: education, theory and practice', *Construction History*, 8 (1992), pp. 89–108. Copyright © 1992 by the Construction History Society, Ascot, Berks.

Chapter 17: Henry J. Cowan, 'Structural design by observation of failures: how Gothic master masons determined the dimensions of their structures', *Architectural Science Review*, 35 (1992), pp. 51–58. Copyright © 1992 by Henry J. Cowan.

Chapter 18: Henry Petroski, 'The Britannia Tubular Bridge: a paradigm of failure-driven design', *Structural Engineering Review*, 5, no. 4 (1994), pp. 259–270.

Copyright © 1994 by Elsevier Science Ltd., Oxford. Reprinted with permission.

Chapter 19: D.I. Blockley and J.R. Henderson, 'Structural failures and the growth of engineering knowledge', *Proceedings of the Institution of Civil Engineers*, 68, pt. 1 (1980), pp. 719–728. Copyright © 1980 by the Institution of Civil Engineers, London.

Chapter 20: William Addis, 'The evolution of structural engineering design procedures: a history of that skill called design', *Transactions of the Newcomen Society*, 61 (1989–90), pp. 51–62. Copyright © 1989–90 by the Newcomen Society, London.

***

Chapters 9, 11, 12 and 20 are reproduced from the *Transactions of the Newcomen Society* by permission of the Society's Council. The Newcomen Society promotes, encourages and co-ordinates the study of history of all types of engineering and technology from the earliest times to the present day. Details from the Executive Secretary, The Newcomen Society, The Science Museum, London SW7 2DD, Tel/ Fax 0171-589-1793.

Every effort has been made to trace all the copyright holders, but if any have been inadvertently overlooked the publishers will be please to make the necessary arrangement at the first opportunity.

General Editor's Preface

Joyce Brown

Civil engineering has a long history. Although practitioners did not begin to describe themselves as 'civil engineers' until the eighteenth century, the origins of their work lie in the construction skills of the ancient world and in the works of military engineers. Shortly after British civil engineers had formed themselves into a professional body, the Institution of Civil Engineers in London, their role was described in 1828 by Thomas Tredgold in these words:

> Civil engineering is the art of directing the great sources of power in Nature for the use and convenience of man.[1]

He thus neatly encapsulated what the civil engineer's job is about – the provision of the means to give society what it wants for its survival.

As time has gone by, the civil engineer's skills have had to diffuse over an ever-widening field and at the same time become more specialised. In all periods he has had to be the ingenious 'fixer', but more than a mere artisan – a point observed in the Institution's motto, *scientia et ingenio*.

The history of civil engineering is a fascinating one. Certainly 'the great sources of power in Nature' are not always directed without a certain amount of opposition by them, and the provision of the basic requirements – habitation, water supply, main drainage, harbours bridges, places of worship, transport systems by land and water – has called for inventiveness and sometimes for daring and courage.

Yet interest in the history of civil engineering is not of long standing, and it seems fair to state that most of the important work has been done in the last forty years. With a few notable exceptions, most of the major books on the subject have appeared in that era; many specialist societies with their own transactions have been founded; and many of the journals in which articles might be published have come into existence. While archaeologists and architectural historians have long made their contribution to this subject, their ranks have been swelled by a large number of serious scholars – including engineers themselves – now interested in recording and evaluating the engineering achievements of earlier times. In particular, there is a lively interest in the study of our industrial past, at least in part because some of the evidence of it are already disappearing from the landscape.

Our intention in creating the series of which this volume forms part is to provide, through the reproduction of important contributions to this subject, an

---

[1] *Minutes of Council, Institution of Civil Engineers*, 4 January 1828.

invaluable reference collection for its study. The series encompasses many different branches of engineering from early times to the beginning of this century; its perspective is global; and the chosen articles have an international authorship, to the extent that this can be achieved for an essentially English-language series. The introduction to each volume has allowed the volume editor to set the pages selected in the context of the whole history of the subject and its historiography, while the provision of references and a bibliography will enable the reader to go further into the study of the topic.

This series will have succeeded if it gives civil engineers some unexpected insights into their own craft and other readers a new way of looking at engineering structures.

***

The volume editors were chosen for their particular knowledge and expertise in the field. They have worked with enthusiasm and good will to put together collections of articles intended to be both informative and stimulating, and I am grateful to them for their efforts. Where possible, they have incorporated the work of the best scholars, while at the same time giving preference to articles often obscurely published and not readily available. The learned journals are represented, as also are conference proceedings and essay chapters contributed to books.

In all of this, I have been much assisted by my Advisory Panel. Emeritus Professor A.W. Skempton (Imperial College) has made himself freely available to me for consultation on many aspects of this venture. Mike Chrimes (Librarian of the Institution of Civil Engineers), Dr. Norman Smith (Emeritus Reader, Imperial College) and James Sutherland (formerly a Senior Partner in Harris and Sutherland, London, consulting engineers) have all contributed valuable advice based on their extensive knowledge of particular fields. The idea for the series was developed in discussion with Dr John Smedley of Ashgate, and Publisher of the Variorum Collected Studies Series, who has given me much personal support and the benefit of his experience.

I am grateful also for the help given by the staff of several libraries, notably Kay Crooks and Susannah Parry in the library of the Department of Civil Engineering, Imperial College, and Mike Chrimes and staff of the library of the Institution of Civil Engineers, London. Many friends have given me encouragement and useful advice; in particular, I want to thank Paula Kahn and Mark Baldwin.

Formerly Department of Civil Engineering,
Imperial College, London

Introduction

W. Addis

It may appear superfluous or likely to duplicate the content of other volumes in this series to devote a separate volume to the theme of engineering design. It has been undertaken, nevertheless, because how engineers designed the artefacts they built tends often to get lost among other features of engineering history. The history of civil and structural engineering is generally addressed in the following ways:[1]

- the development of a type of structure (e.g. suspension bridges, timber roof trusses, dams, masonry domes);
- developments and inventions in engineering technology (especially materials, plant and machinery, and methods of manufacture or construction);
- biographies of engineers;
- the stories of individual projects;
- developments in engineering science and mathematics ('theory' and techniques of analysis);[2]
- various non-technical issues (economic, commercial, contractual, management, social, political, etc.).

Much has been written from all these points of view, but they usually report only on the final state of various works of engineering – their dimensions, materials and construction methods used. Few address the process by which the engineer arrived at this final state, or the problems that face all engineers at the beginning of their projects – what shape? What size? What dimensions? What materials? What joints and connections? What construction methods?

This volume focuses on the historical development of the process which precedes actual construction – the activity we now call 'engineering design', which is so fundamental to our very idea of what modern engineers do. This, of course, begs the question 'what do we mean by design?'.

---

[1] There are few good general histories of civil and structural engineering, and none of these is currently in print. Four I like are Parsons, *Engineers and Engineering in the Renaissance*; Cowan, *The Masterbuilders*; *idem*, *Science and Building*; and Straub, *A History of Civil Engineering* (available in German in more recent and expanded editions as *Die Geschichte der Bauingenieurkunst*).

[2] The best studies in this field are Timoshenko, *History of Strength of Materials*; Benvenuto, *An Introduction...*; Hamilton, *The Development of Structural Science...*; Hamilton, 'The development of structural theory'; and Heyman, *Structural Analysis...* Studies of *how* mathematical and scientific knowledge came to be used in the engineering community are rare. Some of the best are to be found among the score of papers collected in Radelet-de Grave, *Between Mechanics...* I have also discussed this issue in Addis, *Structural Engineering...*

The nature of design in engineering

The word *design* has its origins in the process in which a drawing, plan or scheme is produced to communicate to a client what a proposed work may look like, or to a builder how it is to be constructed. Around the middle of the nineteenth century the word began to denote the visual appearance of something, in the way we might nowadays talk about a dress, a chair or a vacuum cleaner. In this context the word came also to indicate the origination of an idea, and 'designed by' is a phrase now used in a way similar to 'written, composed or painted by'. The creation of an idea is essential in much engineering design, too. The word 'invention' is often used, though I do not like it, for it is too closely linked to the idea of patents to be appropriate to most design engineering. I prefer the word 'conception', which, indeed, is the word the French use.[3]

But in engineering, *design* means far more than appearance, creative inspiration or conception – it embraces the whole process by which an artefact becomes the thing it is – the rationale which underlies the choice of materials, the sizes and relative disposition of components, the methods of manufacture and assembly, the surface treatment given to components, and much more. This process usually results in a set of drawings, but to confuse the final appearance shown in a drawing with the process by which it was created, is to misunderstand the nature of engineering and to devalue the skills of the engineer.[4] I like the word *devise* which used to be widely used to describe what we now call design; it implies not only the origination of an idea, but also a degree of deliberation and effort in working upon the idea to turn it into something practical.

In essence, to design an artefact is to plan what you are intending to do before you start, and to devise a practical and economic way of constructing it. With characteristic simplicity, Ove Arup once wrote that:

> Design is nothing else than indicating a sensible way of building. It includes all drawings, specifications, descriptions and detailed instructions about what should be built and how it should be built.[5]

Even definitions such as this, however, overlook the most important aspect of engineering design – the process of achieving the necessary confidence to begin building. It is one thing to sketch a magnificent bridge or roof for a building,

---

[3] Various aspects of conception in structural design are discussed in Addis, 'Free will and determinism in the conception of structures' and Mainstone, 'Intuition and the springs of structural invention'.

[4] For recognition of this point, see Happold, 'Can you hear me at the back?'. Even today it is not uncommon for non-engineers involved in construction projects to value an engineering drawing according to the time taken to draw it. When, as an old man, Picasso was asked how long it took him to do his simple sketches, he would answer '60 years and 10 minutes'.

[5] Arup, 'The problem of producing quality...', 23.

quite another to be confident that it can be constructed and will safely carry the loads to which it may be subjected. This aspect distinguishes fundamentally the work of those engaged in civil and structural engineering from those engaged in visual design, who never need to address questions as to whether their creation will work, or be safe, or be possible to make.

Design in engineering is, thus, a process whose output, along with sets of drawings and specifications, must also include the necessary degree of confidence in the proposed structure. This confidence must overcome any doubts about success the engineer may harbour, as well, of course, as any doubts felt by the client whose money and reputation are also at risk.

A history of engineering design must address these different aspects of the design process:

- what was the final design for the artefact or structure?
- how was the final design arrived at or devised?
- what means did the design engineer use to increase his confidence, *before commencing construction*, that the proposed structure would work and could be constructed, sufficiently for him to begin building?

A history of confidence among engineering designers would be a challenging subject to undertake since confidence depends very much on the experience and the cultural and intellectual background of the engineer as well as the context in which the engineer is working. There are also important questions about how different engineers in the past might have perceived the gap between what they knew had already been done, and what they might have believed to be possible though they did not yet know by what means. At most periods of history it seems to have presented little problem to build a bridge 5%, or even 10% longer than was considered normal. However, the Conway and Britannia Bridges (discussed in Chapter 18) were not in this category; the spans of these bridges were, respectively, about 450% and 500% (four and a half, and five times) greater than those of any beam bridges previously built, and the story of how Stephenson, Fairbairn and Hodgkinson achieved them is probably the greatest in the history of engineering design.

There are, of course, problems in using modern words like 'engineer', 'design' and 'designer' to refer to the construction of a Gothic cathedral, indeed to any edifice completed before the twentieth century when these words acquired their modern meanings. I use such words to describe the past without implying the professional demarcations, knowledge and working methods of the twentieth century.[6] Rather, and at the risk of being accused of tautology, I use them to

[6] Relatively little has been written about the development of the modern profession of engineering. Examples are Ricken, *Der Bauingenieur...*; Picon, *Architectes et ingénieurs...* and in histories of various professional institutions. An interesting account of how construction

embrace the work that must have been undertaken to complete a cathedral, without specifying precisely what it was or who did it (since we do not know).

It is, indeed, possible to discuss the process of designing a cathedral, for instance, without becoming entangled in questions as to whether it was designed by an architect or an engineer, or whether its structure was designed separately from the architecture. These questions are, literally, meaningless because to use the very words 'engineer' or 'structure' is anachronistic. What happened, over a period of many centuries, was that the activity of designing a large building, whoever did it, was gradually broken down into separate issues. Thus the methods of designing a Greek temple or a medieval cathedral embraced visual appearance, sense of space, function, materials and structure, as well as what we now call the internal environment (lighting, heating, ventilation and acoustics) in a single, holistic design process. The visual appearance, sense of space and function (the 'architecture' in the modern, narrow sense) became a distinct concern during the fifteenth and sixteenth centuries. About a century later, designers first began to think about the load-bearing aspects of buildings in terms of loads (weight), materials and structure.[7] Thinking separately about materials and structures grew during the late-seventeenth and eighteenth centuries, following Galileo's work,[8] and aspects of the internal environment came to be considered independently of the fabric of the building during the late-nineteenth and twentieth centuries.

Thinking about engineering structures

Civil and structural engineering design involves dealing with loads and forces far greater than we can experience directly as human beings. For this reason engineers have needed to devise various ways of representing the real world in a form and at a scale they *can* experience directly, or as a conceptual representation in their minds. Nowadays we refer to such representations as *models* and they may take many forms:

- physical models;
    - *scale models* aiming to represent the relevant engineering behaviour of the full-size structure by reducing dimensions or loads, using model materials, simplifying the structure or types of connection, etc.

---

professionals earned their money is also given in histories of construction contracts e.g. Nisbet, *A Proper Price ...*; and *idem, Fair and Reasonable...*

[7] See chapter 1 for details of Rodrigo Gil's rules which do take some account of the weight of vaults.

[8] In *Dialogues Concerning Two New Sciences*, Galileo distinguishes, for the first time in print, how the strength of a rope is independently affected by the total cross-sectional area of fibres and the strength of fibre (in general).

  - *analogue models* – using one phenomenon to represent another, e.g. water flow to model flow of stress or loads; air flow to model water flow, or vice versa; hanging chain model to find the best form for an arch;
  - prototypes or *full-size models* where as little as possible is different from 'the real thing';

- mathematical models;[9]

  - *geometric models* – two-dimensional plans and sections, perspective drawings, three-dimensional, non-structural scale models;
  - *statical models* – models of loads, models of materials, models of structure in terms of the concepts of statics (elasticity, forces, etc.).[10]

All these models have two distinct uses:

  - to *store information* and to help *communicate* it from one person to another (e.g. designer to client, builder or supplier);
  - to help produce in the engineer and client the necessary level of confidence that the proposed structure will work and can be built, i.e. to justify or substantiate the design choices that have been made. This often embodies an *explanation* and gives the engineer a better *knowledge* and *understanding* of the real world.

Models have long played a fundamental part in engineering design. Physical models have probably been used in the design of buildings, bridges and other structures for thousands of years. The interesting engineering question is 'what was the model used to represent?' Many would have been scale models to show the appearance of a structure or building, perhaps in the context of the surrounding land or townscape. The testing of full-size prototypes of real structures would obviously yield useful results and help raise the engineer's level of confidence, but would always have been very expensive. Some scale models of cathedrals were certainly made, though not deliberately as a means of testing their structural performance under either gravity or wind loads. Nevertheless, the construction of some of these models must have given their builders some

[9] All modern computer models are mathematical models.

[10] *Statics* is the branch of mathematics which deals with forces – how two or more forces can be combined to be equivalent to a single, resultant force, and how we can analyse the equilibrium which is achieved when all the forces (including gravity) acting on an object combine to give an overall resultant of zero. Inside a material we need to consider the force, measured in Newtons (N), that acts on a certain area of that material, say one square millimetre. This is called the stress acting in the material and is usually measured in N/mm^2. The elongation of a piece of material caused by a certain stress is a measure of its *stiffness* and is the measure of the material's elasticity. Loads on structures are usually constant (static) or assumed to be so. When they clearly vary they are called dynamic loads, such as caused by the wind or earthquakes, and design engineers need to consider their effects separately from constant loads.

confidence in the likely stability of the full-size counterpart.[11] The key issue with physical models is the question of scale – if a model is one-tenth the size of the proposed real structure, by how much are other parameters, especially its ability to carry loads, reduced? The density, strength and stiffness of a material will remain unchanged, but self-weight loads will be reduced by one thousand times, and the strength of a beam by one hundred times.

Scale models of masonry structures are very effective since their stability is in direct proportion to their size. If a model worked at one-twentieth scale, there was a good chance the cathedral would work at full size, as long as materials were not changed, for instance by replacing solid stone with rubble. The stability of a masonry structure is not influenced by the elastic properties of the material and so the material of which the model is made is irrelevant. The same is not true of a beam or slender column. In order to scale up results from a model test of these, it is essential to understand the effect of dimensions on structural behaviour – strength, bending, buckling – and this requires evidence from many previous tests of models of different sizes. Galileo was aware of scale effects,[12] but others were not. In his book of 1766, Walter[13] reports that the load carried by one-tenth scale models of several types of stone and timber bridge could be scaled up in direct proportion to predict the capacity of the bridges at full scale. He did not report whether this procedure resulted in any accidents.

The mathematical models of loads, materials and structures that have been developed by mathematicians and scientists during the last five hundred years or so have given us some extraordinarily powerful means to think about engineering structures – it is easy to overlook the intellectual achievement that led to considering loads as a *force* and representing it by a single line on a piece of paper, or the invention of the *second moment of area* of a beam section that reflects its stiffness irrespective of the material of which the beam is made. These and many other concepts have been useful for the designer in two distinct ways. They can provide a theoretical justification of a design and they can be used to generate the dimensions that are needed to specify the sizes and relative dispositions of components that go together to make engineering structures.

Nevertheless, it is worth reflecting that structural understanding is neither a new phenomenon, nor one that *requires* a knowledge of statics and elasticity. It

[11] Having visited and studied the cathedrals at Florence and Milan, Antonio di Vicenzo made an enormous model in brick and plaster of his design for Bologna cathedral in 1390. It was 18.7m long – a scale of about 1:10 – and was large enough to walk inside. Construction lasted from 1390 to about 1430, but the full building was never completed. It would have been one of the largest in Italy and Vicenzo knew he was at the very limit of known precedent. The model was dismantled in 1406. See Frankl, *The Gothic...*, 299.

[12] See Galileo, *Dialogues Concerning Two New Sciences*, Second day.

[13] Walter, *Brückenbau.*

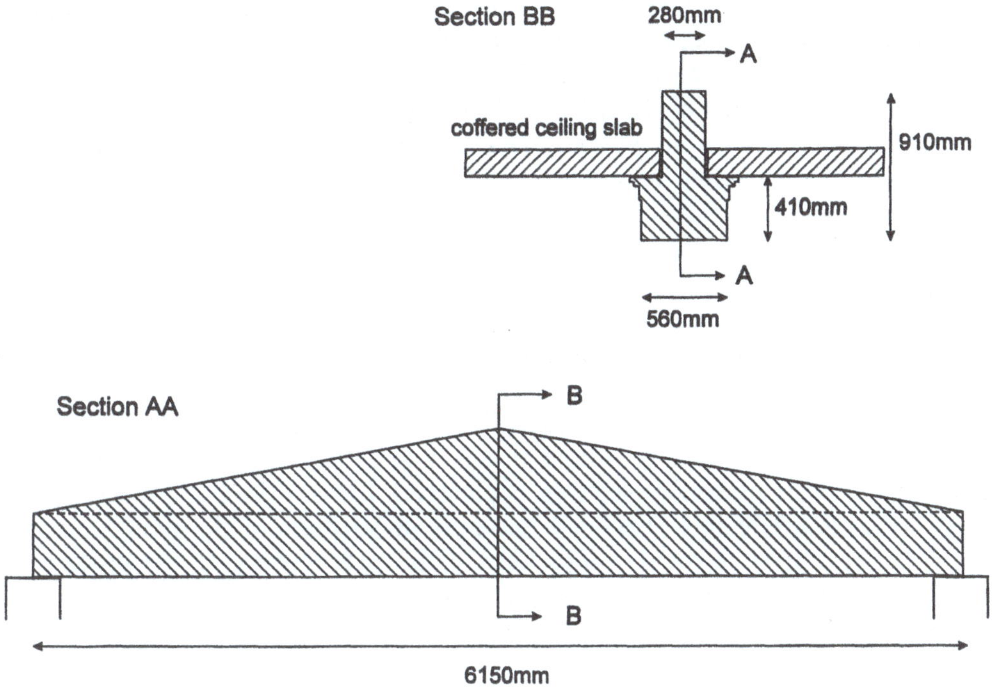

Figure 1: Stone beam supporting coffered marble slabs for ceiling of temple, Greece, *c.*350 BC. Based on information in J.J. Coulton, *Ancient Greek Architects at Work*, 146–7. The shape of this stone beam along its length broadly reflects the bending moment it must provide – being deepest at centre-span. Also the shape of the cross-section (an inverted-T) corresponds to where the material is most needed – being weaker in tension than in compression, more stone is needed at the bottom of the section than the middle or top. Similar forms were used for the earliest cast-iron beams between the early 1790s and the 1820s.

is a commonly held misconception that new types of structure and structural forms were devised first by mathematicians or scientists and later taken up by builders, engineers or other designers of structures. In fact, there is only one example of a practical form of structure being developed or predicted from theoretical ideas – the hyperbolic paraboloid, in the 1930s. Many children and sculptors display a remarkable understanding of materials and structures without a knowledge of engineering statics, and many structures from long ago show their creators understood the essence of all the basic structural actions.[14] The beam shown in Fig. 1 shows that some ancient Greek builders had a good qualitative understanding of bending moments and second moment of area.

The Papers

Although this volume covers such a large field and historical period, making the final selection was made easier by the relatively few papers which focus directly on the subject of engineering design.[15] Generally speaking, the history of engineering design is to be found permeating works written about eminent engineers and significant projects in engineering history. Several papers of this type are included in this volume to serve as illustrations of the design process at work. The book is divided into three parts each looking at the process of engineering design from a different point of view:

- Design rules and methods
- Theoretical justification in design
- Progress in engineering design.

[14] The basic structural actions, and corresponding structural elements, are tension (tie, hanging chain, membrane), compression (column, arch, dome) and bending (beam, slab). The degree of bending in a beam, called the *bending moment*, increases with both the size of load and length of the beam, and varies from low, near the supports, to largest at mid-span. The stresses in the beam also vary across the section – compression at the top, tension at the bottom and zero at a point in between. The sizes of these stresses depend on both the bending moment at that point along the beam, and the shape of the beam cross-section. The effect of the shape varies with *the second moment of area* – a quantity which varies only with the dimensions and geometric shape (it is independent of the material). For instance, using the same area of material in a cross-section, an I-section has a much larger value of second moment of area that a square section. An I-section is a much more efficient use of material. As a mathematical concept it was developed by Euler and others in the early 18th century. See Hamilton, *The Development of Structural Theory* ... and Timoshenko, *History*...

[15] I shall return to address some omissions at the end of this introduction.

Design Rules and Methods

Much guidance for design engineers in the twentieth century is stored in the form of Codes of Design Practice, which embody appropriate design data, criteria and calculation procedures to help ensure the safety of civil and engineering structures. In Britain they now have the same status and rôle as British Standards. Some relate to loads – wind, user-imposed gravity loads; others to types of structure – dams, bridges; others to load-bearing materials – steel, reinforced concrete, timber, masonry. While not restricting what sort of structure may be used, nor prescribing what mathematical technique of structural analysis would be needed, they give guidance as to how an engineer should ensure the safety of a structure or check its stability. They recommend values for working and ultimate stresses that are acceptable for a material, the minimum safety factors that should be used, and the sets of loading conditions which should be considered. Together they represent a condensation of many millions of man-years of engineering experience and are presented in a form that can be easily passed from one generation to the next.

In many ways, modern Codes of Practice resemble the design rules of the medieval mason or the eighteenth-century bridge builder. Like ancient design rules, they give guidance but do not tell the designer exactly what to do; they still require considerable experience to use effectively. In themselves, they do not explain *why* things need to be done in a certain way. However, unlike most ancient design rules, they are written down. They do not, therefore, embody those aspects of engineering knowledge which cannot be written down. As in past times, good engineers today know things they cannot always express, and have ideas they cannot immediately justify. Their professional task is to devise adequate explanation and justification.

The modern codes have their origins around the beginning of the twentieth century, but it is not too difficult to trace their roots further back: through the early use of modern engineering concepts, such as bending theory, stability and factors of safety, to eighteenth-century geometric methods embodying statics, and further back still to the use of geometry alone as a sophisticated means of helping the designer of a wide range of bridges and buildings in the thirteenth and fourteenth centuries. The earliest design rules which survive are for various buildings and civil engineering structures from Classical Greek times. Vitruvius contains precise rules for temples of the different orders and Heron gives some design rules for fortifications.[16] Surprisingly, perhaps, many of these Greek design procedures were algebraic, although often based on simple geometric figures and ratios of simple numbers.

---

[16] See Kerisel in chapter 3, Marsden, *Greek and Roman Artillery...*, and Addis, *Structural Engineering...*

Design rules for entire plans and smaller parts of Gothic cathedrals must have existed from the earliest days,[17] though we get our first glimpse of them only towards the end of the Gothic era in the short booklets by Roriczer and Schmuttermeyer giving design guidance for finials and gablets, neither of which, frustratingly, have much structural significance.[18] In chapter 1 Sanabria presents the most recent, and to my mind, the best discussion of the design rules for vaults, columns and abutments in cathedrals which were written down by Rodrigo Gil de Hontañón in around 1540, and are the most interesting that have survived from the Gothic period.[19] These do seem still to be 'purely' geometric, yet are clearly beginning to embrace the size of the vault – what we might now call the area-loading on a structural bay, in plan view. While the various rules would not satisfy a modern engineer (by using the absolute minimum of material, for instance), they do all generate plausible dimensions for the buildings and, as with all design methods, they are to be judged for their usefulness, not for their ability to reach the absolute limit of what might be possible.

In chapter 2, Yeomans describes the down-to-earth manner in which the sizes of timber beams were chosen before any hint of statics had entered the picture: there were tables and simple formulae. Design methods such as these had probably existed for centuries, even millennia. They were a simple way of storing knowledge gained by experience in a form that allowed the information to be easily retrieved and used again when needed. However, the rules were just formulae and contained no hint of an explanation of the dimensions they yielded, for instance by linking sizes to imposed loads and material properties. But no explanation was needed to justify their use, for they already embodied sufficient power to instil confidence, because in condensed form, they accurately represented established practice.

As the eighteenth century progressed, more and more structural testing was being undertaken, and it was widely recognised by carpenters and engineers that the strength of a rectangular beam varied with its breadth and the square of its depth long before beams were designed using a fully-scientific understanding of elastic bending. The problem with simple design rules was that they were not for general application: different tables would be needed depending on the span, the type of construction, on whether the wood was oak or fir, on whether the

[17] See, for example, Bucher, 'Mediaeval architectural design methods, ...' and *idem*, 'Design in Gothic architecture'. By the late-fifteenth century Italian Renaissance architecture was rivalling the Gothic achievements. For details of their design rules for buildings and bridges, see Ackerman, 'Architectural practice...'; Alberti, *De re aedificatoria*; Betts, 'Structural innovation...'; Parsons, *Engineers...*; Straub, *History....*

[18] See Shelby, *Gothic Design Techniques* for details and discussion of these booklets.

[19] Gil's work was previously discussed in Kubler, 'A late Gothic computation...' It is interesting to compare them with Lorenz Lechler's nearly contemporary design rules discussed in Shelby and Mark, 'Late Gothic structural design...'.

beam was a girder or a joist, and on how heavily the floor might be loaded. Such tables and design rules based on simple ratios and proportion were adequate as long as the use of the beam was within the range of past experience. They enabled the user to calculate the *relative* strength of a beam (as it was called), but they did not permit a calculation of the *absolute* strength.

In the following chapter, Kerisel quickly reviews several millennia of retaining wall history, but not without pointing carefully to evidence of good engineering design thinking on the way. Vitruvius recommends certain proportions and ratios of dimensions in retaining structures – perhaps the earliest examples of design rules which continued to develop throughout the middle ages and reappeared in various Renaissance treatises on buildings and fortifications. The paper then focuses more closely on the design methods that came to embody mathematical models for the behaviour of soil – the nascent science of soil mechanics[20] – in the early and mid-eighteenth century, long in advance of an equivalent use of engineering science by the designers of bridges or the structures of buildings.

As explained above, the use of models is fundamental to the work of the engineer to gain understanding about how structures behave, to study how they might be constructed and to establish suitable geometry and dimensions for components of structures – all roles they sometimes still play today.[21] In chapter 4, Denis Smith considers this important tool in the design engineer's toolkit in the context of design studies for suspension bridges.[22] With its use of contemporary sources, Smith's paper brings out the various functions that the various models were felt to serve. Several writers mentioned explicitly the role of models in persuading people of the viability of a proposed idea. Over the period considered by Smith (*c*.1813–*c*.1867), it is clear that the use of models developed from being largely qualitative in nature – demonstrating a principle – to being an aid to design, for instance in determining the geometry of the structural form,[23] the method of assembly and the relative strengths and stiffnesses of different alternatives.

Porter Goff's paper on Clifton Suspension Bridge (chapter 5) presents a different view of the design process – the story of the design development in

[20] Retaining walls are only one of many structures which rely on soils as load-bearing materials and often involve water. Breakwaters are discussed by Tanimoto and Goda, 'Historical development ...'; canals by Schofield, 'The design and construction...'; earth dams by Skempton, 'Historical development ...'; retaining walls for railway cuttings by Hamilton, 'Sir Benjamin Baker ...'. For good overviews of early developments in soil mechanics, see Skempton, 'Landmarks...', and Heyman, *Structural Analysis...*

[21] For some recent examples of using models in design, see Addis, *The Art...*

[22] Denis Smith has also written about their use during the middle of the 19th century in the design of railway bridges in 'Structural model testing ...', reprinted in volume 9 of this series.

[23] For other early examples of using models for form-finding, see Tomlow, 'Die Kuppel des Giesshauses ...', and Gräfe, 'Zur Formgebung ...' and, of course Gaudí's use of hanging chain models for the vault of the cathedral in Barcelona (see, e.g. Mainstone, *Developments...*).

one project by Brunel, who was undoubtedly one of the most talented engineers of the nineteenth century.[24] The project is especially interesting because Brunel did not live to see the bridge constructed. Among his papers Brunel left a record of different design schemes he undertook at various times when it seemed likely that the bridge project would go ahead: the first schemes were responses to the original design competition, during 1829–35; the second during 1838–39; and his final work in 1850. Together they provide an illuminating picture of how he progressed from conception to scheme design to detail design. During the design development he reworked certain key elements, especially the geometry and construction of the chains again and again, and eventually achieved a weight-saving of over 30% compared to the chain designs of earlier engineers such as Telford at Menai. Interesting, too, were his various ideas for reducing and accommodating the effects of uneven heating of the chains by the sun. This paper presents a rare record of the design engineer at work – a stage in construction projects that all-too-often is rendered invisible when, in the case of nearly all completed projects, all evidence of initial ideas is literally thrown away, leaving posterity with records of only the as-built scheme and details.

Peck's paper on the foundations of buildings in Chicago is among the best on engineering design history that I know.[25] Unlike many that purport to look at design, but in fact look at engineering science and mathematics, his paper looks

[24] For discussion of Telford's Menai suspension bridge which Brunel studied, see Paxton, 'Menai Bridge…'. For a broader look at design methods for suspension bridges in the nineteenth century, see Buonopane and Billington, 'Theory and history…', Peters, *Transitions in Engineering…*, and Kranakis, *Constructing a Bridge…* For a wider discussion of Brunel's understanding and use of engineering theory, see Charlton, 'Theoretical work' in Pugsley, *The Works of I.K. Brunel*, 183–202. For more on design methods used for bridges, see Alberti, *De re aedificatoria*; Parsons, *Engineers…*; Rennison, 'The High Level Bridge …'; Ruddock, *Arch Bridges…*; *idem*, 'The building of the North Bridge …'; Swailes and Marsh, 'Victorian iron arch bridges'; Shirley-Smith, 'Royal Albert Bridge, Saltash', in Pugsley, *The Works of I.K. Brunel*, 163–82; and Shipway, 'The Forth Railway Bridge…'.

[25] In the same series as Peck's is another excellent historical/review paper, on how wind-loading was taken into account when designing early steel-frame buildings (Wilson and Maney, 'Wind stresses …'). Although strictly it is just outside the catchment period for this volume, the paper is noteworthy, as are three short papers on other aspects of early steel frame design – Thomas Sputo, 'History of steel beam-column design', *Journal of Structural Engineering* (American Society of Civil Engineers), 119, no. 2 (February 1993), 547–57; T.V. Galambos, 'History of steel beam design', *American Institute of Steel Construction (AISC) Engineering Journal*, 14, no. 4 (1977), 141–7; and R. Bjorhovde, 'Columns from theory to practice', *AISC Engineering Journal*, 25, no. 1 (1988), 21–34. An excellent book which reviews frame building design methods in the USA in the late-nineteenth and early-twentieth century is Friedman, *Historical Building Construction…*. The same period for Britain is covered briefly in M.N. Bussell, 'The development of reinforced concrete: design theory and practice', *Proceedings of the Institution of Civil Engineers, Structures and Buildings*, 116 (Aug./Nov. 1996), 317–34 and M.N. Bussell, *Appraisal of Existing Iron and Steel Structures* (Steel Construction Institute, Ascot, 1997).

at what practising engineers and architects told us of how they set about their work and what methods they actually employed.[26] Practical design methods, in foundation design and elsewhere, often do not employ the most sophisticated engineering science or mathematical models that might be available. However, they do serve the needs of the designer adequately for the purpose at hand, and that is all that the design engineer needs. Practising engineers always have to balance the pressures of time and cost with the need to be sufficiently confident in the results of their design work. The elegance of an engineering design method can just as well lie in the ease and speed with which it solves the problem at hand, as it might in yielding a structure that is everywhere loaded to the maximum level that the material can safely sustain. Peck communicates this message with great clarity. He also furnishes his argument with a wealth of detail concerning the engineering triumphs behind, or rather beneath, such architectural icons as the Home Insurance Building, the Monadnock Block, and works of architects Adler and Sullivan – except that few now realise that Adler was not an architect, but an engineer of considerable talent.

Theoretical Justification in Design

Before considering the various mathematical tools which can aid the designer, it is important to mention the oldest and most powerful means by which an engineer can be confident that a proposed structure will work: precedent. Most building and civil engineering structures throughout history were designed and constructed the way they were by copying examples that were already known to be successful. The following of precedent is an example of inductive logic: the more often we have experienced an event or a certain course of action – the sun rising tomorrow, an apple falling when dropped, a timber plank or brick arch safely spanning a few metres – the more confident we are that the outcome will be repeated in the future. Induction is also the basis for the increasing level of confidence that accompanies the repeated, successful use of a certain design method.

However, inductive logic always carries the caveat *ceteris paribus*, 'other things being equal'. The next plank will only be as safe as the last so long as the span is the same, and the depth and breadth the same, and the timber of the same quality. The problem is that we can never know in advance which properties are relevant and which not (Is colour important? Yes, if it indicates rotten timber!). Another factor is that humans are wont to experiment – how great a distance can a plank of a certain cross-section span? One of the engineer's most valuable skills is knowing the limits to which certain materials or design methods can be taken, and

---

[26] For a fascinating discussion of the design methods and construction of the foundations of the Rialto bridge in Venice, see Parsons, *Engineers ...*', and for a review of concrete foundations, see Chrimes, 'Concrete foundations...'.

judging by how much they might be exceeded and what sort of additional justification must be sought if a well-established limit is to be passed.

As a means of avoiding the need to proof-test all structures, it has long been seen by man that logical argument can be used to reach conclusions not apparent in certain raw data or assumptions. Geometry is a paradigm of this: simply by drawing and reasoning, it is possible to reach an infinity of conclusions and be entirely confident of their truth.

Justification using Geometry In chapter 7 von Simpson describes the intellectual world of the middle ages and how designers might have thought about their task of designing and constructing a cathedral; it was very different from today. Since Classical Greek times (and probably earlier), people had sought explanations for natural phenomena in order to understand the world they lived in. In every field of what we now call natural science – optics, acoustics, astronomy, music, mechanics, botany – philosophers had explained phenomena in terms of *geometry, harmonics* and *number* which were seen as earthly manifestations of the principles the god(s) had used to create the universe. Euclid's book on geometry had been rediscovered in the late Middle Ages and lifted the status of geometry, and its power to provide explanations, to something not very different from the role of physics today. This practice persisted until well into the eighteenth century in many fields. To a medieval cathedral designer, then, a geometrical model of the building served not only as a mathematical model to describe the proposed construction, but also, being based on geometry, it served in some manner to justify its adequacy, in conjunction with other engineering knowledge, of course. A description of the functions of geometry included in a thirteenth-century text book show an interesting parallel with the notions of theory and practice discussed by Hall, in chapter 14, and myself in chapter 20.[27]

	Geometria theorica	*Geometria practica*
finis (purpose)	to teach something	to do something
officium (duty)	to give reasons and dispel doubt	to give measurements or limits which the work should not surpass

The several functions of mediaeval geometry, according to a twelfth-century treatise.[28]

[27] There is some discussion of these questions in chapter 8 by Mainstone, and in Frankl, *The Gothic ...*; Heyman, 'How to design a Gothic cathedral ...'; Ackerman, *'Ars sine scientia nihil est...'* and in several papers included in, *The Engineering of Mediaeval Cathedrals*, volume 1 of this series.

[28] From Victor, 'Practical geometry ...', 9.

Justification using Geometry and Statics It is particularly intriguing for the modern reader to look for the first hint of using forces and statics in design procedures for cathedrals. In fact, this is not as straightforward a quest as one might imagine. It depends what you mean by 'using forces and statics'. While we might nowadays be clear that using the triangle or parallelogram of forces is an unambiguous use of statics, in 1400, or even as late as 1700, it would not have been so clear. It could equally well be seen as another clear example of using *geometry* in a design procedure. Leonardo da Vinci sketched what looks to us like a triangle of forces for the vault of a cathedral, but we cannot say what he would have called it.[29]

Stresses are seldom very high in masonry structures and their survival depended mainly upon adequate stability, not only for gravity loads, but horizontal loads as well, due to the wind and earthquakes. Mainstone has written widely on the development of design methods[30] and in chapter 8 considers the theme of stability through four centuries as the means of theoretical justification developed from purely geometrical to statical, expressed in a graphical way, to more modern, non-geometric approaches.

There is, in fact a very close link between geometry and statics as ways of modelling a structure. The whole subject of graphical statics, which survives to the present day, is based on the ability of a certain sort of graphical model of a structure to represent the relative proportions of forces within the structure. The size of a stone has a constant ratio to its weight, namely the density of the material; the lateral thrust of an arch varies with the geometric ratio of span to rise, and so on. Indeed, the stability of compression (masonry) structures is principally dependent upon the geometry of the structure, and has led Heyman to discuss this stability in terms of a geometric factor of safety.[31] Although we do not know exactly how learned minds in the late-seventeenth century thought about stability, it is clear that the transition from thinking and designing in terms of 'pure' geometry to using 'pure' statics was not a sudden switch from one to the other.

The case of Christopher Wren is an intriguing one. Far more than most engineers, builders or architects, he knew about forces – he was an astronomer and was familiar with the nascent ideas of universal gravitation first postulated by his close friend Robert Hooke.[32] In chapter 9 Hamilton considers Wren's structural understanding, both in existing cathedrals at Salisbury and Westminster Abbey, and in devising the structure for his own St Paul's. He clearly understood

---

[29] See Mainstone's paper in chapter 8.

[30] See several references in the Bibliography and his forthcoming *Structure in Architecture*, a collection of papers published by Ashgate Publishing Ltd.

[31] This is discussed in Heyman, 'The safety of masonry arches'.

[32] Some of Hooke's work is discussed in Dorn, *The Art of Building...*

the lateral thrust in arches, Gothic vaults and his dome. But Wren was by no means a modern structural engineer. He shows no evidence of 'taking moments' (considering the overturning effect on walls of the horizontal component of thrusts in arches and vaults) in order to investigate the stability of an arch abutment. As Hamilton observes, Wren's attempt to explain the stability of an arch, and to indicate a design method for establishing the necessary width for the abutment, does not consider the lateral thrust of the arch, even though Wren makes it clear elsewhere (both in his writings and his cathedral) that he is well aware of these thrusts. Nevertheless, Wren rightly criticises Blondel's rule[33] by imagining the consequences of adding weights (implying a statical approach), and concludes that this design rule was 'neither true nor universal', partly because it was not based 'upon any sure geometrical theorem'. Wren's understanding of stability seems, then, still to be largely based on geometry, as was his probable use of a hanging chain (on Hooke's suggestion) to derive the best shape for the brick cone in his dome at St Paul's Cathedral.

Justification using the Concepts of Stress and Elasticity The remaining chapters in this part relate mainly to the beam and to bending, and to the use of cast and wrought iron in structures. This is no coincidence. With the introduction of cast and wrought iron, around 1790, there was suddenly a need to be extremely careful about the quantities of materials needed for buildings. Apart from their very high cost, the method of manufacture was also significant. With timber and stone you start out with a large piece of material and cut it down to size. It was, therefore, more expensive to make a beam as small as possible and just-strong-enough than it was to make it larger, and hence stronger than necessary. With cast and wrought iron, the opposite was the case – there were clear benefits in using as little material as possible. In this respect, cast-iron was an ideal material because it could so easily be made into the optimum shape.[34]

As Skempton reports in chapter 10, from their earliest use in the 1790s, the shapes of cast-iron beams reflected their structural duty, being deeper at centre span and with a cross-section in the form of an inverted T. The same was the case for the early cruciform cast-iron columns which were thicker at mid-height than at the ends.[35] Here is ample evidence that Charles Bage and William Strutt, when using a new material, were no less inventive or competent than their modern

[33] See Fig. 4 in Hamilton's paper (chapter 9), and Mainstone's paper chapter 8.

[34] For discussion of other ways in which production methods have influenced building and component design, see Strike, *Construction into Design....*

[35] None of these shapes had been used in timber, masonry or wrought iron building components. For more detail on cast-iron beams and columns see also Fitzgerald, 'The development of the cast-iron frame ...', and Swailes, 'Nineteenth-century cast-iron beams ...'. For other issues relating to the transition from timber to iron see Guillerme, 'Wood versus iron'.

counterparts. In addition to their use of mathematical bending theory, they learnt from the use of iron in other situations such as machinery, undertook experimental testing to establish the limits to strength, and proof-loaded critical components to ensure they contained no invisible manufacturing flaws or inferior material. Within a very short time, the most suitable and economical forms of beams had been developed and Tredgold published design guidance in his *Practical Essay on Cast-Iron* of 1822.

Tredgold deserves a special mention in a history of building design, partly because he happened to be professionally active when design methods were changing so radically, and partly because his two books, devoted to designing structures in timber and in iron, both ran to many editions and were very influential in Britain throughout the century. His book on the principles of carpentry, published in 1820, was one of a long line of practical design handbooks for structural carpentry and much of its style follows these precedents. But it is different from its ancestors in being the first which also introduces a considerable amount of engineering science related to the strength and stiffness of beams and to the design of columns. His *Practical Essay on Cast-Iron*, published two years after his book on timber, was the first book giving design guidance on this relatively new material. His books were the first (in English) which used maximum stress as the limit to acceptable structural behaviour instead of maximum deflection or rupture load, as had been usual during the eighteenth century. He applied this approach to columns also, though he seems not to have been aware of how to calculate Euler's buckling load for a column. Tredgold's work is especially interesting, as Booth and Sutherland elaborate so eloquently in chapters 11 and 12, because it gives us an insight into approaches to structural design at just the period when it was in major transition.

In the final paper of the section, Sutherland not only presents an excellent summary of design methods up to the early use of cast-iron, but takes the story forward into the wrought iron age.[36] By this time all engineers were designing their beams on the basis of a maximum stress rather than a load that would cause collapse, one reason being that wrought iron fails by plastic deformation well before it fractures, whereas cast iron fails by fracture with no plastic deformation. While the elastic theory needed for these new design methods was well known, the confidence with which the methods could be used depended on having reliable data on the strengths and stiffnesses of irons from different manufacturers, not only throughout Britain, but also abroad. These were provided by a rapid development in methods and improved reliability of materials testing.

[36] Sutherland tells more of the story in 'The age of cast iron ...', reproduced in volume 9 of this series.

Progress in Engineering Design

It is evident from the earlier papers that engineering design has changed and improved over the centuries, but it is not at all clear how this change and improvement has comes about. In particular, the role of engineering science in influencing engineering design is not without dispute. The papers in this final part consider the epistemology of engineering design – 'what sort of knowledge is it?' and 'how does it grow?'. These questions are hotly debated among philosophers of engineering and technology.

Hall (chapter 14) considers the theory:practice model of engineering knowledge, which is especially prevalent in Britain and the USA. This presupposes a dependence of engineering practice on a body of theory or scientific knowledge. There are, I believe, problems with this model, as I have argued elsewhere.[37] If 'scientific knowledge' or 'theory' are taken to be what we would expect to find in a modern engineering textbook, then obviously there could have been no design 'practice' of engineering before this theory was devised, at dates, depending on the specialist field of engineering, between about 1700 and the present day. Yet people *did* design and construct buildings, bridges, dams and other structures long before there was any modern engineering theory, and judging by the successes, they did so with considerable skill.

Pippard's paper, in chapter 15, was presented as the Unwin Memorial lecture, which explains its slightly misleading title. Unwin and Pippard were great university academics and clearly saw university-generated research as the main stimulus to progress. The lively exchange of opinions, lasting several years, arose, as Pippard recounts, out of a proposal in 1905 to raise the Aswan Dam by about 20 feet. The story illustrates, in the field of dam design, a similar tension between academic engineering scientists and practising designers that has been encountered in other papers in this collection.[38] But this is not to criticise either – coming to understand, for the first time, the way in which enormous and new types of engineering structure actually work is very difficult. Every type of explanation must be proposed and tested. The various models of dams that were made were both ingenious and illuminating. Often the result was not that an explanation was wrong or right, but that it might be more or less useful to an engineer faced with a real dam to build.

[37] Since the late-nineteenth century this model has often been used by those wishing to justify much technological research in the academic environment, despite several studies which have demonstrated that this strategy seldom results in outcomes that are useful to industry. The model has also been used in the UK to justify the teaching of engineering mathematics and science in universities, to the exclusion of engineering and design practice. I have discussed this and the theory/practice issue in greater depth in Addis, *Structural Engineering...*, where several tens of further references are given.

[38] The design and construction of dams is treated in Binnie, 'Masonry and concrete dams...', Skempton, 'Historical development ...' and in *Dams*, volume 4 of this series.

It is the engineer who constructs the dam who ultimately must be convinced that it will stand; inevitably, different engineers will be impressed to different degrees by the rival arguments of different engineering scientists and the results of various experiments used to test and demonstrate the ideas being put forward. Engineering design decisions were not, and still are not, based on a logic as infallible as that which underpins geometry, algebra and arithmetic.

Stan Smith's paper in chapter 16 presents a very different view of British engineering just a decade or so before the debate about dam design.[39] Even around 1890, most design in engineering firms was still based more upon pragmatism and experience than upon the latest ideas in engineering science that were spreading throughout continental Europe and the USA. But this was not to say the engineering was bad. The British iron industry had gained more experience than any other country in the world during the previous fifty years and this was still largely accessible by word of mouth. Except for extraordinary projects such as the Forth Bridge, most projects would have had little need for innovation or change. On the continent and in the USA, there was no such wealth of experience and they had needed to undertake much research and development to catch up. By the end of the century, however, Britain was poised to be overtaken in virtually every field of construction engineering.

Progress has always been made in engineering by careful analysis of the results and causes of failures. In chapter 17 Cowan discusses how the engineers of Gothic cathedrals might have learnt from failures. The reader is advised, however, not to make the inference, all-too-frequently made by writers about early construction, that such building was a 'trial and error' process similar to that used by a 3-year old playing with bricks. The people who built the Gothic cathedrals, as well as those throughout the Egyptian, Greek and Roman Empires before them, knew a great deal about the properties and behaviour of materials, structures, soils and water. That we now know more, and can store and convey our knowledge in convenient and efficient ways, does not diminish the quality and usefulness of their expertise.

In chapter 18 Petroski, too, has written on the significance of engineering failures and used the Britannia Tubular bridge as an illustration.[40] In keeping with

---

[39] Smith covers similar but not identical ground in 'Craftsmen and contractors...'.

[40] It is a reflection on the state of civil engineering history that this paper on the Britannia Bridge is, as far as I am aware, one of only two devoted to what was probably the greatest structural engineering achievement of the nineteenth century. (The other, looking at the larger picture, is Smith, 'The development and use of the tubular beam', which is reprinted in *Structural Iron and Steel, 1850–1900* volume 10 of this series.) Compare this to the numerous papers on architectural aspects of a myriad relatively insignificant churches.

For more about this extraordinary project, see Beckett, *Stephensons' Britain*; Clark, *The Britannia and Conway Tubular Bridges ...*; Rolt, *George and Robert Stephenson*; and Rosenberg and Vicenti, *The Britannia Bridge...*

the author's fascination with failures in engineering history, he describes the development and final design of the bridge as a series of examples of progress resulting from failure. It is valuable to study and record the way engineers learn from experience and experiment, for this is how they develop their confidence. But, to my mind, it is a little misleading to portray the engineering design process as if it is exclusively 'failure-driven'; it can be just as strongly driven by success – the successful anticipation, avoidance and solving of problems never previously encountered. This was especially true at Menai, and not only for engineering reasons; the unique site conditions and constraints, the political purpose of the bridge, and the attitudes of the appropriate regulatory authorities were all major influences on the final design for the bridge.

All engineering design is a matter of having an idea and testing it – either in the mind, or in discussion between several minds, or using physical or mathematical models, indeed testing it by any and every conceivable means. It is the engineer's responsibility to prevent failure and, in order to do this, he or she must first *imagine* every possible type of failure. A structure of such originality and scale as the Britannia bridge had two requirements: engineers who believed it would *somehow* be possible to build it, and engineers who believed they could, by their ingenuity and rigour, raise the confidence of everyone concerned sufficiently to be allowed to proceed. Fairbairn and Stephenson had gained the initial confidence from their previous work with wrought iron, respectively in shipbuilding and railway locomotives. They also knew that Brunel had, in 1843, already built the steamship *Great Britain* – effectively a tubular, wrought iron beam nearly 100 metres long. The structural form of the bridge was different from both, however, and with advice from the mathematician Eaton Hodgkinson, they devised model tests that would, they believed, replicate all the modes of failure that the full size bridge might suffer, some already imaginable, others not, because the structural form and construction were without precedent. By these means the risk of failure was reduced to a level that still left the engineers confident enough of success. With hindsight it is easy to underestimate the magnitude of the achievement represented by the first successful attempt at a new challenge – both technical skill and the skill needed to persuade a sceptical client;[41] in March 1848 no-

[41] Brunelleschi was faced with a similar problem with his clients and rival competitors for designing and constructing the dome over Florence Cathedral. He refused to divulge his method for building the dome without centring and was dismissed as either mad or a charlatan. To make the point that it would be unwise for him to divulge his highly original idea, he summoned an egg be brought to the room. He threw down the challenge that whoever could balance the egg on its end should build the cupola. All tried and failed. Then Brunelleschi tapped the egg lightly on the marble table to flatten one end, and stood it up with ease. The others complained that they could have done the same if they realised they were allowed to do that. Brunelleschi replied

one in the world knew if a 122-metre tubular iron bridge would work; by May 1848 everyone knew that it did.[42]

In chapter 19 Blockley and Henderson present an account of the growth of engineering knowledge based on the model of scientific progress expounded by philosopher Karl Popper. The authors propose that a construction project can be seen as an experiment, analogous to those of the scientist whose experiments in the laboratory are undertaken to falsify the scientific hypothesis under test. As far as it goes, the model is fine, but it does not seem to me to address precisely the design engineer's predicament, since I do not believe that any engineering projects have been started in the belief that they were experiments which might succeed or might fail. It is, however, entirely proper to regard engineering design as an intellectual process, as Blockley and Henderson do. In chapter 20 I argue that progress in engineering design follows a pattern similar to that proposed for the natural sciences by the philosopher Thomas Kuhn. He uses the word *revolution* to describe the major upheavals in scientific thought that arise when new explanations and theories are proposed which are entirely in conflict with previously accepted beliefs.[43]

Engineering design, too, is based upon beliefs about the world, and 'revolution' is not too strong a word to apply to the many conflicts that have arisen out of different beliefs as to how best to design safe structures. It is a fundamentally different sort of activity from mathematics and engineering science. The goal is not new knowledge or understanding – it is a new building or dam or tunnel. Some knowledge or understanding of the important issues is a prerequisite, but the nature and depth of this knowledge has varied from time to time and place to place, throughout history. The good engineer uses whatever is available to achieve his own aims.

The engineer's knowledge comprises far more than engineering science or theory, in the modern sense, and is gained through various types of experience, both personal and vicarious. Each engineer, for each project, needs to raise his *own* confidence that his plans are viable and safe. It is not enough for there to have been a successful precedent, or a scientist's theory. The engineer needs also

simply that they would also have known how to build the dome themselves if he showed them his model and plans. He got the job. For more about the dome see Giorgio Vasari, *The Lives of Artists*, published in translation by Penguin Books, 1965.

42 Stephenson deliberately planned to use the Conway span as a trial run for the Menai spans which were slightly longer and very much higher above water level (over 30m). Foundations for both bridges were started in early 1846; the first tube at Conway went into service on 1st May 1848 and the first tube of the Menai span was raised in summer 1849.

43 Galileo was imprisoned for suggesting that Jupiter had moons of its own, for this contradicted the 'fact' in Ptolemaic astronomy that all heavenly bodies rotated around the earth. For an exposition of his ideas about progress in science, see Kuhn, *The Structure of Scientific Revolutions*.

to believe that the successful precedent is similar in relevant respects, or that the scientist's theory will be appropriate, to the job at hand. The origins of such beliefs are complex, especially when the nature of the knowledge often seems not to be expressible in mathematical models or even in words. As important as the nature of engineering knowledge are the means by which this body of engineering knowledge can be stored, for instance in design rules, codes of practice and scientific theories, and the means by which it can be learnt by newcomers to the profession – both blacksmiths and materials scientists know a great deal about wrought iron, but each learns it and passes it on to others in very different ways.[44]

Conclusion

At present the written history of engineering design is unevenly covered, even though the source material may exist. As yet, for instance, no-one has traced the history of how timber and, later, wrought iron roof trusses were designed (as opposed to constructed).

Rather more serious is the absence of due attention from historians to probably the two most important subjects in the history of engineering design – the development and use by designers of *graphical statics* and *the factor of safety*. These are themes in need of research.

The mathematical technique called 'graphical statics' embodies a wide variety of geometric representations of forces in structures. Despite its early invention – both Leonardo and Simon Stevin had a hand in this – and its development by mathematicians in the eighteenth century, its use by design engineers came to full fruition only in the second half of the nineteenth century. As a tool to aid the design of structures it is quite brilliant for its simplicity, its power, and for its ability also to convey an understanding of how structures work. Taken together, this explains why most engineers still use graphical methods for understanding the statics of structures from time to time, and why it is still taught in some university engineering courses. Its impact must have been no less dramatic in the last century than that of the personal computer in recent years. In Britain, it is especially remembered for its use in designing statically determinate trusses for roofs and railway bridges, and for the notation of the force diagrams devised by Robert Bow.[45] This use of graphical statics, however, was rudimentary in

[44] For further discussion of these important issues, I recommend Vicenti, *What Engineers Know...* and also a rare paper written about design methods by practising designers, Happold, et al., 'Design towards convergence ...'. The paper is summarised and discussed in Walker and Addis, *Happold...*

[45] A 'statically determinate' structure is one in which the forces in the members are independent of the materials of which they are made, and can be determined using statics alone.

comparison to its use by continental European engineers, where it was developed much further to embrace statically indeterminate structures too.[46] Nor should we forget the widespread use of graphical methods in soil mechanics, and in the design and analysis of both masonry and suspension structures. It is no exaggeration to say that graphical statics was one of the two essential ingredients which enabled the transition to be made from ancient and mediaeval thinking about civil and structural engineering to our own in the twentieth century.

The second essential ingredient was the factor of safety.[47] Its early quantitative use, in the eighteenth century, corresponded to its literal meaning – the multiple by which the maximum load on a bridge could be exceeded without causing collapse, and was typically between four and six. This idea had its origin in the practice of proof-testing and was of more use to the owner of the bridge than its designer. During the late-eighteenth and early-nineteenth century the safety factor came to be adopted as a guide for the designer as to how much safety he should aim to ensure, by design. Safety factors came to reflect the confidence with which the properties of materials could be relied upon, and the need to allow for a certain amount of unforeseen overloading. Their values were based largely on experience and common sense. Only in the middle of the nineteenth century did Professor Rankine formalise the modern concept of the safety factor – a means of being able, when designing structures, to make use of mathematical models of loads, materials and structures that were known to be imperfect representations of the real world. Hitherto, much of the engineering science and mathematics developed during the eighteenth century had remained largely unused by designers who, rightly, pointed out that their real materials and structures did not behave like the idealised mathematical representations. By demonstrating how the so-called 'gap between theory and practice' could, effectively, be quantified,

---

In structures which involve the bending of structural elements, or which contain multiple load paths through the structure, the forces and stresses in the structural elements depend upon the elasticity (stiffness) of the materials themselves and so cannot be determined using statics alone; these are called 'statically indeterminate'.

[46] The classic work on the subject was Culmann's *Die graphische Statik* (1866). It was soon followed by *Le figure recoproche nella statica grafica* (Milan, 1872) by L. Cremona and the first comprehensive text in English by Du Bois, *The Elements*... (first edition 1875). Bow's book *Economics of Construction*... (1873) differs from the other three in being an immensely practical book and aimed at those with a modest understanding of mechanics and mathematics. It comprises two parts: a classification of structures and diagrams of forces, illustrated by many dozens of examples of force diagrams as shown in Fig. 2. The book's elegance lies in applying the very simplest techniques to the design of many common types of roof structure and truss bridge. The work of Culmann is addressed in Maurer, *Karl Culmann*..., and some commentary on the use of statics is presented in Charlton, *A History of Theory of Structures*..., and Timoshenko, *History of Strength*...

[47] For some discussion of the early use of safety factors, see Addis, *Structural Engineering*..., Blockley, *The Nature of Structural Design*...; Heyman, 'The safety of masonry arches'; and Randall, 'Historical notes on structural safety'.

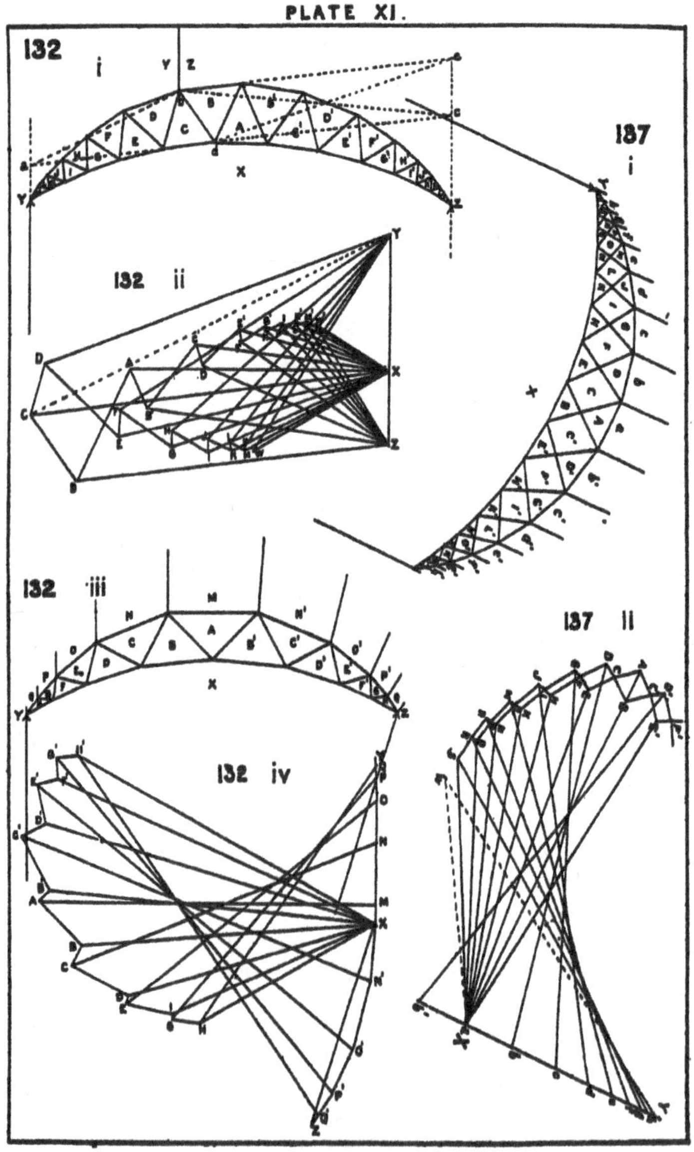

Figure 2: Diagrams of forces from R.H. Bow *Economics of Construction in Relation to Framed Structures* (Plate XI). The experienced user of graphical statics can easily see the inter-relation between the geometry of the structure and the shape of the diagram of the forces in the structure (and hence their magnitudes) and also the loads acting on the structure. Note the inclined imposed loads in Diagram 132iii representing gravity + wind loading. Diagram 137 is Bow's design work for the roof-principal of the Drill Hall at Forrest Road, Edinburgh, completed in 1872.

Rankine showed how design engineers could use the idealised 'theory' without compromising their need to develop sufficient confidence in their schemes for new structures. This conceptual revolution was to change engineering design for ever.[48]

It has not been my intention in this volume to present or even summarise a complete history of design in civil and structural engineering. Rather it offers readers a view of engineering from a somewhat different point of view than is usual, that of the designer of engineering artefacts. This cannot be done merely by looking backwards, for this inevitably focuses on the *outcome* of the design process – rather like reading the last page of a detective story first. It is essential to step backwards and then try to look forwards – to try to look at projects from a point of view when the solution was not yet known, and to trace how engineers managed to set and to attain their objectives. I believe this approach presents a more faithful picture of what engineering, and the work of engineers, is really about. When it is well done, it can be more exciting too, and more likely to inspire the reader.

The perceived value of any human activity depends, to some extent, on people's perceptions of its history, its heritage. And an important part of our understanding of history is an evaluation of what has been achieved – the quality and originality of works of music, sculpture or engineering, and the skills of the people who created them. In engineering history we do not yet have a well-developed discipline of engineering criticism that matches the sophistication of art, music, literary or architectural criticism. Although much has been written over the last century about the significant people and projects in civil and structural engineering history, very little is widely read or easily accessible, and very little of what has been written addresses the important matter of *why* the people and projects are significant. To achieve an understanding and wide recognition of the skills of design engineers from our past is the essential task of historians and a pre-requisite of valuing the same skills in the engineers of today and tomorrow.[49]

---

[48] See Channell, *A Unitary Technology...*

[49] By contrast, the study of military history – from the nineteenth century back to Greek and Roman times – is still seen as vital in the training of today's army and navy. For further discussion of the value of studying engineering history, see Addis, 'Structural criticism ...'; and *idem*, 'Structural engineering history ...'. Four books which stand out for their contribution to analysing and explaining the achievements of engineers are Penfold, *Thomas Telford...*; Pugsley, *The Works of I.K. Brunel*; Ruddock, *Arch Bridges...*; and Skempton, *John Smeaton...*

Bibliography

Ackerman, J., '*Ars sine scientia nihil est* – Gothic theory of architecture at the cathedral of Milan', *Art Bulletin*, 31 (1949), 84–111.

—, 'Architectural practice in the Italian Renaissance', *Journal of the Society of Architectural Historians*, 8 (1954), 3–11.

Addis, Bill [W.], *The Art of the Structural Engineer* (London, 1994).

Addis, W., *Structural Engineering – The Nature of Theory and Design* (Chichester, 1990).

—, 'Structural engineering history and the education of structural engineers', in *Education for Structural Engineering*, ed. D. McKay (Sheffield, 1992), 21–8.

—, 'Structural criticism and the aesthetics of structures', *Congress Report, 15th Congress of International Association for Bridge and Structural Engineering*, Copenhagen, (1996), 121–6.

—, 'Free will and determinism in the conception of structures', *Journal of the International Association for Shell and Spatial Structures*, 38, no. 2 (1997), 83–9.

—, Review of Eda Kranakis, *Constructing a Bridge – an exploration of engineering culture, design and research in 19th century France and America* (MIT Press, Cambridge, Mass. and London, 1997), *Annals of Science*, 56, no. 2 (April 1999), 205–9.

Alberti, L.B., *De re aedificatoria* (Florence, 1485; translated by Leoni, 1775 as *The Ten Books of Architecture*, reprinted Dover Publications, 1986).

Arup, Ove, 'The problem of producing quality in building', *The Arup Journal* (Ove Arup's 90th Birthday Issue), 20, no. 1 (Spring 1985), 23–5.

Beckett, Derek, *Stephensons' Britain* (Newton Abbott, 1984).

Benvenuto, Eduardo, *An Introduction to the History of Structural Mechanics*, 2 vols.: *Part I: Statics and Resistance of Solids; Part II: Vaulted Structures and Elastic Systems* (New York, 1991).

Betts, Richard, 'Structural innovation and structural design in Renaissance architecture', *Journal of the Society of Architectural Historians*, 52 (March 1993), 5–25.

Binnie, G.M., 'Masonry and concrete dams, 1880–1941', *Industrial Archaeology Review*, 10 (Autumn 1987), 41–58.

Blockley, D.I., *The Nature of Structural Design and Safety* (Chichester, 1980).

Bow, R.H., *Economics of Construction in Relation to Framed Structures* (London, 1873).

Bucher, F., 'Design in Gothic architecture', *Journal of the Society of Architectural Historians*, 27 (1968), 49–71.

—, 'Mediaeval architectural design methods, 800–1560', *Gesta*, XI, no. 2 (1972), 37–51.

Buonopane, S.G., and D.P. Billington, 'Theory and history of suspension bridge design, 1823–1940', *Journal of Structural Engineering*, 119 (1993), 954–77.

Channell, D.F., *A Unitary Technology: The Engineering Science of W.J.M. Rankine*. Thesis submitted for the degreee of Doctor of Philosophy: Case Western Reserve University (1975).

Charlton, T.M., *A History of Theory of Structures in the Nineteenth Century* (Cambridge, 1982).

Chrimes, M.M., 'Concrete foundations and substructures: a historical review', *Proceedings of the Institution of Civil Engineers: Structures & Buildings*, 116 (August/November, 1996), 344–72.

Clark, E., *The Britannia and Conway Tubular Bridges with general inquiries on beams and on the properties of materials used in constuction. Published with the sanction, and under the supervision of Robert Stephenson* (London, 1850).

Coulton, J.J., *Ancient Greek Architects at Work: Problems of Structure and Design* (New York, 1977).

Cowan, H.J., *An Historical Outline of Architectural Science* (2nd edn., London, 1977).

—, *The Masterbuilders* (New York, 1977).

—, *Science and Building* (New York, 1977).

Cross, Hardy, 'The relation of analysis to structural design', *Proceedings of American Society of Civil Engineers*, 61 (1935), 1363–1408.

Culmann, C., *Die graphische Statik* (Zürich, 1866).

Dorn, H.I., *The Art of Building and the Science of Mechanics: A Study of the Union of Theory and Practice in the Early History of Structural Analysis in England*. Thesis submitted for the degree of Doctor of Philosophy: Princeton University (1970).

Du Bois, A.J., *The Elements of Graphical Statics and their Application to Framed Structures* (4th edn., New York, 1883).

Fitzgerald, R., 'The development of the cast iron frame in textile mills to 1850', *Industrial Archaeology Review*, 10, no. 2 (Spring 1988), 127–45.

Frankl, P., *The Gothic: Literary Sources and Interpretaton through Eight Centuries* (Princeton, 1960).

Friedman, Donald, *Historical Building Construction: Design, Materials and Technology* (New York and London, 1995).

Galileo, G., *Discorsi e Dimonstrazioni matematiche, intorno a due nuove scienze attenenti alla mecanica e i movimenti locali* (Leida, 1638). Translated by H. Crew and A. de Salvio as *Dialogues Concerning Two New Sciences* (New York, 1914).

Gräfe, Rainer, 'Zur Formgebung von Bögen und Gewölben', *Architectura* (Journal of the History of Architecture, Munich), 16, no. 1 (1986), 50–67.

—, *Zur Geschichte des Konstruierens* (Stuttgart, 1989).

Guillerme, André, 'Wood versus iron', *History and Technology*, 6, no. 3 (1988), 239–50.

Hamilton, S.B., *The Development of Structural Science in the 17th and 18th Centuries and its Relation to Contemporary Practice in Engineering and Architecture*. Thesis submitted for the degree of Doctor of Philosophy: London University (1951).

—, 'The development of structural theory', *Proceedings of the Institution of Civil Engineers*, 1 (1952), 374–419.

—, 'Sir Benjamin Baker: with particular reference to his contribution to the study of earth pressures', *Géotechnique*, 8 (1958), 104–12.

Happold, E., Presidential Address: 'Can you hear me at the back?', *The Structural Engineer*, 64A, no. 12 (December 1986), 367–78.

Happold, E., Liddell, I., and Dickson, M., 'Design towards convergence: a discussion', *Architectural Design*, 46, no. 7 (1976), 430–35.

Heyman, J., 'The safety of masonry arches', *International Journal of Mechanical Sciences*, 11 (1969), 363–85.

—, 'How to design a Gothic cathedral: some fragments of the history of structural engineering', *Proceedings of the Institution of Civil Engineers: Civil Engineering*, 92 (February, 1992), 24–9.

—, *Structural Analysis: An Historical Approach* (Cambridge, 1998).

Kranakis, E., *Constructing a Bridge – An Exploration of Engineering Culture, Design and Research in 19th Century France and America* (Cambridge, Mass. and London, 1997).

Kubler, G., 'A late Gothic computation of rib-vault thrusts, *Gazette des Beaux Arts*, 26 (1944), 135–48.

Kuhn, T.S., *The Structure of Scientific Revolutions* (2nd edn., Chicago, 1970).

Mainstone, R.J., 'Structural theory and design: 1. Early proportional theories', *Architecture & Building*, 34 (1959), 106–13.

—, 'Structural theory and design: 2. Basic concepts of modern theory', *Architecture & Building*, 34 (1959), 186–95.

—, 'Structural theory and design: 3. Modern theory and design', *Architecture & Building*, 34 (1959), 214–21.

—, 'Structural theory and design before 1742', *Architectural Review*, 143 (1968), 303–10.

—, 'Intuition and the springs of structural invention' in *Structures Implicit and Explicit*, ed. J. Bryan and R. Sauer (University of Pennsylvania, 1973), 47–67.

—, 'The springs of invention revisited', *Structural Engineer*, 73, no. 22 (November, 1995), 392–9.

—, *Developments in Structural Form* (2nd edn., Oxford, 1998).

Marsden, E.W., *Greek and Roman Artillery – Technical Treatises* (Oxford, 1971).

Maurer, B., *Karl Culmann und die graphische Statik* (Diepholz, Germany, 1999).

Nisbet, James, *Fair and Reasonable: Building Contracts from 1550 – A Synopsis* (London, 1993).

—, *A Proper Price: Quantity Surveying in London, 1650–1940* (London, 1997).

Parsons, W.B., *Engineers and Engineering in the Renaissance* (Baltimore, 1939).

Paxton, Roland A., 'Menai Bridge, 1818–26 and its influence on suspension bridge development', *Transactions of the Newcomen Society*, 49 (1977–78), 87–110.

Penfold, Alistair, ed., *Thomas Telford: Engineer* (London, 1980).

Peters, Tom F., *Transitions in Engineering: Dufour and the Early 19th Century Cable Suspension Bridges* (Basel and Boston, 1987).

Picon, Antoine, *Architectes et ingénieurs au siècle des lumières* (Marseilles, 1988).

Pugsley, A., ed., *The Works of I.K. Brunel – An Engineering Appreciation* (Institution of Civil Engineers, London, and University of Bristol, 1976).

Radelet-de Grave, P., and E. Benvenuto, *Between Mechanics and Architecture* (Basel, 1995).

Randall, F.R., 'Historical notes on structural safety', *American Concrete Institute Journal* (October 1973), 669–79.

Rennison, R.W., 'The High Level Bridge, Newcastle: its evolution, design and construction', *Transactions of the Newcomen Society*, 52 (1980–81), 180–207.

Ricken, Herbert, *Der Bauingenieur: Geschichte eines berufes* (Berlin, 1994).

Rolt, L.T.C., *George and Robert Stephenson* (London, 1960).

Rosenberg, N., and W.G. Vicenti, *The Britannia Bridge: The Generation and Diffusion of Technical Knowledge* (Cambridge, Mass., 1978).

Ruddock, E.C., 'The building of the North Bridge, Edinburgh, 1763–75', *Transactions of the Newcomen Society*, 47 (1974–75), 9–33.

Ruddock, Ted [E.C.], *Arch Bridges and their Builders, 1735–1835* (Cambridge, 1979).

Schofield, R.B., 'The design and construction of the Cromford Canal, 1788–1794', *Transactions of the Newcomen Society*, 57 (1985–86), 101–23.

Shelby, L.R., *Gothic Design Techniques* (Southern Illinois University Press, 1977).

—, and R. Mark, 'Late Gothic structural design in the "Instructions" of Lorenz Lechler', *Architectura* (Journal of the History of Architecture, Munich), 9 (1979), 113–31.

Shipway, J.S., 'The Forth Railway Bridge centenary 1890–1990: some notes on its design', *Proceedings of the Institution of Civil Engineers, Pt. 1*, 88 (December, 1990), 1079–1107.

Skempton, A.W., 'Landmarks in early soil mechanics', *Proceedings of 7th European Conference on Soil Mechanics & Foundation Engineering*, Brighton, 5 (1979), 1–26.

—, ed., *John Smeaton, FRS* (London, 1981).

—, 'Historical development of British embankment dams to 1960', in *Clay barriers for embankment dams*, Proceedings of the conference organised by the Institution of Civil Engineers and held in London on 18 October 1989 (London, 1990), 15–52. Reproduced in vol. 4 of this series.

Smith, Denis, 'Structural model testing and the design of British railway bridges in the nineteenth century' *Transactions of the Newcomen Society*, 48 (1976–77), 73–90.

Smith, Stanley, 'The development and use of the tubular beam, 1830–1880', *History of Technology*, 14 (1992), 100–34.

—, 'Craftsmen and contractors: a re-appraisal of their role in the design and fabrication of mid-nineteenth century iron structures in Great Britain', *Journal of Architectural and Planning Research*, 11, no. 4 (Winter 1994), 311–38.

Straub, Hans, *A History of Civil Engineering* (London, 1952).

Strike, James, *Construction into Design: The Influence of New Methods of Construction on Architectural Design, 1690–1990* (Oxford, 1991).

Stussi, F., 'Karl Culmann und die graphische Statik', *Schweizerische Bauzeitung*, 69, no. 1 (1951), 1–3.

Sutherland, R.J.M., 'The age of cast-iron: who sized the beams?', *The Iron Revolution: Architects, Engineers and Structural Innovation, 1780–1880*, ed. R.Thorne (Royal Institution of British Architects, London, 1990), 25–34.

Swailes, T., 'Nineteenth-century cast-iron beams: their design, manufacture and reliability, *Proceedings of the Institution of Civil Engineers, Structure & Buildings*, 114 (February 1995), 25–35.

—, and J.O. Marsh, 'Victorian iron arch bridges', in *Arch Bridges*, ed. C. Melbourne, Proceedings of the First Conference on Arch Bridges held at Bolton, UK on 3–6 September 1995 (London, 1995), 65–74.

Tanimoto, K., and Y. Goda, 'Historical development of breakwater structures in the world', in *Coastal structures and breakwaters*, Proceedings of the conference organized by the Institution of Civil Engineers, and held in London on 6–8 November 1991 (London, 1992), 193–206.

Timoshenko, S.P., *History of Strength of Materials* (London, 1953, reprinted by Dover Publications, 1983).

Tomlinson, M.J., 'Telford and soil mechanics', *Géotechnique*, 6, no. 3 (1956), 98–105.

Tomlow, Jos, 'Die Kuppel des Giesshauses der Firma Henschel in Kassel 1837: eine frühe Anwendung des Entworfsverfahrens mit Hängemodellen', *Architectura* (Journal of the History of Architecture, Munich), 23, no. 2 (1993) 151–72.

Vicenti W.G., *What Engineers Know and How They Know it* (Baltimore, 1990).

Victor, S.K., 'Practical geometry in the High Middle Ages', *Memoirs of the American Philosophical Society*, 134 (1979).

Walker, D., and B. Addis, *Happold: The Confidence to Build* (Bath, 1997; London, 1998).

Walter, C., *Brückenbau* (Augsburg, 1766).

Werner, E., *Technisierung des Bauens* (Dusseldorf, 1980).
Wilson, W.M., and G.A. Maney, 'Wind stresses in the steel frames of office buildings', *Bulletin No. 80, Engineering Experiment Station*, University of Illinois (June 1915).
Yeomans, David T., *The Trussed Roof: Its History and Development* (Aldershot, 1992).

1
The mechanization of design in the 16th century: the structural formulae of Rodrigo Gil de Hontañón

Sergio Luis Sanabria

The existing fragments of an architectural booklet by the 16th-century Spanish architect Rodrigo Gil de Hontañón reveal an ingenious attempt to systematize the design process by creating a sequence of formulaic procedures to be followed in ecclesiastical projects. The formulae are addressed to two more or less separate issues. The first is to synthesize Gothic and Classic proportioning methods, and demonstrate their fundamental identity. The second is to establish an independent "science" of structural design. Aside from the more theoretical writings of Leonardo da Vinci, the work of Rodrigo Gil is the principal evidence extant for the development of structural thinking among 16th-century master masons.

Seven formulae discussed here are concerned with the correct depth of a buttress to support an arch or a rib vault. The formulae do not seem to have been derived through theoretical analysis, using the medieval Scientia de Ponderibus. *Rather they are the result of new experimentation and traditional Gothic geometric thinking applied to classical arches, and of new arithmetic procedures applied to Gothic rib vaults.*

Introduction

ITALIAN ARCHITECTS of the 16th century developed the expressive range of classical vocabulary far beyond its tentative appearances in the 15th century. This dazzling artistic development has masked the contributions made to the building arts by Late Gothic architects, so that the entire range of progressive ideas of the 16th century often is ascribed mindlessly to the Renaissance movement. This is true of technical developments in structural analysis. The standard story, as told by W. B. Parsons, begins with the precocious but uninfluential work of Leonardo ca. 1500 on the bending of beams, the geometric resolution of force diagrams, and the analysis of arches. It then leaps ahead to the 17th century and Galileo's publication in 1638 of *Dialoghi delle nuove scienze*, with its still incorrect analysis of a cantilevered beam. The inventions of the triangulated truss, exemplified in the work of Vasari and Palladio, and of laminated beams by Philibert de L'Orme are treated as practical inventions of the second half of the 16th century, of limited theoretical importance.[1]

Even Robert Mark, a staunch admirer of Gothic structural achievements, views Late Gothic with its codification of design rules as having stifled structural experimentation.[2] Certainly the peak of intuitive structural experimentation was reached earlier, with the construction of the great cathedrals of the Ile-de-France in the 13th century and of Catalonia in the 14th. At this time scholastics such as Jordanus Nemorarius (fl. 1220–1230), who rekindled the theoretical study of mechanics (*scientia de ponderibus*), were apparently unable or unwilling to make any contribution to the analysis of the vast and complex buildings then going up, often in the midst of the schools where they taught their Archimedean doctrines.[3]

Thus, according to a widespread viewpoint, the 15th and 16th centuries saw a decline in structural inventiveness, with no significant experimentation other than the introduction of some new forms, and no progress in theoretical analysis. I believe that this overly-negative evaluation of Late Gothic achievements is premature, due to insufficient studies. The writings of the Spanish architect Rodrigo Gil de Hontañón illuminate this obscure subject and show the transitional development of structural

I would like to thank Professor Robert Mark for his helpful technical criticism of this paper.

1. William Barclay Parsons, *Engineers and Engineering in the Renaissance*, 1939, 67–73, 481–492. More recent publications are Bertrand Gille, *Les ingénieurs de la renaissance*, Paris, 1964; and Rowland Mainstone, "Structural Theory and Design Before 1742," *The Architectural Review*, April 1968, 303–310.

2. Robert Mark, "Structural Experimentation in Gothic Architecture," *American Scientist*, 6, 548–549.

3. See Guy Beaujouan, *L'interdépendance entre la science scolastique et les techniques utilitaires*, Paris, 1957; Guy Beaujouan, *Manuscrits scientifiques médiévaux de l'université de Salamanque et de ses "Colegios Mayores,"* Bordeaux, 1962; Marshall Clagett, *The Science of Mechanics in the Middle Ages*, Madison, Wisconsin, 1959; Joseph Brown, "The *Scientia de Ponderibus* in the Later Middle Ages," dissertation, University of Wisconsin, 1967; Marshall Clagett, *Archimedes in the Middle Ages*, 3 vols., Philadelphia, 1978.

Fig. 1. Rodrigo Gil de Hontañón. Choir of Cathedral of Astorga. Late 1530s–1559 (author).

Fig. 2. Rodrigo Gil de Hontañón. Alcalá de Henares. Universidad Complutense, Colegio de San Ildefonso. ca. 1538 (source unknown).

thinking in the 16th century, stimulated by Gothic tradition, humanist ideas, new mathematical tools, and an incipient experimental approach to theory.

Rodrigo Gil de Hontañón and his treatise

Rodrigo Gil de Hontañón (ca. 1500/1510–1577) was perhaps the most prolific and versatile Spanish architect of the 16th century.[4] His career spans the last flowering of the Gothic and the simultaneous development of the Renaissance in Spain. As a Gothic master mason, the director of a thriving construction firm, Rodrigo Gil conducted the works at the Cathedral of Astorga from their inception in the 1530s to 1559 (Fig. 1). He was responsible from 1538 for the two largest cathedrals under construction in Spain, Salamanca and Segovia, both already begun under his father, Juan Gil de Hontañón. Rodrigo also inherited two major works by Juan de Alava, the cloister of the Cathedral of Santiago de Compostela in 1538, and the soaring Cathedral of Plasencia in 1544. These buildings rank Rodrigo as the most important Late Gothic master in Spain.

At the same time, Rodrigo was very aware of the new classical vocabulary spilling from Italy and was one of the foremost masters of the Plateresque Renaissance in Spain. The façade of the College of San Ildefonso at the University in Alcalá de Henares, begun about 1538, is perhaps his best known work, where he combines Late Gothic geometric proportioning methods with Renaissance forms (Fig. 2).

Despite an extremely busy and peripatetic career, Rodrigo found time to jot down notes for an unfinished treatise on architecture, attempting a synthesis of Classicism and Gothic. Although his original writings have not survived, parts of his text and illustrations, which apparently were kept at the *fábrica* of the Cathedral of Salamanca, were copied before 1681 by Simón García, a Salmantine architect born ca. 1651, as part of a Baroque compendium on a variety of architectural subjects (Fig. 3). Simón García's *Compendio de Arquitectura y Simetría de los Templos*, now at the Biblioteca Nacional de Madrid, MS. 8884, is a handwritten codex of 141 folio pages, with 77 chapters.[5] Chapters 1 through 6 are based mostly on the notes of Rodrigo. Chapter 75 plus an illustration at the end of Chapter 16 can also be attributed to him.[6]

4. The best and most comprehensive study of the work of Rodrigo Gil is John Hoag, "Rodrigo Gil de Hontañón, His Work and Writings," dissertation, Yale University, 1958, with complete bibliographies to date. Later works on Rodrigo's architecture are Marqués de Lozoya, *Rodrigo Gil de Hontañón en Segovia*, Santander, 1962; Pedro Navascués Palacio, "Rodrigo Gil y los entalladores de la fachada de la Universidad de Alcalá," *Archivo Español de Arte*, XLV, 1972, 103–117; A. C. Casaseca, "Trazas para la Catedral de Segovia," *Archivo Español de Arte*, LI, 1978, 29–51; J. García-Murga, "La intervención de Rodrigo Gil de Hontañón en la iglesia de Santa Maria de Guareña," *Goya*, 144, 1978, 314–323.

5. The manuscript was first published partially by Eduardo Mariátegui in *El Arte en España*, VII, 1868, 113–127, 154–184, 193–215. A more complete edition by José Camón Aznar was published in Salamanca in 1941. The latest edition of the *Compendio* is by Carlos Chanfón Olmos, Mexico, 1980, which I have not seen.

6. For Chapters 1–6 see José Camón Aznar, "La intervención de Rodrigo Gil de Hontañón en el manuscrito de Simón García," *Archivo Español de Arte*, XIV, 1941, 300–305; and Hoag, "Rodrigo Gil," 405–

Fig. 3. Simón García. *Compendio de Arquitectura y Simetría de los Templos.* Title page. 1681 (Biblioteca Nacional de Madrid).

It is almost impossible to distinguish between the rough and unpolished writing styles of the two men, which suggests that Simón rewrote much of Rodrigo's text. The only clue is that Simón seems more prolix and prone to insert learned references in inappropriate settings. Some caution in attribution of ideas is therefore essential while reading Rodrigo's chapters.

Dating of the original manuscript of Rodrigo Gil can be based only on internal evidence. The earliest notes on human proportioning in the book must be dated after 1544, by a reference in Chapter 2, 2r, to Guillaume Philandrier's *In Decem Libros Vitruvii de Architectura Annotationes*, published in Rome in 1544. In Chapter 6, 22r, there is a reference to what may be a personal discussion with Pedro Sañchez Ciruelo, Darocensis, polymathic professor of mathematics, theology, philosophy, logic, and music at Paris, Alcalá and Salamanca, tutor of Philip II, who died in 1554. Thus much of Chapter 6 must date before this time.

In his booklet Rodrigo describes a nearly mechanical design process for church architecture guided by strictly predetermined routes, but with many free choices along the way.[7] Although Rodrigo probably never employed this methodology in his own designs, the process interested him because it would have simplified the repetitive tasks of design in a large architectural firm. The first step is to determine the size of the church, and this is done with a formula using the parish population. The formula, in Chapter 2, 3v, yields 28 square feet per inhabitant, based on the average size of a burial plot. This area is then increased by 30 percent to $36^{2}/_{5}$ square feet (2.85 square meters) to allow for future population increases.[8] Given the area, an appropriate plan type can be selected with as many aisles as may be useful. Then the interior proportions of the building are obtained by either of two methods, human proportionality or geometry.

The first method, proportionality, arises from a Vitruvian tradition that included, among others, the works of Francesco di Giorgio and especially Albert Dürer.[9] Following the common inspiration of Vitruvius, both Francesco di Giorgio in the 15th century and Rodrigo in the 16th drew the analogy between the

409. Chapter 18 has been incorrectly attributed to Rodrigo by Marcelino Menéndez y Pelayo, *Historia de las ideas estéticas en España*, 1883, I, 475. It is a short introduction to the principles of geometry, and is the first of several chapters on mathematics extracted by Simón Garcia from standard texts. Chapter 75 is a set of specifications for the construction of the tower illustrated in 10r. The conditions seem to correspond to those in Salamanca in the 16th century, and probably refer to a project by Rodrigo. See Hoag, "Rodrigo Gil," 381–387; José Camón Aznar, "Sobre la torre de la Catedral Nueva de Salamanca," *Archivo Español de Arte*, XIV, 1940–1941.

7. A related idea for such a science of building had occurred at about the same time to Philibert de L'Orme, who in his *Le premier tome de l'architecture*, ed. 1568, III, 6, 62r, proposes to reduce the very disordered and confused book of Vitruvius to the theory and methodology of Euclid.

8. A Spanish foot is approximately .28 meters. This rule can be contrasted with the slightly later one of Carlo Borromeo in Chapter 1 of his *Instructiones Fabricae et Supellectilis Ecclesiasticae* of 1577, where the proposed area per parishioner is a square 1 braccio 8 onze per side. If these are Milanese braccia of .595 meters the area is 0.98 square meters, or slightly more than one third of Rodrigo's area. Borromeo's intention must have been to provide standing and kneeling room only, and a more crowded, urban feeling to the sanctuaries.

9. Vitruvius, *The Ten Books on Architecture*, trans. M. H. Morgan, ed. New York, 1914, I, ii, 4–5; III, i; IV, i, 6–8, discusses human proportioning and suggests architectural applications. This idea was adopted and extended by Francesco di Giorgio in the late 15th century; see Richard Betts, "The Architectural Theories of Francesco di Giorgio," dissertation, Princeton University, 1971. The ideas of Vitruvius were made available for the first time in Spanish in Diego de Sagredo, *Medidas del Romano*, Toledo, 1526. This was the first text on classical architecture published outside Italy, and soon was translated into French. Spanish translations of Vitruvius circulated in manuscript form no later than the mid-16th century. Pedro Navascués Palacio, *El libro de arquitectura de Hernán Ruiz, el Joven*, Madrid, 1974, 4ff, gives a comprehensive list of known translations.

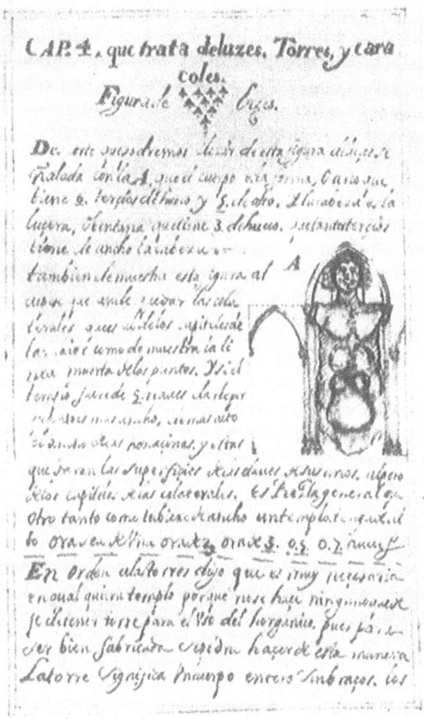

Fig. 4. Simón García. *Compendio*. Chapter 4, 9r. Human Body Proportioning a Church Section (Biblioteca Nacional de Madrid).

Fig. 5. Simón García. *Compendio*. Chapter 4, 9v. Human Body Proportioning a Tower (Biblioteca Nacional de Madrid).

human body and the fabric of a church. Both plan and elevations can be established from the proportions of the ideal human body (Fig. 4). Rodrigo advocates the proportions given in Albert Dürer's *Vier Bücher von Menschlicher Proportion*, first published in Nuremberg in 1528.[10] Smaller parts such as towers, staircases and pediments can be proportioned using smaller *homunculi* (Fig. 5). The human dimensioning system is extended to all elements visible in the interior of the building, including rib vaults, which are proportioned according to the thickness of the fingers of the hand. It is interesting to note that although the human dimensioning system is based on Renaissance thinking, the proportions are applied to buildings in the Gothic style.

The second method of proportioning buildings is by geometry, and is founded on older Gothic principles. The plan layout is constructed starting with a square which is superimposed upon the entire building or on a major part, such as the crossing (Fig. 6). The square is then manipulated geometrically with the aid of compass and straight edge. Whenever favorable accidents occur, such as two distances being nearly equal, or two arbitrary lines intersecting at a useful location, the fortuitous circumstance is exploited to yield dimensions of lesser parts, such as nave bays and side aisles. Although the system appears rigorous, it is as convoluted a mode of thinking as reading the Tarot or casting a horoscope. It is useful only because it allows decisions to be made where criteria for deciding are neither clear nor exact.

The most surprising aspect of all this is that both proportioning methods yield similar results. Plans by proportionality are often nearly identical to those by geometry. This is explained in Chapter 1 by invoking Vitruvius III, 1, and his generation of the circle and square from the canonical human body. Proportions based on man, the "Lesser World" or Microcosm, must be in harmony with those of the Universe or Macrocosm ruled by the

10. This is Rodrigo's evidence for and contribution to a dispute about correct human proportions which raged in Spain in the 16th century. Further contemporary evidence for it is found in Juan de Arfe y Villafañe, *De varia commensuración para la escultura y arquitectura*, Seville, 1585, Book 2.

Fig. 6. Simón García. *Compendio.* Chapter 5, 12r. Second Temple Plan Designed by Geometry (Biblioteca Nacional de Madrid).

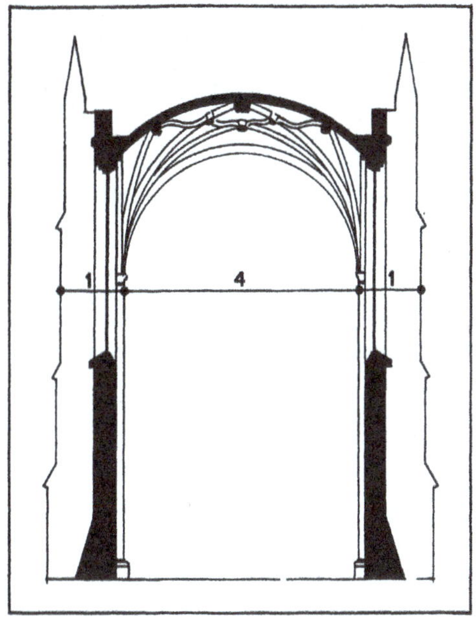

Fig. 7. First Structural Rule. 1:4 Buttress Depth to Span Ratio (drawing by author).

geometry of the Ptolemaic system. For Rodrigo there is a natural step from the older geometric method to the newer and richer human proportioning system. The two systems are complementary and there can be no contradiction between the two.

The structural formulae

All interior dimensions of a church are set out according to these philosophically controlled aesthetic systematizations. However, structural members are not sized according to spatial proportioning formulations. In Rodrigo's notes all problems of support are solved with independent rules.[11] The discussion of these rules begins in Chapter 6, 18v–19r, where Rodrigo professes to a long puzzlement about the problem of the correct depth of a buttress to support an arch, such that it is neither more nor less than needed for stability. He has asked many Spanish and foreign masters if they know any such rule, and has gotten only rules of thumb with no good explanations. Hence he has undertaken to find some rules that will accomplish this.[12] Rodrigo addresses a variety of structural problems; of these I would like to discuss seven formulae for dimensioning piers and buttresses.[13]

Structural Rule No. 1
1:4 Rule of Thumb

One rule that Rodrigo offers is no different from the unexplained rules of thumb he obtained from his inquiries. This first

11. The independence of the spatial ordering system and structural sizing is an old dichotomy in Gothic buildings. I was able to establish the existence of two such separate systems of measurement at the late 13th-century choir of St. Bénigne in Dijon. See "Metrics and Geometry of Romanesque and Gothic St. Bénigne, Dijon," *The Art Bulletin*, LXII, 1980, 510–532.

12. Is it Rodrigo or Simón Garcia who is asking this question and proposing solutions? This passage, as well as the following structural formulae have been attributed to Rodrigo by Camón Aznar, Kubler, and Hoag. I am convinced of the essential correctness of this attribution, although it is possible to argue for a 17th-century date by contrasting the three geometric formulae in the *Compendio* with the more primitive formula published by Derand in 1643: see discussion in Conclusion and footnote 32 below. The counterargument, based on internal textual evidence, is too complex to pursue here.

13. The structural formulae were first discussed by George Kubler, "A Late Gothic Computation of Rib Vault Thrusts," *Gazette des Beaux-Arts*, s.6, 25, 1944, 135–148. However, this important publication is marred by computational errors and some misreadings of the text. Hoag, "Rodrigo Gil," 426–435, 441, gives a good general discussion of the structural formulae, although again there are some new misreadings.

rule, given twice in the text, in Chapter 2, 2v, and Chapter 6, 19r, requires that the buttress depth to support a vault should be one-fourth of the span (Fig. 7). Buttress depth for Rodrigo includes the wall thickness and wall shaft, and so consists of the measurement from the interior of the church to the outermost surface of the buttress. This measurement is not to be taken at ground level, but at the *tas de charge* or higher.[14] The rule can be compared to an earlier buttress depth formula given by Lorenz Lechler in his *Instructions to His Son* of 1516, which specifies an average buttress depth at ground level of 3/10 of the span.[15] This is 1/20 more than Rodrigo's rule, but would be reduced by the talus setbacks as the pier rises. Further, Lechler allows a variation of +1.4% −2.4% depending on the stone quality. Hence these two rule-of-thumb formulae are in very close agreement.

Three Arithmetic Formulae For Various Vault and Arch Supports

Structural Rule No. 2
Hall Church Pier Diameter

A structural formula discussed by Rodrigo in Chapter 6, 17r, gives the diameter of the interior piers of hall churches, that is, churches in which central nave and side aisles have equal height. Piers of 16th-century Spanish hall churches were usually cylindrical, thus the diameter is the essential dimension. The formula given is:

$$\text{Diameter} = \tfrac{1}{2}\sqrt{H+W+L},$$

where H is the height of the pier, and W and L are width and length of a bay of the central nave. This and formulae 3 and 4 are of special interest since they are the earliest masonic rules in which the square root is extracted arithmetically rather than geometrically.

The pier diameters obtained with the formula vary over a rather narrow range, if dimensions for both large and small hall churches are used to obtain extreme values. Given a very large building, such as the Cathedral of Plasencia, with rectangular bays approximately 16×12 meters, and a springing height of about 20 meters, the required pier diameter would be 3.5 meters. The actual pier diameter at Plasencia is approximately 2.3 meters. The Cathedral of Barbastro, 1500–1533, by Juan Segura, with its 10.5 meter square bays and 21 meter high springing, has piers only 1.3 meters wide—far slenderer than the 3.2 meters Rodrigo's formula would require. By contrast a small church such as the Colegiata of Lerma in the province of Burgos, with bays 6.5 by 6.7 meters, and a pier height of about 11 meters, has piers 1.76 meters in diameter. Rodrigo's formula would require piers 2.5 meters wide, only 28 percent less than the value for Plasencia. In all cases the formula is very conservative, and has a substantial safety factor built-in.[16]

Structural Rule No. 3
Outer Wall Buttress Depth

Another arithmetic formula is found first in Chapter 2, 5r–5v, where it is unintelligibly applied to a church design, and again in Chapter 6, 17v, where it is fully explained. This formula is intended to determine the size of a wall buttress for a hall church. The computation involves finding the width W of the buttress at the springing of the vault. The plan of the buttress is preordained to be a double square, so its depth is 2W. Rodrigo specifies a complex operation to find the buttress width:

$$W = \tfrac{1}{3}\sqrt{H + \tfrac{2}{3}\Sigma R},$$

where H is the height of the buttress to the springing and ΣR is the sum of the perimeters of all ribs coverging on the buttress, whether transverse, diagonal or tiercerons, measured from the springing to their respective keystones, usually a quarter-circle.[17]

Rodrigo offers a strange reason for taking two-thirds of the sum of the ribs: approximately one-third of the volume of the stone of the ribs is removed from the originally squared voussoirs while cutting the moldings. This subtraction of one-third is an attempt to refine a formula which is not even grossly adjusted to the relevant parameters it controls, since it makes no provision for the greater thrust of the transverse ribs, and the lesser of the diagonals and tiercerons. Later in Chapter 6, 21v, Rodrigo himself states that the transverse ribs cause the greatest injury to buttresses because they are widest and thrust directly, while diagonal ribs and tiercerons cause little harm because they are thin and thrust diagonally. Thus indirectly Rodrigo casts doubt on the validity of his own formula.

Structural Rule No. 4
Semicircular Arch Buttress Depth

The last of the arithmetic buttress depth formulae, in Chapter 6, 20v–21v, is intended for a simple semicircular arch. The

14. See Rodrigo's comments in 5r and 18r where he discusses the arithmetic formula for the depth of an exterior buttress.

15. Lon Shelby and Robert Mark, "Late Gothic Structural Design in the 'Instructions' of Lorenz Lechler," *Architectura*, 9, 2, 1979, 113–132.

16. For the dimensions of Plasencia and Barbastro see plans in Fernando Chueca, *Historia de la arquitectura española. Edad antiqua. Edad media*, Madrid, 1965, 579, 634–5. For Lerma see Castor de Uriarte, *Las iglesias "salón" Vascas del último período del gótico*, Labastida, Alava, 1978, 19.

17. The quarter-circle profile of the ribs results from the typically domical rib vaults, *bóvedas baidas*, constructed in Spain in the 16th century. Kubler's discussion of this formula, "Late Gothic Computation," 141–142, is clear and perceptive. He emphasizes the fact that web thrusts are ignored by this formula and explores the impact of this omission on the Gothic Rationalism dispute between Pol Abraham and the followers of Viollet-le-Duc.

buttress depth must equal the square root of the sum of the buttress height and half the arch intrados perimeter, or:

$$\text{Depth} = \sqrt{H+P/2}.$$ [18]

This rule is similar in format to formulae 2 and 3 for rib vaults. The fascination with the square root in all these arithmetic formulae indicates it is a novel mathematical tool for master masons, and is being indiscriminately used. A powerful stimulus for its use was undoubtedly its connection with the rotation of squares, the old geometric proportioning method of the master masons.

All three arithmetic formulae handle linear units inconsistently: to find the pier depth or diameter, which is in feet, the square root of a sum of feet is taken, resulting in a value no longer in feet but in square-root feet. The inclusion of pier height in the formulae is irrelevant to static load conditions. Clearly these Late Gothic formulations are not based on an understanding of theoretical mechanics, but remain empirical guides yielding conservative values over a limited range, and with garbled units of measurement.

Three Geometric Formulae for Semicircular Arch Buttresses

The next three rules are perhaps the most interesting ones in the booklet. All three are geometric constructions, and share with the fourth formula just discussed a different context from that of the other formulae for rib vaults. These three geometric rules establish the depth of a buttress to support a semicircular arch with a flat wall above of a specified height. Thus these rules are intended for an arcade or aqueduct, or for a single free-standing arch, perhaps a barrel vault, any of which suggests a Classical rather than a Gothic style.[19] These formulae are still in the Gothic geometric spirit, but are applied to a non-Gothic technical problem, the correct width of a Classical arch wall. This was a new question for Classical architecture, which used modular proportioning methods rather than structural ones.

Structural Rule No. 5
First Geometric Formula

The first geometric rule, in Chapter 6, 19r–19v, is presented as a recipe to be followed strictly step-by-step (Figs. 8 and 9). The steps of the construction are shown in Fig. 9, I–VII. The object is to find two points, R and F, which can then be used to define the structurally correct proportions of this arch (9, VIII).

It can be shown easily that if the arch diameter, GE, is handled

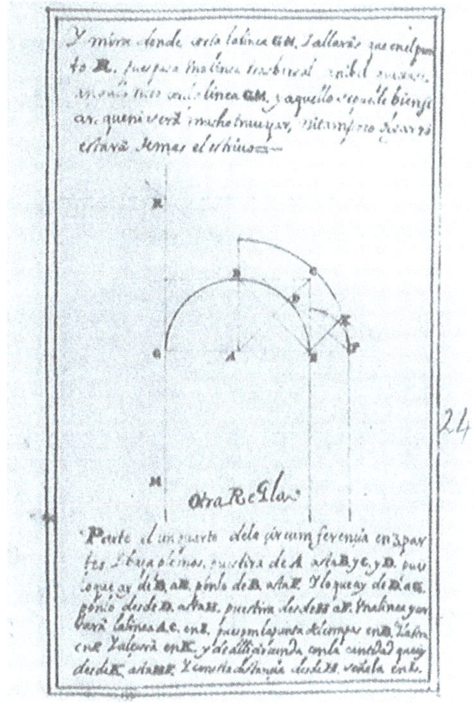

Fig. 8. Simón García. *Compendio*. Chapter 6, 19v. First Geometric Formula (Biblioteca Nacional de Madrid).

as 2 radii, the buttress depth EF is $2 - \sqrt{2}$, or 0.586 radii. The buttress depth to span ratio is 1:3.414, which is more conservative than the 1:4 ratio of the first structural rule.[20] Under this formula the arch can be loaded to a higher level than would be required in a vault, since the distance GR is $2/3 + \sqrt{2}$, or 2.081 radii. The recognition that buttress depth and loading height are related is important, but the rule is inflexible as it is, since the two parameters are fixed and cannot be altered.

Structural Rule No. 6
Second Geometric Formula

The second geometric rule, presented in Chapter 6, 19v–20r, is also a set of step-by-step instructions (Figs. 10 and 11).[21] First the arch intrados is drawn, with one springing point labelled A, the center E, the crown B, and the quarter circle opposite A divided into three equal parts at points C and D (11, I). From C

18. Kubler, "Late Gothic Computation," 142–143, read the term *circumferencia* as radius, so he incorrectly believed the formula to be $\sqrt{\text{radius} + \text{height}}$.

19. Kubler, "Late Gothic Computation," 142, believes the first geometric formula is for a barrel vault. Hoag, "Rodrigo Gil," 430–435, argues convincingly through contemporary examples that these arch formulae are not intended merely for barrel vaults.

20. Kubler, "Late Gothic Computation," 142, states the opposite, claiming that the 1:4 formula has a higher safety factor.

21. The second and third geometric formulae have not been discussed in print so far to the best of my knowledge.

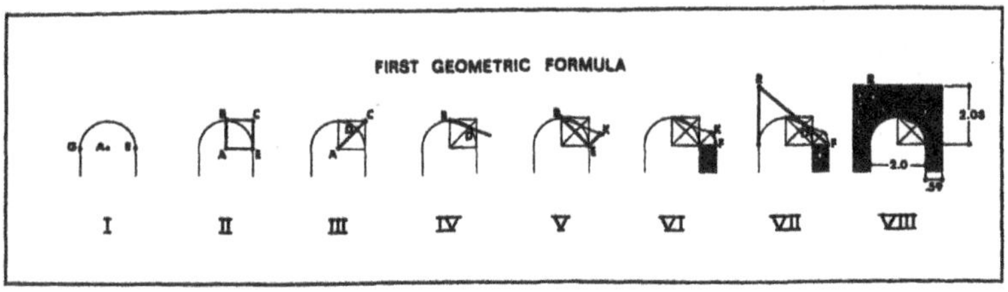

Fig. 9. First Geometric Formula (drawing by author).

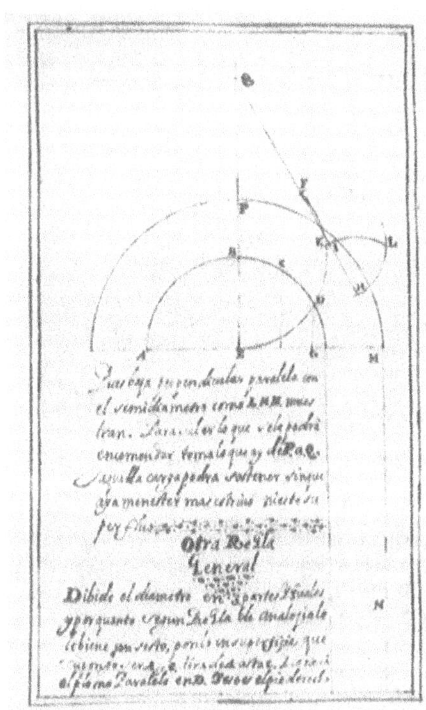

Fig. 10. Simón García. *Compendio*. Chapter 6, 20r. Second Geometric Formula (Biblioteca Nacional de Madrid).

and D plumb lines are dropped to AE (11, II).[22] Three lines are drawn from A to intersect and extend beyond B, C and D (11, III).[23] With the centerpoint at B an arc ED is drawn intersecting line AC at K and line AB at F; with another centerpoint at D arc GH is drawn intersecting AD at H (11, IV). Two of the lines defined, AF and AH, are nearly the same but not equal.[24] Rodrigo evidently believed they were equal. Points F and H are now connected with a line intersecting the central axis EB at Q, and this defines the permissible loading height (11, V). The next step demands a theoretically impossible construction. Using K as the centerpoint, arc FH is to be drawn (11, VI). Since AF is not equal to AH, K is not equidistant to F and H.[25] The error is so slight (0.95%) it would be unnoticeable even in a large drawing. The final step in the construction calls for an arc KL with the centerpoint at H (11, VII). Point L defines the buttress projection. Using points L and Q the arch now can be completed with its buttresses and high wall loading (11, VIII).

Through a tedious computation it can be established that the permissible loading height above the springing is 2.894 arch

22. The specified dropping of the plumb line from C is an unnecessary operation since it is not used in the subsequent construction.

23. Although Rodrigo does not mention it, angles EAD, DAC, and CAB are equal, each subtending 1/24 of a circle, or 15 degrees.

24. If the diameter of the arch is 2 radii, distance AF is $1 + \sqrt{2}$, or 2.414 radii, and distance AH is $1/2 + \sqrt{2 + \sqrt{3}}$, or 2.432 radii.

25. Distance KH is .632 radii and KF is .626.

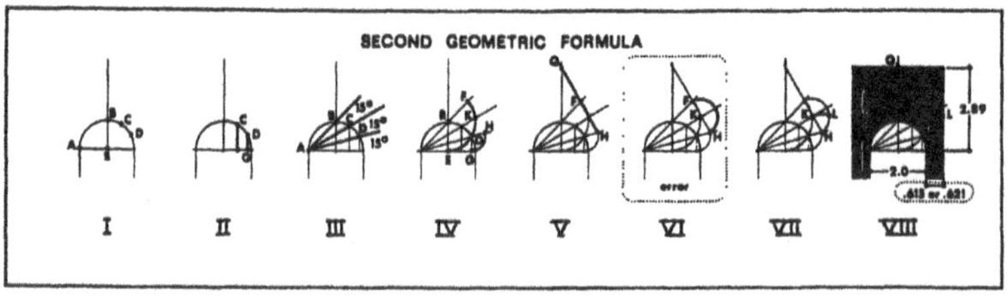

Fig. 11. Second Geometric Formula (drawing by author).

radii, and the buttress depth is either 0.615 or 0.626.[26] The buttress depth to span ratio is either 1:3.22 or 1:3.25. This is deeper than the first geometric formula, which allowed a ratio of 1:3.414 but the permissible load is also greater, 2.894 against 2.081.

This gradual adjustment of two variables with respect to one another is very interesting. It appears as if two experiments were conducted in which two arches of a given span and slightly different buttress depth were loaded until they somehow failed. The height of the load was then reduced to a safe level, and two geometric constructions invented to allow the duplication of these safe levels. Rodrigo was without question familiar with the concept of a safety factor, as he discusses it in 18r, at the end of his description of the Third Structural Formula. He points out that although his formula gives the right size to hold the thrust of the arches, the workmen can add somewhat more, because it is better to have too much than too little. However, he insists that the size obtained with his formula is quite sufficient, implying that a margin was already provided. If Rodrigo had conducted experiments to determine the failure point of arch-buttress systems, he would have known he had to provide generous safety margins, as is evident in all his conservative rules.

The very eccentricity of the two geometric constructions also argues for independent, externally predetermined values for loading height and buttress depth which had to be accommodated. A series of experiments, no matter how primitive, seem a reasonable explanation for the existence of such predetermined values. Certainly Rodrigo's position as master of the *fábricas* of Salamanca and Segovia put him in a position where he could use the nearly free labor of apprentices to conduct experiments under the guise of a training program. If this interpetation is correct, the experiments are the earliest recorded instances of any such investigation, and should be seen as landmarks in the development of experimental mechanics.

Structural Rule No. 7
Third Geometric Formula

The third geometric rule, exhibited in Chapter 6, 20r–21r, and in a drawing appended to Chapter 16, 59r, is a remarkably imaginative attempt to establish a general structural rule for buttress depths (Figs. 12 and 13). It is at once the most important and yet the most garbled and obscure passage in the entire *Compendio*. As a result it has been brushed aside in all previous discussions.[27] Careful reading shows the text to be divided in three distinct parts, and only in the third part is inarticulateness and confusion severe enough to render that section meaningless.

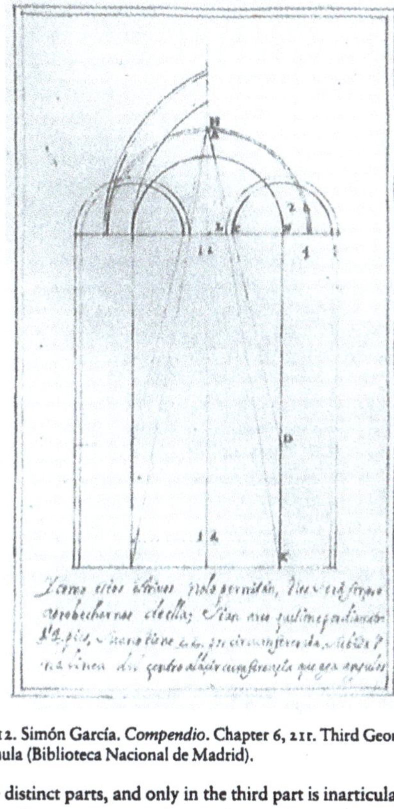

Fig. 12. Simón García. *Compendio*. Chapter 6, 21r. Third Geometric Formula (Biblioteca Nacional de Madrid).

In the first part Rodrigo designs a buttress for an arch with a voussoir depth one-sixth of its intrados diameter (Fig. 14, I).[28] This voussoir proportion is established "by analogy," which must refer to human proportioning, although which body parts are being used is unclear. In this example, given the voussoir depth and a preselected buttress depth, the buttress height is obtained. The construction requires that the intrados diameter be divided into three parts. The buttress depth is given by distance CE, which is 1/3. Thus the buttress depth to span ratio is 1:3. Next a line is drawn from the extrados crown at A through one of the two dividing points on the diameter to the intersection with the base of the pier at D. D defines the height of the pier, which will be 1 1/3 times the opening. Although at first this appears to be another specific formula applying to a special case,

26. The two buttress depths are obtained by assuming that the arc centered at K has either a radius KH, which gives the greater depth, or KF. Again the differences in buttress depth are nearly unnoticeable in a drawing, 1.8%.

27. Kubler, "Late Gothic Computation," footnote 19. Hoag, "Rodrigo Gil," 433.

28. Part I reads: Divide el diametro en 3 partes iguales y por quanto segun regla de analogia le biene un sesto, ponlo en superficie que supongo ser A. y C. tira de A. asta C. Y cortará el plomo paralelo en D. Y eso es el pie derecho. Y lo que ay de C. asta E. es la magnitud.

290 JSAH, XLI:4, DECEMBER 1982

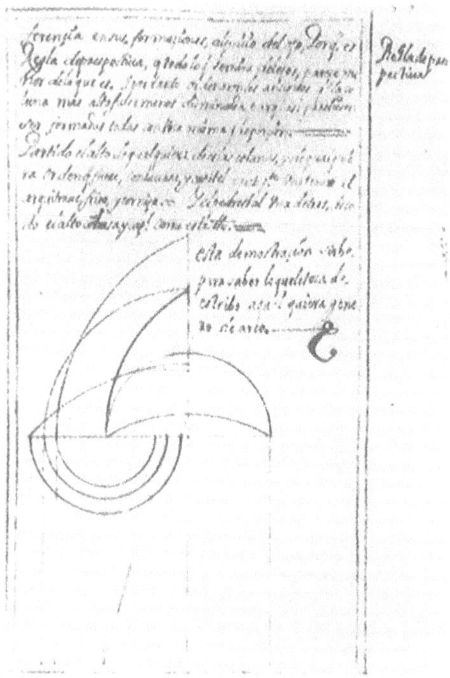

Fig. 13. Simón García. *Compendio.* Chapter 16, 59r. Generalization of Third Geometric Formula (Biblioteca Nacional de Madrid).

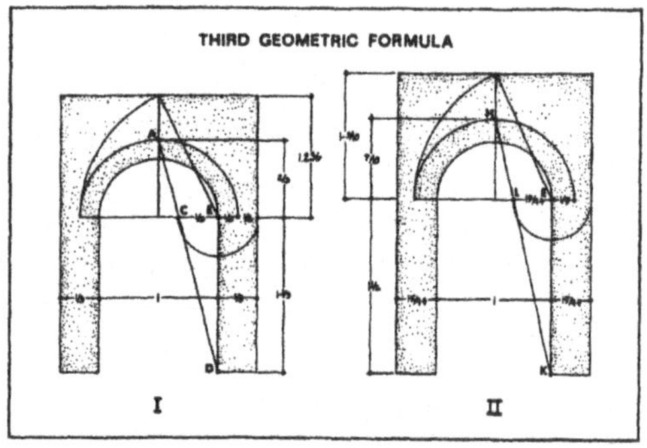

Fig. 14. Third Geometric Formula. Parts I and II (drawing by author).

the construction procedure can be generalized to other pier heights or voussoir depths, as is done in the second part. The allowable loading is established by drawing an arc with the center at an impost and the free end at the opposite extrados springing point. The intersection of this arc with the vertical axis of the arch gives the maximum loading line at a height of 1.236 arch openings above the impost (2.472 arch radii).

In the second part a different set of conditions is established (Fig. 14, II).[29] The proportions of the opening below the arch are arbitrarily fixed at a ratio of 2 parts to the diameter and 3 parts to the height. In the first part this proportion was established by the geometric construction. The voussoir thickness also is established arbitrarily, this time one-fifth of the arch opening. The buttress depth then is established by the same construction as in the first part. A line HK is drawn from the extrados crown to the pier base, intersecting the impost line at L. The distance from L to the nearer impost E is the buttress depth, which is $^{15}/_{44}$ of the arch opening. The loading height is established in the same way as in the previous example, and is 1.3 arch openings (2.6 radii) above the impost. In Simón García's diagram this second construction is superimposed upon the first one, and proportions are elongated (Fig. 12).

It is important to note that four parameters are related by this formula. These are 1) arch diameter, 2) voussoir depth, 3) buttress depth, and 4) buttress height. In order to find the value of one of these the other three must be given. However, if only two parameters are known, the formula gives a curve relating the other two uniquely. Thus, knowing arch diameter and voussoir depth, the formula gives a hyperbolic curve relating buttress height to buttress depth (Fig. 15). This curve is obtained by generalizing Rodrigo's method to every point along the outer face of the pier. The buttress depth approaches an asymptotic value of A/2 as the buttress height increases indefinitely. Although this hyperbolic profile is not explicitly discussed by Rodrigo, its generalization is implicit in the geometric construction.

In the third and last part of the discussion a very confused attempt is made to explain how the geometric rule exemplified in parts I and II could be generalized arithmetically to any buttress.[30] Here it is clear that neither Rodrigo nor Simón García fully understand the hyperbolic generalization discussed above.

29. Part II reads: Y si a de subir tanto y medio que es de 2 de hueco, 3 de alto, se le de la una quinta parte a la rosca que sera en H. pues tira desde H. hasta K. y cortará el diametro en L. Y lo que ay de L. a E. es lo que le cabe.

30. Part III reads: i) Y si mas sube se le de por la razón de la regla de 3 diciendo si doce pies que este arco tiene de hueco bienen a la rosca 2 de tanto que le bendrá ni mas ni menos será en la salida de este mismo estrivo,

ii) será por regla de 3 diciendo si de 12 de hueco bienen 4 y un ochavo de salida de estrivo, de tanto que bendrá esto es subiendo su cuadrado de pie derecho,

iii) mas si es que sube tanto y medio, como si tiene 12 de hueco 18 de pie derecho, tenga de rosca la una quinta parte que serán 2 pies y mas $^{2}/_{5}$, que será haciendo un pie 5 partes, y tomar las 2 de manera que le biene de los dichos 12 pies, 2 pies y $^{2}/_{5}$, esto es de rosca, y de aqui se sacará para otro y sabida la rosca será sabida la grandeza del estrivo, usando la regla de 3, y sabido el hueco y alto de pie derecho, por la dicha regla de 3 se sabrá la rosca, y estrivo.

This is undoubtedly the most problematical text encountered in the *Compendio*, and may be very corrupt, as Kubler and Hoag believe, although it may consist of very brief notes jotted down by Rodrigo as his ideas developed. Simón García in later chapters exhibits enough mathematical fluency to demonstrate that this very confused passage cannot be his own. The lack of an algebraic system is sorely missed throughout this section, as the rule of three (proportionality) is invoked continuously but weakly and without any understanding of the basic rules for balancing algebraic equations, in the intuitive suspicion that something like it may indeed work. This is not unreasonable, since algebraic knowledge still was rather esoteric in the 16th century. Had Rodrigo been able to make a successful translation of his geometric vision into algebraic language, he would have foreshadowed René Descartes' later invention of analytic geometry. However, such a level of mathematical sophistication was not found amongst architects at this time.

The third geometric formula is illustrated one more time by Simón García at the end of Chapter 16, 59r (Fig. 13). It is out of context and is obviously an afterthought. There is no descriptive text, but a caption under the figure states: "This demonstration serves to know the buttress depth of any kind of arch." Three constructions are superimposed upon one another for pointed, semicircular and low segmental arches. The constructions are in every case identical to those already discussed, except that the sloping construction line is drawn from the crown to the intrados rather than the extrados. This must be an error, although perhaps done intentionally to avoid a maze of criss-cross lines. The constructions to establish the height of the safe loading line are different from one another in that the arcs are centered about different points, although they are conceptually the same as those in parts 1 and 2.

It should be made clear that the third geometric formula is still far from being a modern structural formula. There is no implicit analysis of forces or strains, and the formula remains an almost magical recipe. Yet its remarkable sophistication lies in identifying all the major design variables in an arch-buttress system and integrating them in a bold although unfortunately still incorrect set of interrelations. According to the formula, increasing the voussoir depth is a major contribution to the stability of an arch, since it allows a thinner buttress. In fact broader voussoirs exert a greater thrust and would require a deeper buttress. The formula makes buttress depth directly related to buttress height as well as to arch span, similarly to the arithmetical rules 2, 3 and 4. But under static loading conditions the correct depth of a stable arch-pier system is a constant regardless of pier height, if only arch thrust and pier weight are considered. Despite these criticisms, the implied hyperbolic relationship between height and depth is an intelligent approximation to the correct catenary path of the centroid of forces of an arch. Since ideal static loading conditions are not likely to be encountered in real structures, the

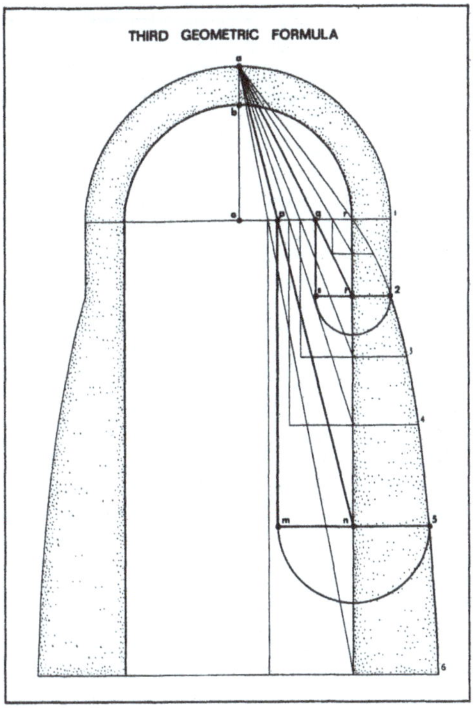

Fig. 15. Third Geometric Formula. Hyperbolic Generalization (drawing by author).

formula apparently treats the pier as an extension of the arch, which must respond to some bending forces, and thus must become broader towards its base.

The Third Geometric Formula can be viewed as an important speculative development, representing the most advanced synthesis of the formulaic tradition of the Late Gothic, as it merged with the Classical forms of the Renaissance. Even though many of its assumptions are wrong, the search for a practical, general solution to a difficult structural problem reveals an incipient attempt to postulate theoretical relationships between significant variables, and to integrate these relationships in a coherent, even eloquent, manner.

Conclusions

Rodrigo's formulae are clearly classifiable into two major formats, geometric and arithmetic. The geometric formulae fit in a medieval tradition that can be traced back continuously to the 11th century, which has antique sources of immemorial age. The arithmetic formulae are the result of the slow assimilation of a new mathematical vocabulary made available to masons and other craftsmen for the first time in the 14th-century Italian abacus schools.[31] The use of advanced arithmetic in masonry appears to spread outside Italy only in the 16th century, along with the taste for Classical forms. The formulae of Rodrigo are also divisible into those that apply to Gothic problems, characterized by rib vaults, and those that apply to a simplified version of the Renaissance semicircular arch.

One exception to both these distinctions is the 1:4 rule of thumb, which may be viewed as an old conservative verity that would have survived through centuries by word of mouth in the masonic circles. Its limited capacity to deal with a variety of essential variables, such as buttress height or voussoir depth, would make it no less a valuable friend when making preliminary design decisions on a floor plan.

It is interesting that formulae 2 and 3 which apply to Gothic forms are set in the more progressive arithmetic-algebraic format. It suggests Gothic builders were interested in keeping up with technical consequences of the new cultural exports from Italy. Formulae 5, 6 and 7, that apply to Renaissance forms, by contrast have a geometric format which is an advanced development of the old geometric Gothic tradition known through Magister II, the minutes of the Milan controversy, Roriczer, Schmuttermayer, and Lechler. This tradition of geometric structural formulae can be followed into the 17th century, when in the famous treatise on stereotomy of François Derand a similar formula was included.[32] Although elegant and apparently very adaptable, Derand's formula reveals a stunted development of geometric structural thinking, reducing as it does to the 1:4 rule of thumb, while ignoring the variables of voussoir depth and loading height. In Rodrigo's treatise only one formula, the fourth, follows at once Renaissance forms and the more progressive arithmetic format.

In the transitional world of Rodrigo Gil a need must have been felt to systematize an architectural art that was operating with two independent sets of rules, one Gothic and the other Classical. Rodrigo's systematic eclecticism allowed the two styles to be used together, and the mathematical technologies that accompanied them were shared just as easily.

We have seen that for Rodrigo there is a natural separation between aesthetic decisions, which are connected to theoretical and philosophical issues, and structural decisions, which are pragmatic engineering problems. The structural formulae are a clear expression of this separation, which may be more fundamental to the character of late medieval masonic tradition than the stylistic distinction between Gothic and Renaissance forms.

31. See Lon Shelby, "The Geometric Knowledge of the Medieval Master Masons," *Speculum*, XLVII, 1972, 395–421.

32. See R. P. François Derand, S.I., *L'architecture des voûtes ou l'art des traits et coupe des voûtes*, Paris, 1643, 16–17. Brief notices can be found in Hoag, "Rodrigo Gil," 433–434; and Mainstone, "Structural Theory and Design," 305–306, and fn. 39.

Nonetheless the structural rules coexist easily with the aesthetic ones in a system which encompasses the entire design process. Such a synthetic, comprehensive aim is Scholastic in spirit, akin to the grand systematizations of a philosopher like Francisco Suárez, or the erudition of Pedro Sánchez Ciruelo, and thus exhibits the Gothic drive for unity rather than the more modern fragmentation of knowledge and faith. Yet, as the philosophy of Suárez was at once rooted in the Middle Ages and able to inspire the late 17th-century rationalism of Leibniz, so the synthetic Late Gothic writings of Rodrigo reveal an incipient drive towards late 17th-century mechanistic rationalism.

One sign is the theoretical formulation of the problem of structural sufficiency, in Chapter 6, 19r–19v. Two conditions are imposed on the search for a rule giving buttress depth. The first is that the rule must give the correct buttress depth for any arch, which must mean any size or shape. Thus the first requirement is that the rule must be general in its application. The second condition is that the rule give a sufficient depth, that is, as much as is necessary to support the arch and no more. Even though each of these requirements remains unanalyzed and poorly articulated, the very expression of the technical ideal of structural sufficiency is part of the necessary groundwork for the development of a rigorous study of mechanics in engineering. The formulae themselves certainly do not meet the stated requirement of structural sufficiency, and on one occasion we have seen Rodrigo's implicit criticism of his own work.

A second, more controversial indication of an independent, progressive technical outlook is the possible experimentation which the analysis of the formulae has suggested. If specific structural experiments were indeed performed, they are among the first instances of such behavior. The very nature of the suggested experimental setup, simple free-standing semicircular arches, indicates a rather modern attempt to isolate a few variables to investigate their mutual relationships. This is similar to a procedure already present in Leonardo's own thought experiments.

Even the format of the resulting formulations, with their utilization of square roots and summations, their geometric complexity, and the multiplicity of variables being interrelated, points towards an increasing mathematical sophistication and learning. By recognizing that a larger variety of interior dimensions ranging from bay proportions to voussoir depth are significant to the sizing of structural members, Rodrigo's system exerts a more subtle control over the designer than earlier Classical or Gothic proportioning schemes. The designer in turn has more options with which to contend so that the range of theoretical design tools is appreciably expanded.

Rodrigo attempted a broad synthesis of the proportioning systems of the past. The Classical methods of Vitruvius, using fixed ratios, were extended by grafting onto them Dürer's comprehensive measurements of the ideal human body. Gothic geometric constructivism also was absorbed and related to Vitruvian human design. In addition, distinct technical formulations were introduced in an attempt to bridge the gap between the intuitive practice of Gothic construction and the structurally inflexible traditional proportioning systems.

For an ongoing enterprise such as Rodrigo's this effort to unify Gothic, Humanist and technical design methods had a practical purpose. The largest, most important buildings that Rodrigo was working on were begun in the Gothic style, which clearly was falling gradually out of favor. In order to complete these buildings harmoniously future architects would have to understand the possibility of an abstract common ground between the Gothic and the Classic styles. Rodrigo must have hoped that buildings such as the Cathedral of Salamanca could be completed by classicizing architects in full awareness of aesthetic and technical principles which he believed to be common to all architectural proportioning systems.

2

Designing the beam: from rules of thumb to calculations

Summary

Rules of thumb were used before 'scientific' methods of design and have been recorded in early carpenter's manuals. The sizes recommended for floor joists and beams present a diversity of rules and, during the change from oak to fir, show a poor understanding of structural behaviour, but comparison of the various tables suggests that the intention behind them all was to limit the liveliness of the floor. Experiments at the beginning of the nineteenth century provided the means for a more rational approach to the same problem, enabling designers to use different timbers and to size members in order to limit the deflection for given loads.

There is never an exact correspondence between the state of a technology and the science that it draws upon. The designer needs to be sure that what is built will perform adequately in service, and sometimes scientific understanding is sufficiently advanced to enable such predictions to be made. There are times when technology is developing in a way that shows our understanding to be inadequate, producing a pressure for an improved science. But there are times when there is sufficient informal understanding for designers' needs: an informal understanding that is often encapsulated in what are called 'rules of thumb'. Rules of thumb are used in two ways: when there is no scientific basis for design, they embody the collected experience of previous generations of craftsmen or designers and they are also used alongside formal methods of calculation, providing an "adequate" answer to a problem much more quickly than formal analysis. In this last form, the rule of thumb persists because it is prescriptive rather than descriptive. It tells the designer what size to use, rather than checking a guess, so that such rules remain an essential aid to design because they avoid excessive iteration. Indeed their ease of use ensures that they will survive unless proved either inadequate or uneconomic. We can therefore distinguish two sets of questions about the use of rules of thumb:–

i) what were the rules that were used by early designers, and is it possible to deduce from these the nature of the experience that generated them?
ii) what were the factors that led to their replacement by more formal methods, and what was the process by which this replacement came about?

The simple beam is a good example with which to explore these questions because, apart from the wall, it is the most basic structural device and the early carpenters' manuals provide a useful source as an indication of the way that its behaviour was understood. Moreover, it was a device that attracted the attention of scientists some time before any use was made of their findings by designers but, in spite of this, it can be argued that it was the needs of design that led to a complete understanding of beam behaviour.

In floor structures a series of beams must be arranged to provide a level surface within a limited depth. Boards rest on joists and joists span either between the walls or onto 'summers' or 'girders'. Until the 17th century traditional modest sized dwellings had used joists perpendicular to the main walls of the house, spanning onto summers that spanned longitudinally between end walls or intermediate frames (Fig. 1). This allowed the jettied

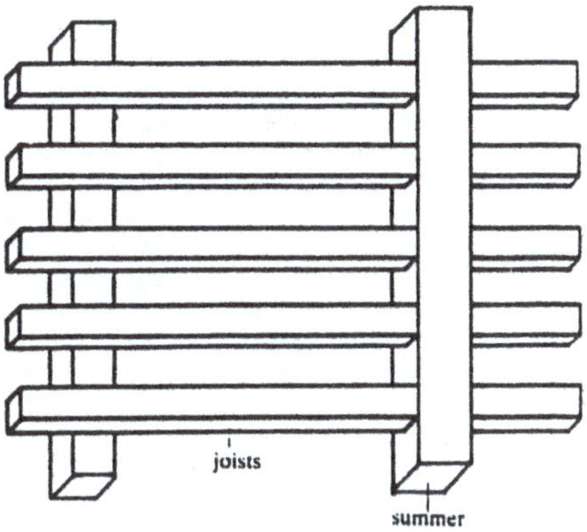

Fig. 1. Early floor construction showing jettied floor joists jettied beyond the head palte of the ground floor wall frame and connected into a 'summer' beam.

form of construction often associated with early timber frames because the joists, set above the head plate of the wall frame, could easily cantilever beyond it. With a central chimney this layout of timbers often resulted in one end of the summer being carried on the chimney breast, an arrangement that was eventually forbidden by the Acts for the rebuilding of the City of London after the Fire. Subsequent carpenters manuals show what appears to be more sensible arrangements with the girders (the 18th century term) which support the joists avoiding the chimney and with the joists trimmed round fire-place openings.

Common experience of timber members is that, when loaded, they deflect considerably before breaking and design normally seeks to limit deflection in order to prevent such effects as the cracking of finishes. Experience also shows that a rectangular beam better resists bending when set upright rather than when laid flat so that deep narrow members are the best shape to adopt for our

Fig. 2. Conversion of timber: the result of splitting a log along a straight line from which a shallow rectangular joist can be made by trimming off the excess.

floor structures. Perhaps surprisingly then the earliest floor joists were laid the other way round. During the medieval period the form used for floor joists and for the summers and girders that supported them was limited by the nature of the process of converting timbers from the round log. For larger beams the whole of the log was used in which case it would be trimmed with an axe or an adze to an approximately square section. For smaller beams, ie for joists, the log would be split with a beetle and wedge, and so halved. Since the log would normally have had some curve to it, the carpenter would turn it to select a straight line along which it could be split (Fig 2). Being curved in the other direction the split surface would form the top (or bottom) of the joist so that the floor was level, with the result that its proportions were roughly 2:1, the joist broad and shallow rather than deep and narrow.

Of course, for many applications it was necessary to saw timbers: floor boards are the obvious example in this context. For the larger spans that were needed in the great country houses of the Tudor and Elizabethan period, narrow and deep joists were made by sawing from larger sections. At Woolaton, Northamptonshire, for example, joists 3" × 18" deep span 12ft. Sawn timbers became common for joists after imported fir began to be used in large quantities following the Fire of London but before that the choice between halved and sawn timbers presumably reflected the relative costs of conversion.

There is some evidence that sawn joists were coming into common use by the 17th century. Moxon (1677), who in his *Mechanicks Exercises* described the work of several trades, discussed the carpentry of house construction and offered an argument for deep rather than shallow joists:-

> *"The joists are framed so as to lie with one of their narrowest sides uppermost. . . the reason is because the stuff of the bressumers and girders are less weakened by cutting the mortices in them in this position" and "The tennants (he means mortices) are cut through the two narrowest sides rather than between the two broadest sides because the stuff of the girders retains more strength when least of the grain of the stuff is cut"*

Clearly he would not have been able to make this recommendation if sawn timbers had not been available, but it is interesting to note that the reasons given for his preference are because of the weakening effect of mortices in the supporting girder, and not because of the more efficient shape of the joist – indeed for the same consideration he recommends that the girders be broad and shallow.

Moxon was writing shortly after the Fire of London and the Acts to control the rebuilding of the city after this disaster (the embrionic London Building Regulations) included rules for the sizing of structural members and are the first record that we have of the relation between the size and the span of a beam. Of the different classes of building covered by the acts, for the smaller and least important the table of sizes only limited the span of the girders and joists, and distinguished between normal floors and those in garrets. A single table of sizes covered the two larger classes of building, providing for a range of spans in the girders, but assuming that the joists would all span 10ft (Table 1). The scientist/

TABLE 1 The Act for the Rebuilding of London

Scantlings of girders		Joists bearing 10 feet
from 10 ft to 15 ft	11 × 8 ins	3 × 6 ins
15 ft to 18 ft	13 × 9 ins	3 × 7 ins
18 ft to 21 ft	14 × 10 ins	3 × 7 ins
21 ft to 24 ft	16 × 12 ins	3 × 8 ins
24 ft to 26 ft	17 × 14 ins	3 × 8 ins

architects Hooke and Wren were among the Commissioners responsible for the drafting of the Acts but it would be unreasonable to expect a scientific approach to the problem of member sizing given the pressures of the occasion; more likely, they simply adopted whatever was the accepted practice of the time and this seems to be confirmed by Moxon, who recorded that they were *"well consulted by able workmen before being reduced into an Act"* These tables of scantlings for the second and third kinds of building were then reproduced by Moxon, Leybourne (1668), a writer of several books for surveyors, and Neve (1703), who produced the first dictionary of building. These figures may of course simply represent London practice while other rules may have been used elsewhere, but both the attraction to London of carpenters for the rebuilding work after the Fire and the wide publication of Neve's book, would have helped to disseminate the London methods to other parts of the country during the early part of the 18th century.

It was not until the mid-18th century that building manuals which included structural details began to appear in large numbers. These included tables of scantlings for a much greater range of spans than those covered by the London rebuilding acts. However, some care needs to be exercised in their interpretation. Nicholson (1812) was eventually to point out the errors in many of these books and I have discussed elsewhere their varying reliability (Yeomans, 1986). Even Francis Price (1733), who wrote one of the earlier and better manuals, has to be approached with caution because, although he was a trained builder, his table of structural sizes has its faults. He distinguished two types of building, large and small, although we may perhaps assume that the large buildings would be of higher quality. He also distinguished between the two structural materials, oak and fir, that were in common use by that time, but seems to have misunderstood the effect of using the different timbers (Table 2).

TABLE 2 **Francis Price, Carpenters' Companion, 1733**

	Small Buildings		Large Buildings	
Girders	Fir ins	Oak ins	Fir ins	Oak ins
16 ft span	8 × 11	10 × 13	9·5 × 13	12 × 14
20 ft span	10 × 12·5	12 × 14	12 × 14	15 × 15
24 ft span	12 × 14	14 × 15	13·5 × 15	18 × 16
Common joists				
6 ft span	6 × 3	5 × 3	5 × 3	6 × 3
9 ft span	9 × 3	7·5 × 3	7·5 × 3	9 × 3
12 ft span	12 × 3	10 × 3	10 × 3	12 × 3
Bridging joists				
6 ft span	4 × 2·5	4 × 3	4 × 3	5 × 3·5
8 ft span	5 × 2·25	5·5 × 3	5·5 × 3	6·5 × 3·5
10 ft span	6 × 3	7 × 3	7 × 3	8 × 3·5

The sizes that he gives for fir members are smaller than those for oak whereas, given the relative strengths of the two materials, we should expect the reverse to be true. But at the time it was their relative weights that were regarded as more significant, oak being heavier than fir – Batty Langley (1736) remarking that;-

> *"altho' oak is much stronger than fir, yet it is of greater specific gravity, it must therefore have larger scantlings for the same purposes than fir which is weaker . . .".*

Taking the recommended sizes for joists spanning 20ft (6m), the self weight of the floor, including the boarding (but without any allowance for the weight of the ceiling below it) amounts to something like 18 lbs/sq ft. Although this compares with the present 30 lbs/sq ft design load for domestic buildings, in practice the imposed load would have rarely exceeded 7 lbs/sq ft so that self weight would have been the principal load. This would have been particularly significant for the girders.

By this time there were some improvements in floor construction. The simple arrangement of girders and

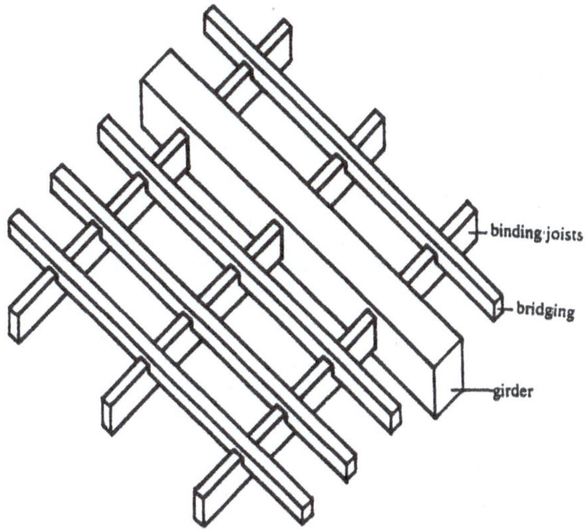

Fig. 3. Eighteenth century floor of binding joists and bridgings.

common joists was still being used but for the better quality buildings *bridging floors* might be used instead (Fig 3). These used larger joists (Price says that they should be 1½ times the width of common joists although of the same depth) but they were placed wider apart with smaller bridgings spanning across them. The effect would have been to spread the loads over more members and so provide a stiffer floor which was presumably the reason for its adoption.

Batty Langley was one of the less reliable authors of the period if we are to judge by some of his curious ideas about roof construction and the scarfing of timbers. In his *Builders' Jewel* (1741) he included an apparently realistic description of floors with a table of sizes with recommended dimensions slightly smaller throughout than those given by the Rebuilding Acts for equivalent spans. This was a pocket sized work and was likely to have been more influential than his encyclopaedic and much larger *Ancient Masonry,* both because of its size and relative cost, and appeared in a great many editions until late in the century. The most difficult author of the period to deal with is William Pain who advertised himself as a builder and carpenter and, like Francis Price, was involved in the daily problems of building. He produced several books between 1758 and 1793 which, while generally reliable, contain tables for structural timbers that differ from each other. Like Price he also gives sizes for both oak and fir in some of his books although not in others. The question is whether or not the differences in the tables in the various books indicate a basic uncertainty on Pain's part, or whether increasing experience over this long period caused him to reconsider some of the sizes that he was using.

What kind of rules might we expect to see? Before metrication the rule used in Britain for sizing the depth of 2" wide domestic floor joists was "half the span in inches plus two" – so for a 10 foot span, half of 10 is five plus two gave a 7" × 2" joist. Clearly this is the kind of easy to remember rule that we would be looking for in the tables and some of them do indeed show this kind of pattern. We should look for it in vain in the Rebuilding Acts. There is a proprotionally greater increase in the scantlings for larger spans. The curious feature of the tables in the act is that the scantling of the joists is increased with the size of the girder even though their span was assumed to remain constant at 10ft. It was this feature of the tables that Leybourne altered. For oak joists, Francis Price has a simple rule making the depth in inches equal to the span in feet for larger buildings, while for smaller buildings he has a difference of 2½" for every 3ft difference in span for oak and 1½" for a 3ft difference for fir. The pattern was completed by making oak joists of small buildings the same as fir joists of larger buildings. For oak girders there was a similar pattern with a 2" difference in one dimension and 3" in the other for every 4ft difference in span although one cannot find any such simple rule in the table for fir girders.

Pain's *Builders' Companion* (1758) has only one class of building but does show both timbers. The tables also have a similar logic to those of Price with improvements that the relative sizes for the two timbers are now the right way round (Table 3). Presumably by the time that

TABLE 3 **William Pain, Builders' Companion 1758, 1765, 1768**

Girders	Fir	Oak
16 ft span	11 × 13 ins	10 × 13 ins
20 ft span	13 × 15 ins	12 × 14 ins
24 ft span	15 × 17 ins	14 × 15 ins
Common joists		
6 ft span	6 × 3 ins	5 × 3 ins
9 ft span	9 × 3 ins	7·5 × 3 ins
12 ft span	12 × 3 ins	10 × 3 ins
Binding joists		
6 ft span	5 × 3.5 ins	4 × 3 ins
8 ft span	6 × 3·5 ins	5·5 × 3 ins
10 ft span	8 × 3·5 ins	7 × 3 ins

this was published a greater experience in the use of fir would have led to an appreciation that it was not as stiff as oak, indeed by 1774 the distinction between the two timbers had been dropped (Pain, 1774) – perhaps because of a reduction in the use of oak in building – and the obvious logic of the tables has also been lost. Table 4 shows the increments in girder sizes for 2ft increments in span up to 20ft. This table was repeated until 1793 but there is some confusion created by a table of 1781 (Table 5) with smaller sizes given for both girders and joists (Pain 1781).

The impression given by these tables is that, although their authors may have wished to have some easy to remember pattern of sizes, they were still going through a process of trial and error; a process made more difficult by the variability of the material. This is suggested particularly by the riders that both Price and Pain added to

TABLE 4 **William Pain, Practical Builder, 1774, 1789, 1793**

Girders		
10 ft span	10 × 9 ins	
12 ft span	11 × 10 ins	
14 ft span	12 × 11 ins	
16 ft span	12 × 13 ins	
18 ft span	13 × 14 ins	
20 ft span	15 × 14 ins	
24 ft span	15 × 16 ins	
30 ft span	16 × 18 ins	
Joists	**Common**	**Binding**
6 ft span	6 × 3 ins	6 × 4 ins
8 ft span		7 × 4 ins
9 ft span	9 × 3 ins	8 × 4 ins
12 ft span	12 × 4 ins	10 × 5 ins

TABLE 5 **William Pain, Builders' Pocket Directory, 1781**

Girders		
10 ft span	10 × 9 ins	
12 ft span	10 × 8 ins	
16 ft span	11 × 10 ins	
20 ft span	13 × 12 ins	
24 ft span	14 × 13 ins	
32 ft span	18 × 16 ins	
Joists	**Common**	**Binding**
6 ft span	6 × 2·5 ins	6 × 4 ins
8 ft span	7 × 2·5 ins	7 × 4 ins
10 ft span	8 × 2·5 ins	8 × 4·5 ins
12 ft span	10 × 2·5 ins	10 × 5 ins
14 ft span	10 × 3 ins	12 × 5 ins

their tables declaring that they did not "*insist that the scantlings of timber ought to be exactly as by this table is expressed*" *but that* "*they may be varied in some respects, as the workmen shall see fit*" (Price 1753 & Pain 1758) Batty Langley (1736) again explains this pointing out that:-

> "*. . . as timber is very different in its strength it is not possible to assign certain scantlings, and therefore they are to be varied at the direction of the carpenter*".

Both the assimilation of fir as a structural material and the appearance of new types of building would have required some learning by carpenters and an adjustment of their ideas. In the early 18th century carpenters would seldom have been concerned with other than domestic floors and occasional public buildings but by the end of the century more might be dealing with such buildings as warehouses and mills with their greater floor loadings.

Apart from a rule to increase the size of timbers with the span, carpenters would have needed some means of judging the overall strength of their timbers. What then was this basis upon which they were assessing the adequacy of the sizes that they were using? Clearly it could not have been the stress level in the timbers because there was no general understanding of bending stresses. At the time the only observable effect on a loaded span would have been the deflection which we know today is governed by the second moment of area for a given span. If these are plotted for the 18th century tables (Fig 4), the stiffnesses are seen to be greater than for present day joists with some trend towards lighter sections. However, the stiffness was increased rather more rapidly with span than our modern rule and the deflections that would have occured under normal loading would have been very small for joists although perhaps noticeable in girders.

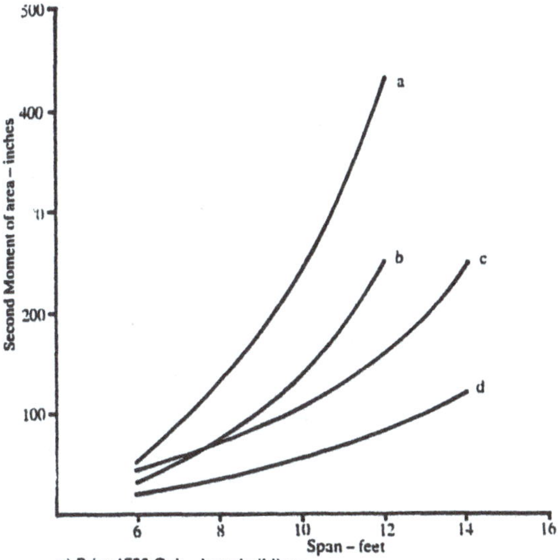

a) Price 1733 Oak – large buildings
b) Price 1733 Fir – small buildings
c) Pain 1781
d) Modern rule

Fig. 4. Graph of second moments of area.

It seems to have been common practice to cut girders to a camber, ie higher at their centre. Langley, for example, recommended 1" for every 20 ft span (Langley, 1741). Long spanning girders were also trussed for the same purpose, ie they had trussing pieces inserted into them and tightened with wedges (Fig 5) to give them some precamber (Dawes and Yeomans, 1985). For the sizes of girders recommended the elastic dead load deflection of a 20 ft span girder would be of the order of 1/3", hardly noticeable one might suppose. However, one must remember that at the time the process of construction involved building the timbers into the walls as construction progressed so that when they were eventually boarded the moisture content of the timbers might still be relatively high. Because of this the deflection might well have been doubled by creep deflection in the timbers, particularly if they were loaded with building materials during construction. The loading of the timbers at that time would comprise their own weight, the weight of the boarding and ceiling and, in some buildings, sound proofing materials placed in the space between the ceiling and floor.

While limiting observed deflection might possibly explain the sizing of girders the deflection of a joist under load would have been much too small to have been noticed even though the trend of the graphs still suggests this as the criterion used. But it is not the deflection under the static live load that we notice in a floor, rather it is its 'liveliness', its spring as we walk across it. Moxon, in discussing the cutting of mortices in a girder, is anxious that they should not unduly weaken the member *'lest the whole floor dance'* (Moxon, 1677) and we may surely assume that it was this effect that determined the carpenters' choice of sizes. This would account for larger sections being used in the better quality buildings, since better quality would be associated with a less lively floor, and it would also explain the move towards bridging floors in the better quality 18th century buildings. Today sizes are governed by limiting deflection to a fixed ratio of the span which is sufficiently ensured by the rule of thumb quoted above. This will result in increasingly noticeable deflections and an increasing liveliness as the span is increased. It is not surprising, thefore, that a comparison of the modern rule with the earlier sizes shows that while they are comparable at the smallest span, the stiffness was greater in the past for the longer spans, the tendency toward a reduction of sizes during the 18th century possibly indicating a search for economies.

The first attempt to introduce some scientific approach to the sizing of beams in a carpenters' manual was made by Peter Nicholson (1792) at the end of the 18th century although the basic principles upon which he drew had been published some time before. In Britain, an account of the relative strength of beams first appeared in the *Builder's Dictionary* in 1734 which reported on a paper by Parent in *The Memoirs of the French Academy* of 1708 and which correctly stated that the load carrying capacity of a beam was in direct proportion to the width and the square of its depth. However, there is no indication that these ideas were ever put into practice and they were completely ignored by the authors of subsequent builders' manuals in England

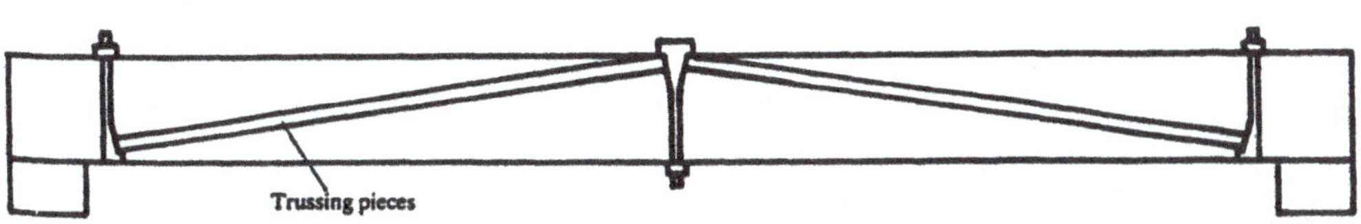

Fig. 5. Trussed girder.

until Peter Nicholson's *Carpenters' New Guide* of 1792. There is nothing to suggest that at this time Nicholson was responding to any need by builders for a more scientific approach. He was widely read in architecture, science and mathematics and made original contributions to the geometry of building, so that his approach to the sizing of timbers was probably just an example of his general interest in science and mathematics and perhaps fortuitous that this was shortly to become of interest to others.

Instead of producing a table of scantlings, Nicholson showed how to calculate the relative strength of beams and gave the results of some of the early experimental work that had been carried out on the Continent, but his calculations would have been of little help to the practicing carpenter, who needed rules for detemining sizes in real situations rather than an academic discourse on relative strengths. What prevented Nicholson from producing a more satisfactory treatment of the beam in the early editions of his books was the paucity of experimental data on which designs could be based. The relatively small number of experiments and the small sizes of the test pieces that had been used by early workers hardly provided a sound basis for extrapolation to real timbers with their natural defects as Nicholson, with his practical training, must have been aware.

This was solved by Barlow who published his *Essay on the strength of timber* (Barlow, 1817) providing figures for the deflections and breaking loads of cantilevers and simply supported beams for a wide variety of timbers based upon experiments carried out on realistic member sizes. He carried out these experiments with the express intention of being able to 'furnish . . . practical rules' for 'practical men'. His specimens were supported either as cantilevers or as beams, both simply supported and encastre beams. Using firstly Riga Fir he tested a range of sizes, up to two inches square, over a 6 ft span and loaded at the centre, measuring both the deflection at various loads and the breaking load. A careful selection of the timbers for the experiments was made so that they should be free of defects and Barlow was aware that these would, therefore, be stronger than timbers used in practice. By weighing his specimens he was able to deduce from the experiments that the strength of the timber was proportional to its density but was able to offer no explanation for this, particularly as it was contrary to his expectations. He then carried out similar experiments on a range of other timbers, principally hardwoods, both home grown and imported. Experiments of interest to the Admiralty were also carried out on triangular section timber and on bent timbers but as these were related to problems of ship-building they are outside my concern here.

The effect of Barlow's work was immediate and the editions of Peter Nicholson's books that followed drew extensively on Barlow's results (Nicholson, 1823) and included similar tables, whilst Tredgold (1820), whose standard work on carpentry first appeared a couple of years after Barlow's, included additional experimental results some of which were taken from his own tests. How then were these experimental results to be used? The relationship between the deflection and breaking load, the span, and the dimensions of the beam were already known. The importance of these early 19th century experiments was that they established the constants of proportionality for different timbers. When a simply supported beam was loaded in the middle the breaking load was given by $W=4ad^2S/l$ and the deflection by $\delta=Wl^3/16ad^3E$, where 'a' and 'd' are the breadth and depth of the beam. The experimental results gave the values of S and E for different timbers.

Manuals clearly show the significance of deflection in design with Tredgold suggesting a limit of 1/40th of an inch for every foot of span, and although both Nicholson and Tredgold, having dealt with deflection, go on to consider breaking loads, Tredgold in particular presents this very much as an after-thought. The rules that he gives for deflection are for finding the size of the beam when either the breadth or the depth is first given and so are in a form that a designer would wish to use. The rules for calculating absolute strength are presented in a form for calculating the breaking load once the size of the beam is known – implicitly it is deflection that is to govern design. Nicholson and Tredgold presented their tables slightly differently, the former reducing all his dimensions to inches, while the latter assumed that the spans would be measured in feet. Thus the coefficients given by each have different values, with Nicholson's strictly the more mathematically correct, while Tredgold's were better suited to the carpenter's measurement (Table 6).

TABLE 6

Peter Nicholson, 1823	
For ultimate deflection	$U = l^2/d\Delta$
Deflection given for a given load	$E = l^3W/\delta bd^2$
Ultimate load	$S = lW/4bd^2$
(where S is resistance for 1" sq)	
Thomas Tredgold, 1840	
Depth of girders and joists	
$= 3\sqrt{\{(\text{length in feet}) \times 2/(\text{breadth in inches})\}} \times K$	
Where K = 2·2 for fir joists	
2·3 for oak joists	
4·2 for fir girders	
4·3 for oak girders	

Parent's work, referred to above, was part of the scientific attempts to understand the behaviour of beams that began with Gallileo in 1638 and were not completed until Navier in 1826. The story of these attempts has been well covered elsewhere (Heyman, 1972) and it is sufficient to note here that, from a practical point of view, the weakness of early experimental work was that it did not deal with timber as it was used but tended to look at small samples of a variety of materials. Since almost all timber beams and joists were rectangular all that was required from a practical point of view was to know how sizes related to span and load for a variety of timbers and this knowledge was adequately provided by the published tables. When Barlow carried out his experiments, the relationship between the behaviour of materials in direct tension and compression, and their behaviour in bending, was still not well understood and Tredgold's 1840 edition of his *Elementary Principles of Carpentry* is still vague about the position of the neutral axis in a beam. But a complete understanding was not important. Barlow's experiments were on pieces of timber sufficiently large and of sufficient span to be capable of direct application to practical problems. The design of beams had moved from a craft base to an empirical base which was clearly sufficient for timber design and, given the persistence of rules of thumb and the tables of scantling which are provided in our current building regulations, is still quite adequate for the craft based building industry. A full scientific understanding, relating bending behaviour to the properties of a material in direct strain, only became necessary with the use of iron as a structural material and the wide variety of shapes that became possible by casting.

REFERENCES

1. *Builders Dictionary*, 1734, London, under Beam.
2. Barlow, Peter, 1817, *An Essay on the Strength of Timber*, London.
3. Dawes, M & Yeomans, D.T., 1985, 'Timbers trussed girders' *The Structural Engineer* **63a:** No. 5 (May), 147-154.
4. Heyman, Jaques, 1972, *Coulomb's Memoir on Statics*, Cambridge, 89-109.
5. Langley, Batty, 1736, *Ancient Masonry*, London.
6. Langley, Batty, 1741, *The Builders' Jewel*, London.
7. Leybourne, William, 1668, *A Platform for Purchasers*, London.
8. Moxon, 1677, *Mechanics Exercises*, London.
9. Neve, Richard, 1703, *The City and Country Purchaser and Builders' Dictionary*, London.
10. Nicholson, Peter, 1792, *New Carpenter's Guide*, London.
11. Nicholson, Peter, 1812, *Architectural Dictionary*, London.
12. Nicholson, Peter, 1823, *Builders' New Director*, London.
13. Pain, William, 1758, *Builders Companion*, London.
14. Pain, William, 1774, *Practical Builder or Workman's General Assistant*, London.
15. Pain, William, 1781, *Builders' Pocket Directory*,
16. Price, Francis, 1733, *A treatise on carpentry*, London.
17. Price, Francis, 1753, *A British Carpenter*, London.
18. Tredgold, Thomas, 1820, *Elementary Principles of Carpentry*, London.
19. Yeomans, D.T., 1986, 'Early Carpenters' Manuals' *Construction History*, **2:** 13-33.

3
History of retaining wall design

J. Kerisel

Being always at the mercy of a new discovery, we shall never know for sure when the first retaining wall was constructed. As soon as man became sedentary, in fact, which was towards the end of the tenth millennium B.C., he set about building defences by raising obstacles that were difficult to climb. These defensive constructions were followed by monuments with a religious purpose inspired by the myth of the ascension towards the heavens.

It would appear that the first examples of the idea of providing a vertical support for a mass of materials are to be found among the megalithic monuments on the Atlantic coast of Europe: a typical example is the Newgrange Cairn in Ireland, erected in the fourth millennium B.C. (Refs. 1 - 3). It was shaped like a drum with a diameter of 82 m (Fig. 1) and vertical sides of 4.20 m high formed of upright raised stones planted side by side in the earth up to a height of 1.20 m and surmounted by a masonry without mortar. This retaining structure enclosed a mass of rounded stones taken from the bed of a nearby river and resisted for just over 500 years.

Builders were quickly made aware of the pressures at work on their walls and sought ways to reduce them. I see evidence of such efforts in two very different techniques to be found in monuments constructed thousands of kilometres apart as early as the fourth and third millennia B.C.: one is the insertion of horizontal mats and the other the technique of compartments.

Insertion of horizontal mats

I see the first use of this technique in the already mentioned Newgrange Cairn (Fig. 1). Between the layers of rounded stones the Irish builders laid mats of grass turf side by side. As we all know, pieces of turf like this weld together rapidly. Some experiments in traction (Fig. 2) based on mats of this kind and some rapid calculations have shown that they are capable of considerably reducing the pressure on the wall of the circular retaining structure which, without them, would have quickly collapsed under the lateral thrust of the rounded stones (Ref. 3).

Similar mats were also used in a few Mesopotamian ziggurats which time has spared. Ziggurats are heavy monuments shaped like superimposed coaxial truncated pyramids (Fig. 6) and are composed of successive courses of raw bricks made of silty clay and partially dried in the sun. The first edifices of this type were built at Sumer on the soft soil of the delta; they punched the

RETAINING STRUCTURES

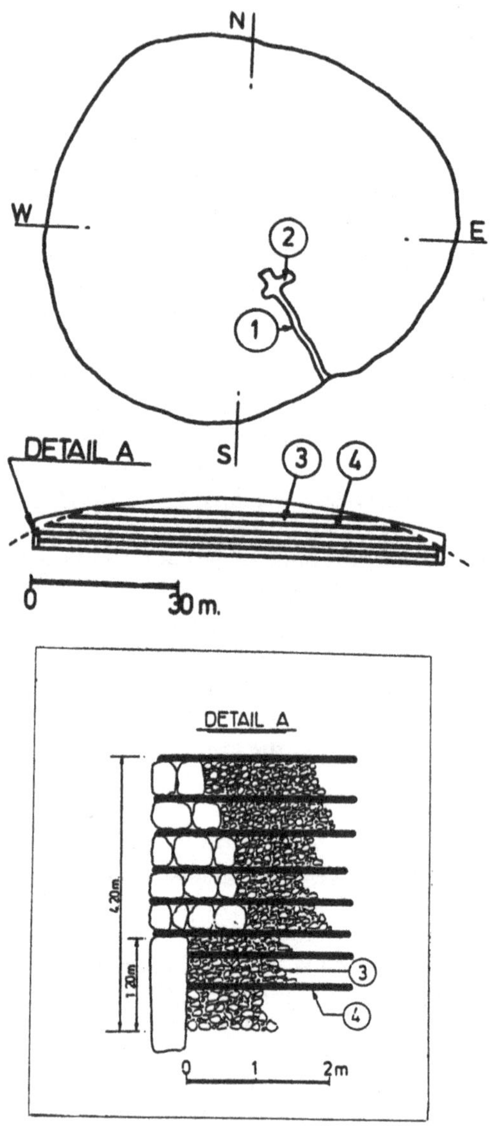

Fig. 1. Plan and cross-section of the Newgrange Cairn (3200 BC): 1 grave passage; 2 chamber (dotted line shows final shape after eventual overturning of retaining wall shown by detail A; 3 small rounded stones; 4 layers of turf

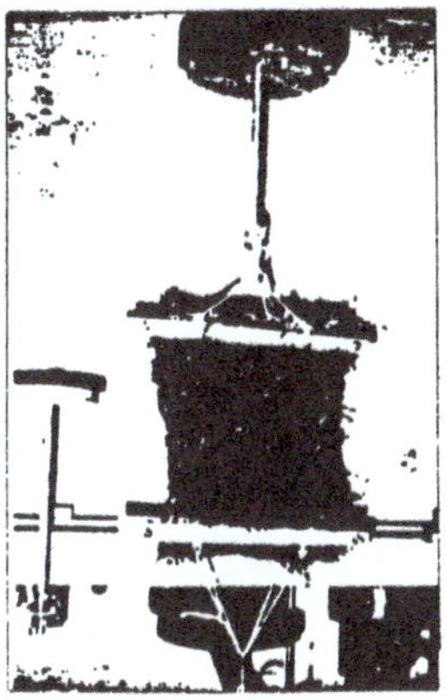

Fig. 2. Laboratory test to measure tensile strength of a square of turf (Simecsol, Paris)

soil and collapsed outwards. Someone then had the idea of inserting a mat of woven reeds laid in a bed of sand every five or six courses of brickwork. The mat neutralised the outward thrust and the sand drained away the water exuded by the raw bricks as the structure rose higher and higher. Cables made from woven plant fibres have also been found in the beds of sand.

System of compartments

The materials liable to exert outward pressure are placed in 'compartments', the walls of which form a grid. An example of this approach is the cairn of Gavrinis in the south of Brittany: the walls of the compartments are made of stones which, unlike the stones in the compartments, appear to have

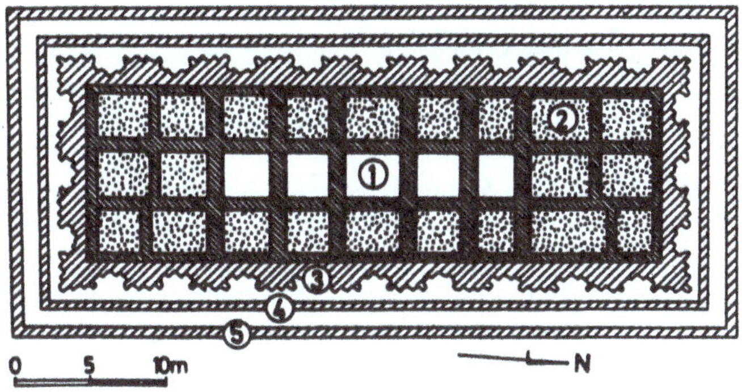

Fig. 3. Plan of tomb 3357 at Saqqarah (after Emery), dating from about 3000 BC: 1 subterranean burial chamber; 2 sand and small stones; 3 thick retaining walls; 4, 5 steps

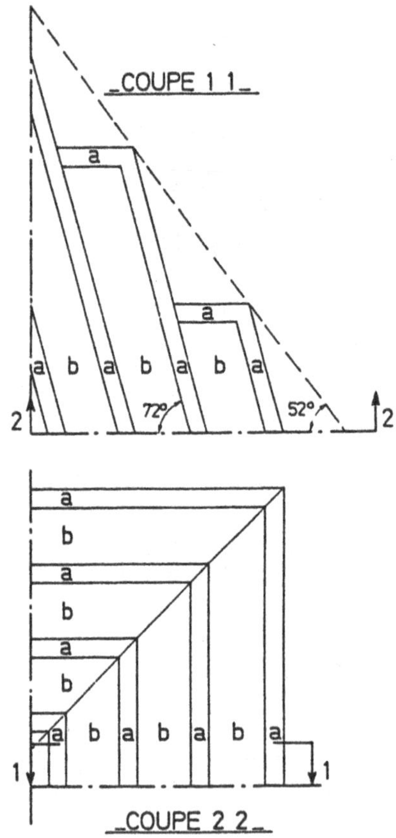

Fig. 4. First stone pyramids; principle of internal retaining walls invented by Imhotep at about 2700 BC: a walls dressed in stone; b chippings of stonecutters

been shaped with some care. We find the same technique in the mastabas, the tombs of important Egyptian dignitaries of the 1st and 2nd dynasties (Ref. 4): here the walls are of brick and the compartments filled with rubble and sand (Fig. 3). The height of these structures is still modest, only a few metres, but it was clearly this idea of compartments that later inspired Imhotep, the renowned builder of the pyramid of Saqqarah, the first major stone pyramid in history and about 60 m high. From the system of compartments Imhotep kept only the idea of walls that formed squares on the ground oriented NS and EW. These squares are centred on the centre of the pyramid, one inside the other, with their walls 5 Egyptian cubits (2.60 m) apart (Ref. 3 and Fig. 4).

Fig. 5. Step pyramid of Saqqarah; countermarches are simply tops of retaining walls

Fig. 6. The ziggurat of Ur, built about 2100 BC; ground level dimensions 62m x 42m; probable height 21m; the third level probably bore a temple

RETAINING STRUCTURES

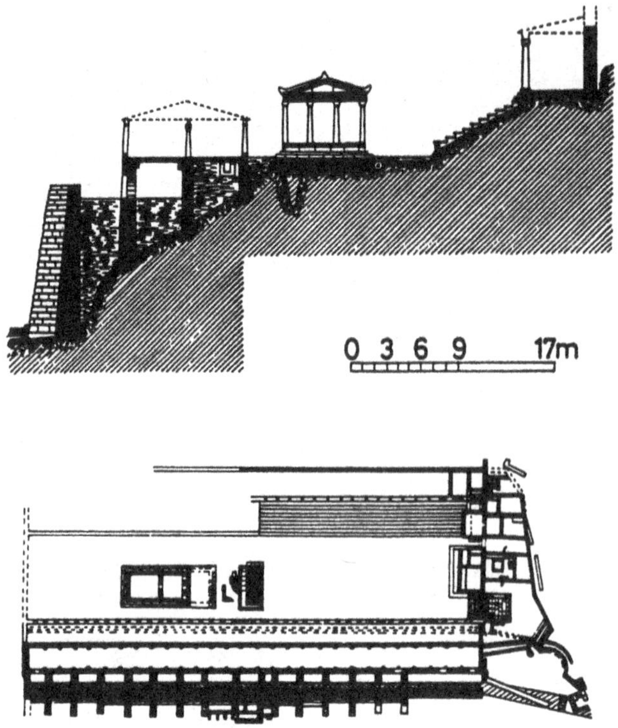

Fig. 7. Vertical and horizontal sections of retaining wall for the terrace of the Temple of Demester at Pergamon, 2nd century BC

We may note in passing that the intervals used for Meidum, a later pyramid, were 10 cubits (5.20 m). To prevent overturning the walls were inclined at 72° inwards; they were constructed of dressed stone and the spaces between them filled with the chippings of the stone cutters. The use of waste materials also economised on manpower (at a time when stone cutting tools were still very primitive). The pyramid (Fig. 5) has survived for nearly five thousand years and now stands as evidence of the intelligence of the first Egyptian builders.

Despite its ingeniousness, the technique of internal parallel walls imagined by Imhotep was used only in a few Egyptian pyramids. Quite quickly builders thought of constructing external buttresses as a means of preventing the retaining walls from overturning. Of this type is the wall of the terrace supporting the Temple of Demeter in Pergamon (2nd century B.C.), one of the finest retaining structures of the Ancient World (Fig. 7). The wall was 14 m high, 80 m long and supported by 15 external buttresses set just over 5 m apart.

KERISEL

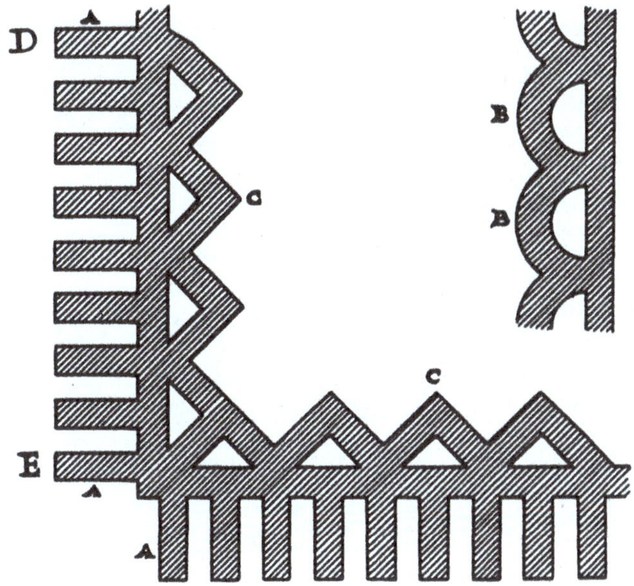

Fig. 8. The teaching of Vitruvius on the structure of walls; illustration by Perrault of the combs proposed by Vitruvius

As civilisations developed, more and more retaining walls were needed for fortifications. To withstand the onslaught of increasingly powerful weapons (catapults, battering-rams, etc) the walls were built higher and higher. The biggest mobile siege towers were about 60 metres high. The very vulnerable external counterforts were then replaced by internal ones or by much thicker walls. The treatises on fortifications of the School of Alexandria (Philo of Byzantium) lay a great deal of emphasis on the need for the thickness of walls to be in proportion to their height.

However, the first important book on retaining structures remains Vitruvius' De Re Architectura (Refs 5 and 6). The use of internal buttresses is commented on at length: 'A series of supplementary walls should be built... to form the shape of the teeth of a saw or of a comb: by this means the earth is broken up into compartments and cannot push on the wall with such a great force' (Fig. 8). The ancients had also found ways of solving other problems relating to temporary retaining walls. We find evidence of this in De Re Architectura. Chapter 12 of Book V is entitled 'De portibus et struturis in aqua faciendis' - 'On building ports and other structures in the water'. For the construction of moles Vitruvius proposed alternative methods, one of which consisted of the construction of a double-walled cofferdam. Between the two walls, reed sacks filled with clay were let down to the bottom and the

RETAINING STRUCTURES

Fig. 9. Illustration for Vitruvius' treatise. Upper scene shows how the seabed near the shore was pumped dry inside a double-walled cofferdam; the space between the two timbered walls was filled with sacks of clay. In the lower scene the cavity is being filled with stone and pozzolana

inside of the cofferdam was then pumped dry with wheels and drums on the screw principle (Fig. 9).

In the middle ages and during the Renaissance there were a great many treatises on fortifications, one example being the study by Francisco di Giorgo in the 15th century. In 1684 Marshal de Vauban published (Ref. 7) a work entitled 'Table explaining the measures ... for building walls 10 to 80 feet high...based on the experience of the construction of 500,000 cubic fathoms (3,700,000 m^3) stonework in 150 forts, built on the orders of the King'. Vauban was in mid-career and, having so far encountered but a fairly limited range of soils, believed that standardisation was possible. He recommended the use of walls with internal counterforts with a centre to centre spacing of 5.40 m. The base of the external facade of the wall itself was slightly set back from the front of the foundations, an arrangement that provided a more effective resistance to stresses. The thickness of the walls at the base, as a function of the height H, was 0.20H + 1.48 (in metres) but the length of the counterforts added to the thickness of the wall came to almost half the height H.

The depth of the foundations is not given. Later, after having constructed fortifications in the north of France where the ditches in front of the walls cut into Sparnacian clay (from a geological level corresponding to London clay

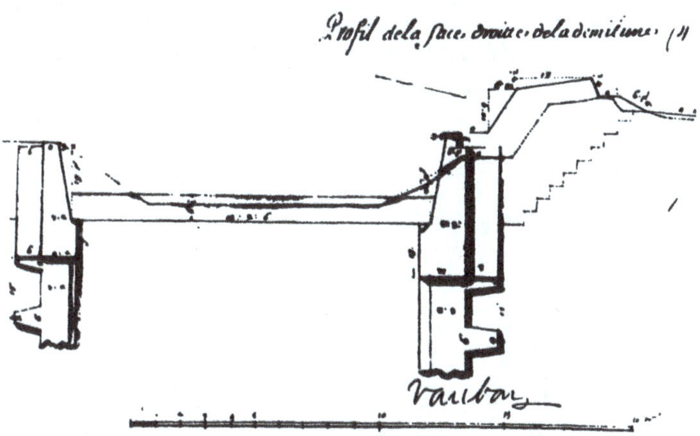

Fig. 10. Two retaining walls with internal buttresses founded in clay, designed by Vauban for the fortification of Ypres in 1699

but less heavily overconsolidated), Vauban was obliged to supplement his tables.

In an entry in his diary dated 25 June 1699 he wrote: 'On this fat soil slippery like soap..., the ground at the bottom of the ditch must be capable of containing the thrust of the covering when it is pushed by the soil of the ramparts'. Here we find the fundamental idea of using the passive resistance of soil by increasing the depth of the wall. In keeping with this idea, Vauban underpinned the walls of the fortification at Ypres which had been constructed before his time in 1678 (Fig. 11). Their foundations had settled and slipped forward quite a lot, with the result that the wall had rotated backwards under the action of its own weight and the friction force on the back face; the counterforts, moreover, had sheared off at their connection with the masonry. Where the wall had not moved too much, Vauban reduced the load on the back of the wall and then used temporary shoring to permit the addition of new masonry underneath.

All this is on the eve of the 18th century. That and the following century would see a great many studies on the question of retaining walls. Professor A. W. Skempton (Ref. 8) gave us a clear account of the subject at the Eleventh International Congress at San Francisco in 1985 in his History of Soil Properties 1717-1927. By observing the resistance of retaining walls and by subjecting scale models to experiments based on back analysis, a start was made on the development of a theory of earth pressure. The publications of Gauthier (1717), Belidor (1729), Sallonyer (1767), Coulomb (1773), Papocino (1781), Prony (1802), Mayniel (1808), Français (1820), Navier (1833), Poncelet (1840), Hope (1845), Rankine (1862), Darwin (1883), Boussinesq (1876, 1882 and 1883) are the main landmarks of the long history of research on retaining

RETAINING STRUCTURES

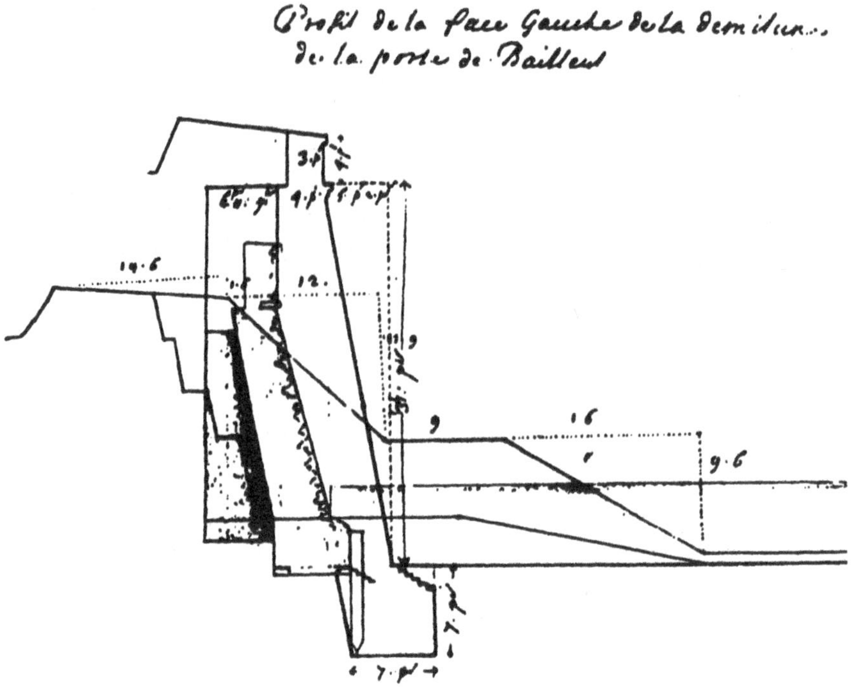

Fig. 11. Under pinning by Vauban of the wall insufficiently embedded in clay at Ypres

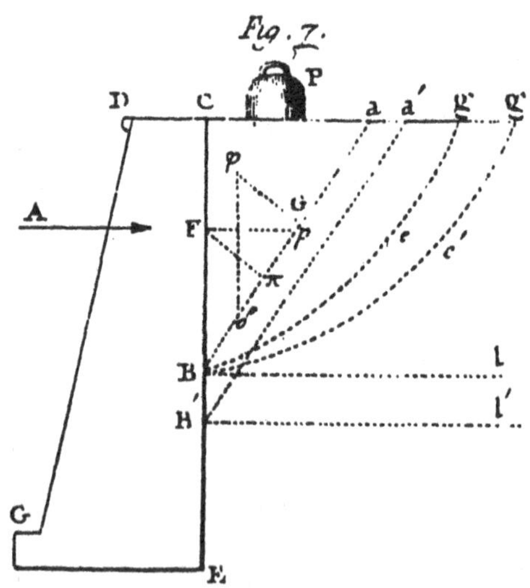

Fig. 12. The basic figure of Coulomb's memoir

walls (for references, see publications of Skempton and Heyman). I should like to make a few comments on just three of these authors - Coulomb, Rankine and Boussinesq.

The bicentenary of Coulomb's essay was celebrated in 1973 at the Seventh International Congress of Soil Mechanics in Moscow (Ref. 9). His life and work are described in detail in the book by Gillmore (Ref. 10), but it is in the book by Professor Heyman (Ref. 11) that the full significance of that essay, which goes far beyond the question of retaining walls, is brought out.

In your call for papers you reproduced one of the major figures from this essay (Fig. 12). 'The force A', writes Coulomb 'must be capable of supporting... all the surfaces CBeg bounded by any curve Beg, which, supported by its cohesion and its friction and acted upon by its own weight, would exert the largest thrust. It will be clear from this statement that if a force were applied at F infinitesimally smaller than that required to support the surface of greatest thrust, the mass of soil could cleave only along that curve, while all the other portions remained united by cohesion and by friction. Thus, to find the force A sufficient to support the whole mass, that surface must be sought, among all the sources CBeg, for which the thrust on the line CB is a maximum. Similarly, to find the greatest force which can act through F without disturbing equilibrium, another curve Be'g' is a minimum. The limits of the horizontal force that can be applied at F without moving the soil will be bounded by the limits A and A', where A will be a maximum and A' a minimum'. (We have used the translation by Professor Heyman). Here Coulomb opposes the active and passive cases and shows that the limits for the latter (far greater) are derived from a study of the minimum, and vice versa.

Coulomb assumes that the curve which gives A is a straight line, experience showing that the surface which breaks is very close to triangular. He does not make the same assumption for the passive case: the curves Beg and Be'g' which he draws by way of example do indeed bend slightly. In fact, the curve for the usual values of ϕ is even more rounded than the ones drawn.

Coulomb draws attention to the fact that the friction created upon the retaining wall has been neglected, and then writes: 'At the moment of rupture the soil is about to slide along CB, which diminishes A and at the same time increases the moment of the retaining wall'.

This problem of friction on the wall occupied researchers for the next two hundred years. Rankine (Ref. 12) applied his stress-field theory for cohesionless soils which leads to the familiar expression

$$Ka = tg^2(45 - \phi /2) \text{ and } Kp = tg^2(45 + \phi /2)$$

for a vertical wall and horizontal soil, confirming Coulomb's value for the active state.

RETAINING STRUCTURES

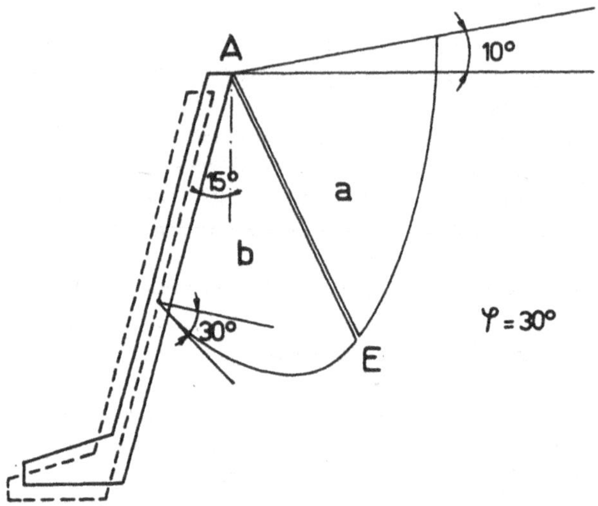

Fig. 13. Boussinesq introduces a discontinuity of the stress field through separating the Rankine equilibrium and the Boussinesq one; case of a wall that has settled

But the theory of Rankine (1857) is restrictive in that the angle of friction on the wall face is established a priori at failure being equal to slope of the free surface of the ground for a vertical wall.

The merit of Boussinesq (1876 and 1882) is to have clearly stated the equations governing the problem of active pressure (Refs 13 and 14) whatever the friction on the wall. The importance of his analysis lies in the introduction of a discontinuity within the stress field along a certain lane AE separating Rankine equilibrium and another one, his own (Fig 13). Moreover, in his note on 'Mr. Darwin's careful and extremely interesting experiments' (Darwin Ref. 15), Boussinesq (Ref. 16) accepts the concept that the angle ϕ within the mass of the sand is not necessarily the same all along the rupture line. As underlined by Professor Skempton, 'this analysis is one of the brilliant steps which, once they are made, seem simple and obvious'.

Starting from the (non-integrable) equations of Boussinesq we can now use computers to draw up numerical tables giving the value of the limiting states of passive and active pressures whatever the angle of friction on the wall, the slope of the ground or the inclination of the wall (Ref. 17).

Some 70 years later, Sokolowski (Ref. 18) made a little addition to the solution of Boussinesq: instead of having a plane of discontinuity between two equilibria they are separated by a fan OQR (Fig. 14). Nevertheless, as Professor Heyman points out, Sokolowski's solution only works if the surface is subjected to a uniform over burden.

KERISEL

In 1885, Boussinesq (Ref. 19) published his celebrated 'Application des potentiels' giving stresses and displacements in elastic foundations. Today the calculation of horizontal pressures on a retaining wall due to a concentrated surface load is still based on this work.

The 20th century put an end to the somewhat abstract research of the 18th and 19th centuries. As Professor Heyman emphasises, these cohesionless theories have been attacked repeatedly as poor models for the behaviour of soils. Professor Peck (Ref. 20) has even spoken of the 'numerical calculation subculture' to which too many theories lead.

The fact is that at the end of the 19th century, in the field of retaining walls with which we are concerned, mathematics was well in advance of physics. The 20th century saw an effort to catch up and there was big progress in our knowledge of the mechanical properties of soils. But the mathematical tool was not challenged for all that. Occasionally, however, certain factors were overlooked: for example, in the problem of horizontal pressures on a retaining wall due to a concentrated surface load, people took from Boussinesq only the stresses and did not examine whether the displacements of the wall at the corresponding point matched those given by that author, or, if they did so, they did not draw the consequences. As for cohesion, that was easy to take it into account as Caquot (1934) showed (Ref. 21) by introducing a uniform external stress $C/Tg\phi$ on the free surface and behind the wall.

In fact, though progress was made, physics did not entirely catch up. In undisturbed sand said to be without cohesion, a low level of cohesion may develop quite rapidly as an effect of ageing. It plays a considerable role in the

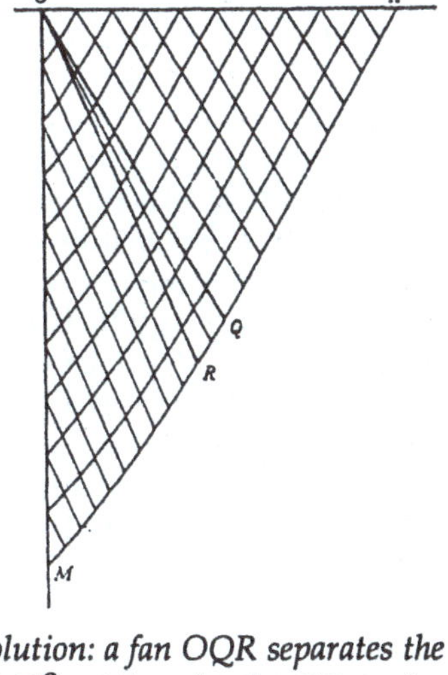

Fig. 14. Sokolowski's solution: a fan OQR separates the two regions; here $\phi = 30^{\circ}$, angle of friction on wall 15°, uniform burden 5C on the ground

calculation of active or passive pressures owing to the size of the multiplier of C/Tgϕ though it is not at all easy to determine in the laboratory or in situ the value of this slight cohesion. Moreover many factors concerning the shear strength of clays are still unknown today.

With the emergence of concrete in the 20th century, gravity masonry walls gave place to concrete walls and reinforced cantilever walls, but there soon appeared a range of new retaining structures such as gabions, cribwalls, cantilevered or anchored steel walls and so on. Nor must we forget reinforced soils: the variety of these structures and the differences in flexibility led engineers to study the extent to which active or passive pressures are strain-dependent and the distribution of lateral pressures. This was the starting point for a considerable number of studies. Limiting myself to the main references prior to 1965, I wold begin by citing the work of Terzaghi (1934) and Rowe (1965) based on models incorporating large rigid screens. This research was extended to flexible models by Tschebotarioff (1949), Rowe (1952) and Terzaghi (1953), and to temporary retaining structures for deep excavations. This then provided an opportunity for Professor Peck (1943) and Bjerrum (1956) to make numerous earth pressure measurements in braced excavations and show their very large dependency upon the method of installation of struts or bracing frames.

Lastly Rowe (1965) (Ref. 23) described the mechanism of progressive rupture about which Boussinesq had had a certain intuition.

In spite of all this knowledge and further progress in the last twenty-five years, I note that your draft Code of Practice for earth retaining structures, prepared under the authority of the British Standard Institution, observes that 'there is still a considerable lack of reliable information upon the actual earth pressures encountered on walls in practice', and that the code is still 'limited to small and medium sized retaining walls which have a free standing height not in excess of approximately 15 m'. This gives some idea of the work before you at this Congress. However, in order not to end on a pessimistic note, I should like to stress that all this research now going on has at least made it possible to build in safety a large variety of structures which melt much better into the landscape through their architecture and perhaps through their decorative plant cover than the mass gravity walls they replace.

I shall bring to an end here this backward glance, which began so deep in the past at a time when our eyes are turned towards the future. It is never without interest to know what has already been accomplished and I have endeavoured to remind you of the most essential achievements.

References

1. O'KELLY M.M. *Newgrange, Archaeology, Art and Legend*, Thames and Hudson, London, 1982.

KERISEL

2. KERISEL J. A glimpse of geotechical engineering during the fourth and third millenia B.C. Proceedings of an international symposium organised by the Greek national group of IAEG. Balkema, Rotterdam, 1988, Vol 3, 1641-1653.
3. KERISEL J. La pyramide à travers les âges. *Mythes et religions.* Presses de l'Ecole des Ponts et Chaussées, Paris 1991.
4. EMERY W. B. *Archaic Egypt.* Penguin Books, London 1961, p 55.
5. VITRUVIUS. *De Re Architecutura.* 1st century B.C. Translated into French and illustrated by Perrault: the ten books on architecture. Coignard, Paris, 1674. English translation by Harper, Cambridge, Mass. and Heineman, London, 1955.
6. KERISEL J. The golden Jubilee book. *Proceedings of the Eleventh International Congress on Soil Mechanics and Foundation Engineering.* Part One: the history of geotechnical engineering up until 1700. Balkema, Rotterdam, 1985, pp 3-95.
7. VAUBAN. Tables publiées dans 'La manière de fortifier selon la méhode du Maréchal de Vauban'; par l'abbé du Roi, 1684.
8. SKEMPTON A.W. *Golden Jubilee Book* (see Ref. 6). A history of soil properties 1717-1927, pp 95-123.
9. KERISEL J. Bicentenary of the 1773 paper of Charles Augustin Coulomb. *Eighth International Conference on Soil Mechanics and Foundaiton Engineering,* Special Lecture, Moscow 1973.
10. GILLMOR C.S. *Charles Augusin Coulomb: Physics and Engineering in 18th century France.* PhD thesis, Princeton University, May, 1968.
11. HEYMAN J. *Coulomb's Memoir on Statics.* Cambridge University Press, 1972.
12. RANKINE W.J.M. On the stability of loose earth. *Phil. Roy. Soc.* 147, 9, 1857.
13. BOUSSINESQ J. Essai théorique sur l'élasticité des massifs pulvérulents et sur la poussée des terres sans cohésion. Paris 1876.
14. BOUSSINESQ J. Sur la détermination de l'épaisseur minimum que doit avoir un mur vertical ... pour contenir un massif terreux sans cohésion dont la surface supérieure est horizontale. *Annales des Ponts et Chaussées,* 6th series, 1882, pp 494-510.
15. DARWIN G.H. On the horizontal thrust of a mass of sand. *Minutes of Proceedings, Inst. Civ. Engrs,* 71, 1883, pp 350-378.
16. BOUSSINESQ J. Note on Mr. Darwins paper 'On the horizontal thrust of a mass of sand'. *Minutes of Proceedings, Inst. Civ. Engrs,* 72, 1883, p. 262.
17. KERISEL J. AND ABSI E. *Active and passive earth pressure tables.* 3rd Edition. Balkema, Rotterdam, Brookfield, 1990.
18. SOLOLOWSKI V.V. *Statics of Solid Media.* English translation: Butterworth Scientific Publications, London 1954.

RETAINING STRUCTURES

19. BOUSSINESQ J. Applications des potentiels à l'étude de l'équilibre et du mouvement des solides élastiques, Paris, 1885.
20. PECK R. *Golden Jubilee Book* (see Ref. 6), The last sixty years, part 3, pp 123-136.
21. CAQUOT A. *Equilibre des massifs à frottement interne*, Gauthier-Villars, Paris, 1934, pp 10-13.
22. TERZAGHI K. Large retaining wall tests. *Engineering news record*, September 1932, February, March and April, 1934.
23. ROWE P. W. and PEAKER R. Passive earth pressure measurements, *Geotechnique*, Vol 15, No. 1, March 1965.
24. TSCHEBOTARIOFF G. P. *Large scale earth pressure tests with model flexible bulkheads*. Princeton University, 1949.
25. TERZAHI K. Anchored bulkheads. *Proc ASCE*, Vol 119, 1953, pp 1243-1324.
26. ROWE P.W. Ancored sheet pile walls. *Proc Inst. Civ. Engnrs*, 1952, Vol 1, p. 27.
27. PECK R. Earth pressure measurements in open cuts Chicago subways. *ASCE*, Vol 108, 1943, p. 1008.
28. BJERRUM L. AND EIDE O. Stability of strutted excavations in clay. *Geotechnique* 6, 1956, pp 32-47.

4

The use of models in nineteenth-century British suspension bridge design

Denis Smith

Introduction

In the nineteenth century the suspension bridge became a prominent feature of bridge-building on both sides of the Atlantic, especially in Great Britain, France and the United States. In Great Britain the first half of the century promised more, both in terms of theoretical ideas and practical techniques, than was to be sustained after about 1870 by which time the suspension bridge had failed to establish itself in the British bridge builders' repertoire.

France was a prolific builder of suspended bridges in the 1830s and 1840s but the effort petered out after 1850 following the collapse of the Pont de la Basse-Chaîne at Angers which killed two hundred and twenty-three people.

By comparison in the United States suspension-bridge building went from strength to strength in the hands of such able practitioners as Charles Ellet and John Roebling, although precisely why American technology succeeded so fully where its European counterpart failed so utterly remains unexplained. The fact is though that by the twentieth century American engineers had established a pre-eminence which has yet to be challenged.

This paper is concerned only with early suspension bridges in Great Britain and with but one, and a very distinctive, facet at that. For the first two-thirds of the nineteenth-century British bridge building in general was characterized by the use of models as an aid to design and as a basis for evaluating new ideas. This is not the place to examine the technical, scientific or even the psychological reasons as to why model techniques endeared themselves so thoroughly to British engineers but it does seem that this was the most distinctive contribution to bridge designing techniques of British origin. The characteristic French approach was through mathematical analysis while in the United States the prevailing philosophy was essentially that of 'trial-and-error'.

Perhaps it is no accident that the demise of the suspension bridge in Britain coincided with a rapid stagnation of the use of models (although one should not overlook the significance of a shift in the application of scale-model techniques typified by the work of Robert Manning, Horatio Philips, William Froude and

Osborne Reynolds). Nevertheless for half a century suspension bridge design by means of models was a matter of interest to some of the most eminent civil engineers (and some of the least) in Britain and what follows is an assessment of their work.

Suspension bridges

The use of suspension structures as footbridges, of modest span, has a long history which need not be repeated here. However, in the early part of the nineteenth century many bridges were built in America, France and Britain designed for the passage of horse-drawn vehicles. Compared with footbridges such structures required a greater degree of design skill to ensure stability. Eventually the demands of the railways were to pose problems for suspension bridge designers, for which there was no precedent, and which were never to be solved in Europe.

The suspension bridge has several advantages over the traditional masonry arch structure. Its relatively light-weight construction made large spans possible, and the structural system being above the deck left unobstructed space below. This was an important factor leading to its adoption in cases where headroom was of prime importance. As a tension structure problems of lateral instability, later to be encountered in the compression zones of metal girders, did not arise. Under the action of dead load only tension structures are inherently stable. However, the price paid for structures with such favourable strength/weight ratios is lack of stiffness and it was undoubtedly the suspension bridge which first made designers face the distinction between *strength* and *stiffness*. The calculation of the necessary sizes of suspension links and hanger rods was, after all, an easy matter after the materials strength experiments of Telford, Barlow, and others had established the tensile properties of wrought iron. But to predict the response of the structure to *live* load was a more complicated matter. This first became a problem with the dynamic effects of wind loading on the earlier road bridges and was the cause of several bridge and pier failures. Methods of making the deck sufficiently stiff to maintain the chain geometry under the passage of moving loads were increasingly discussed and became the prime consideration in the design of railway suspension bridges.

British model experimental work in this field in the nineteenth century can be divided into two phases. Firstly, the exploitation of wrought iron in suspension bridges capable of carrying vehicular traffic. This was in response to the growth of planned road systems in the early part of the century. In this period, model experimenters were preoccupied with the properties of the suspension chain itself. Models were used to help resolve problems arising from variations in geometry, link material, and cross-sectional area of the chain. Road bridge models were concerned, quantitatively, with *strength*, and the *stiffening* effect of the road deck was considered only in a qualitative manner. Secondly, models were resorted to by designers in their attempts to adapt the suspension bridge to railway purposes. Here the preoccupation was with overall *stiffness*. Models were built to investigate the effect of combining a girder with a suspension chain. These were amongst the last model experiments recorded in Britain in the nineteenth century. The models that were made to help achieve all these objectives will be discussed here in chronological order.

THE PROPOSED SUSPENSION BRIDGE AT RUNCORN

The earliest reference to a model test in connection with suspension bridge design relates to the proposed bridge over the Mersey at Runcorn undertaken by Thomas Telford in 1814. A petition to Parliament was drawn up in 1813 to bring in a Bill to establish the River Weaver Fishery Company, 'to preserve the fry in the said River Weaver, and in certain parts of the River Mersey . . . and also to erect a Bridge over the River Mersey at or near a place called Runcorn Gap'.[1] Telford was approached during 1814 and in September a newspaper said 'A magnificent bridge is intended to be erected over the Mersey at Runcorn, which unite Lancaster and Chester . . . A plan worthy this great object, has been produced by Mr. Telford'.[2] Some years later Telford said 'the crossing of the River Mersey at Runcorn has always been the most important consideration whenever shortening the communication between London and Liverpool has come under discussion. In the year 1814 I had my attention particularly drawn to the subject, and was employed to investigate it'.[3] Few details of his work in 1814 survive and the principal source is Telford's Report to the Runcorn Bridge Committee of 1817. However, three of his pocket notebooks survive in the library of the Institution of Civil Engineers, one of which is titled, in his own hand, 'Runcorn Bridge 1814'.[4] This contains river survey notes at Runcorn and details of various iron chains seen in his travels, recording the size of links and the loads carried. In addition, two pages marked 'model' give details of wire joints and incremental loading up to 3,033 lbs. This does not tell us much of the model except that the tests were made on 8 July. Another part of the notebook mentions 'Experiments at Brunton's' made from 5 April, and these he described in his Report to the Committee three years later:

> on considering the matter maturely, it occured to me, that on the suspending principle, chains are not the most perfect means to be employed, but that the metal should be kept, as far as practicable, in straight lines, and also have few joinings. In order to proceed with due caution upon this principle, I

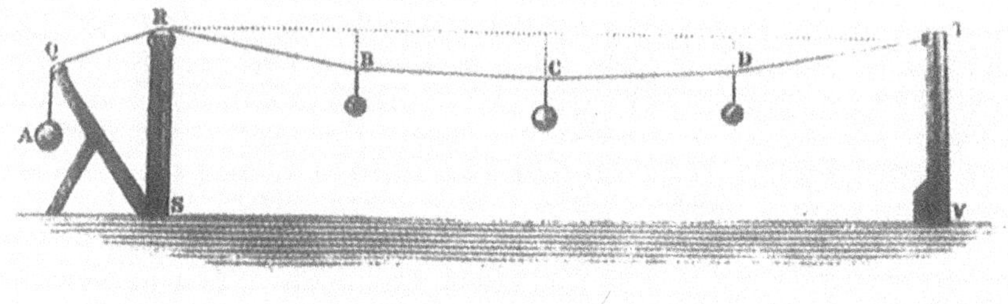

Figure 1. Telford's wire test apparatus.

'Here RS, TV, represent the supporting pillars upon which the wire extended; QS, another prop, over which the wire passed; being placed at such an angle as made it coincide with the direction of the resultant of the vertical and horizontal tensions, in order to prevent any strains upon the other support, RS.

A, B, C, D, represent the places of the several weights with which the wire was loaded; C being in the centre of the length, and B and D at one quarter of the length from each end; and the deflections from the horizontal line RT were measured at these points, as the different weights were applied.'

Life of Thomas Telford, Ed. Rickman, 1838. p. 555.

> made during the summer of 1814 above 200 experiments upon malleable iron, of from one twentieth to one half inch diameter, and on lengths varying from 31 to 900 feet. These experiments were made perpendicularly, horizontally, and with different degrees of curvature.[5]

These tests on the tensile strength of wrought iron were made 'both in regard to elementary parts, and also when combined, partly by welding, and partly by jointing in a model'.[6] Those relating to the 'elementary parts were carried out with a hydrostatic press at 'Mr. Brunton's patent chain-cable manufactury, Commercial-Road, London'.[7] The tests made with 'different degrees of curvature' were conducted with the link-polygonal apparatus shown in Fig. 1.[8] The tests sought to establish the suitability of the material for suspension bridges and he subsequently proceeded with a design and a model — but whether the model was made at Brunton's works is not clear. Telford was assisted by James Rendel who, 'At an early age . . . went to London, and obtained an engagement, as a surveyor, under . . . Mr. Telford, by whom he was employed on the surveys and experiments for the proposed suspension bridge across the Mersey at Runcorn'.[9]

Telford's crucial statement about the model test is the prime source of evidence:

> I constructed a model of 50 feet in length, and one twelve hundredth part of the strength of the intended bridge, and although composed of wire only, and of course, without proper Joints or Braces, it bore 3000 lbs. without any symptom of its parts being deranged.[10]

Several questions arise from this; firstly that of the geometric scale of the model. Telford's design was for a bridge having a central span of 1,000 feet together with two side spans of 500 feet each — making it a bridge of unprecedented span. When he said 'a model of 50 feet in length' we cannot be certain if this included all three spans; although W. A. Provis was more specific saying 'a model of the central opening was constructed'[11] and this is certainly more likely and logical. If so, the geometric scale factor was $50/1000 = 1/20$ or 20:1 so far as the span was concerned, but unfortunately we have no further details of the model. Secondly, as Telford loaded this model it indicates that it was not merely of the visual type, and indeed we know that another, smaller, model was made presumably for this purpose.[12] Lacking details of the model chains it is difficult to understamd what Telford meant by describing the model as being 'one twelve hundredth part of the strength of the intended bridge'. In using the word *strength* we may reasonably assume that he is concerned with the breaking *stress* of a member subjected to pure tension, in which case a square law relates the tensile stress to the geometric scale factor. If we assume Telford's span scale factor to be 1/20 and that this factor was also used to scale down the dimension of the proposed chain then the model would have been 1/400 (i.e. $1/20^2$) the strength of the full-sized structure. The chains of this model were, however, 'composed of wire only' and 'without proper joints' and unless the geometric scale factor for these was $1/\sqrt{1200} = 1/34.6$ it is difficult to add anything to Telford's statement relating the strength of his model to that of the full-sized structure. The model was loaded up to *c.* 3,000 lbs, 'without any symptom of its parts being deranged'; in other words it was not tested to destruction, and indeed Provis added that it 'would have carried considerably more without being injured'.[13] After the tests Telford said: 'This far I proceeded in 1814. I was perfectly convinced of the practicability of the scheme, and am now satisfied I went further than necessary in the quantity of material, and also as to sundry precautions in forming the plan, but I was willing to err on the safe side.'[14] Telford's design for the chains exist; they were to consist of '16 iron cables, each formed of 36 square half-inch iron bars; and of the segments of cylinders proper for forming them into one immense cylindrical iron cable, which was to be nearly half a mile long (including the fixings on shore), and about $4\frac{1}{4}$ inches diameter'.[15]

The bridge Committee met several times in the spring of 1817 and Telford was again consulted to help them decide on various proposals. He was convinced that a suspension bridge was the only possible solution and reminded them that he had 'already prepared a Design and Model',[16] and offered his Report gratuitously. Telford mentioned only one other scheme submitted:

> it is that produced by CAPTAIN BROWN only which corresponds to the principle adopted by me in 1814; that is to say, Suspension by malleable Charcoal Iron, preserved as nearly as possible in straight lines . . . His perpendicular Suspenders, Cross Ties, and diagonal braces, correspond precisely with mine. His main suspending lines of the upper Curve, instead of being composed of a number of flexible rods, as in mine, consist each of one Rod or Bar of malleable Iron united by Forelocks. From four of these it was proposed that the whole structure should be suspended.[17]

It is interesting to note Telford's preference for *charcoal* smelted iron. Captain (later Sir) Samuel Brown's interest in the applications of wrought iron chains led to his patent of 1817,[18] relating to suspension bridges. During 1808 he had made a series of experiments on the 'comparative strength of bolts and bars and chains of different descriptions'[19] at his works on the Isle-of-Dogs in London. His catenary design involved replacing ordinary chain links with long bars having forged eyes at their ends, connected by pins in shear, reducing both the weight of the chain and the number of connections. After the 1808 experiments he made 'drawings and calculations of the strength of bridges of suspension' but was too busy to test the design experimentally 'on a scale commensurate to its importance' until 1813;

> when I constructed a bridge of straight bars for this purpose on my own premises . . . The span or extent of this bridge is one hundred and five feet, and although the whole of the ironwork weighs only thirty-seven hundred weight, it has supported loaded carts and carriages of various descriptions.[20]

No further details of this small-scale bridge have emerged but we may reasonably assume that its chain configuration accorded with his recommendation that they should 'rise or have a versed sine, in the proportion of twenty feet in a thousand feet of length'.[21] This gives a very shallow curve producing a large horizontal force component on the tops of the supporting towers. Brown claimed a favourable strength/weight ratio for his bridge and it is worth noting that he never used the word *model* in describing it. Both Telford and John Rennie visited Brown's works and use the model in describing the bridge. Mentioning it to the Runcorn Bridge Committee in 1817 Telford said 'Your Solicitor and I examined this model and drove a Hackney Coach over it'.[22] John Rennie, when questioned on iron suspension bridges in 1819, said 'the only thing of that kind that I have seen is a model that was made by Captain Brown . . . over which I was drawn in a carriage, and found myself perfectly safe and easy'.[23] In addition he noticed 'very little vibration'.

Telford and Brown met in London and exchanged ideas on suspension bridge design. Both considered the Runcorn scheme practicable but 'although the Committee reported favourably upon the scheme, and subscribed liberally to a guarantee fund, nothing further was done'.[24] A thousand-foot span bridge was obviously ahead of its time and John Nicholson said; 'The size, and I may say, the daring character of this project frightened the capitalists to whom it was proposed; but it had, at least, the advantage of drawing public attention to this new species of construction.'[25] Other bridge and tunnel schemes were subsequently proposed but the Mersey was not bridged at Runcorn Gap until the transporter bridge was built in the first decade of this century.[26]

THE PROPOSED SUSPENSION BRIDGE AT MONTROSE

The next suspension bridge design containing a brief reference to a model is that of George Buchanan — a civil engineer born in Montrose and a lecturer on mechanics at the School of Arts, Edinburgh. In 1821 he published a 'Report on the present State of the Wooden Bridge at Montrose, and the practicability of Erecting a Suspended Bridge of Iron in its stead'.[27] The Report contains full details of the proposed bridge over the Esk with a single span of 420 feet and a versed sine of 60 feet — much steeper than that advocated by Samuel Brown. Buchanan was particularly concerned with achieving adequate *stiffness* in his bridge and said: 'Besides that motion which takes place in suspended roadways from side to side, we must equally guard against any vertical undulation to which such a roadway may be subject.'[28] He attempted to achieve this by having vertical longitudinal iron plates beneath the decking, using the side railings, and multi-strand chains comprising 1½-inch diameter rods 15 feet long. In addition, between the vertical suspender rods he proposed to:

> extend a similar series of rods horizontally . . . so that these horizontal rods, interweaving with those which hang vertically, the whole space directly under the chains, and between them and the roadway, will be filled up by a kind of net or wire work, on a large scale, and which, without interrupting the view, will yet communicate a remarkable firmness and steadiness to the roadway, the reason of which may be easily shown, and has indeed been experienced in the model which I have constructed.[29]

This is the only reference to the model and we therefore have no knowledge of the scale factor, whether a load test was made, or indeed whether this was anything more than a visual model. Even a small-scale model would demonstrate a qualitative increase in stiffness which is all he is claiming for his chain configuration. Buchanan's bridge was not built and the first suspension bridge at Montrose was that of Samuel Brown built in 1829 and destroyed in 1838.

THE MENAI SUSPENSION BRIDGE — CHAIN GEOMETRY MODEL

There is no evidence to suggest that Telford made a structural load test on a model as a preliminary to his Menai Bridge design. He would have gained valuable information relating to *strength* from his Runcorn model and tests on wrought iron a few years earlier, indeed these results were repeated in evidence to the Select Committee which recommended the building of the Menai Bridge.[30] Nevertheless, he did make a quarter-scale model in 1823 to help resolve problems of the chain geometry. W. A. Provis, Telford's resident engineer on the bridge, said:

> In order to get the lengths of the ordinates or vertical rods which were to suspend the roadway from the main chains, a model was made of a single chain to a scale of one-fourth of the real one, and a set of boards on edge, curved to the intended rise of the roadway, were placed below it. Between these the perpendiculars were measured at every joint of the chain, which became the standards for making the rods of the vertical suspenders.[31]

The model has a chord line (or span) of 142 feet 6 inches, a versed sine of 10 feet 9 inches, and the chain comprised links 2 feet $6\frac{3}{4}$ inches long. Provis published a 'Table showing the lengths of the vertical suspending rods and their horizontal distances from each other'.[32] In this measuring exercise no load was applied to the chain to simulate the decking and hence the 'intended rise of the roadway' did not take into account change of chain profile due to its weight. However, any differences due to this interaction could most likely be accommodated by screw thread and nut adjustment at the ends of the suspender rods. Provis explained the reason for using this type of model very succinctly when he said; 'It is true their ordinates might have been determined by calculation, but with a practical man an experiment is always more simple and satisfactory than theoretical deductions'.[33] Whereas Drewry, in his *Memoir* of 1832 commented;

> In reality, those lengths cannot be calculated with perfect accuracy, because, . . . the chains of a bridge are neither a pure catenary nor a pure parabola, and their curve varies in every case; hence, when it is possible, the lengths of the vertical rods should be determined by experiment.[34]

TABLE I

From Charles Stewart Drewry, A Memoir on Suspension Bridges *(London, 1832) p. 158*

TABLE OF EXPERIMENTS ON WOOD CHAINS

Number of Experiments	*Span*	*Deflection*	*Load in Dead Weight*	*Load falling on the Platform*	*Momentum of the falling Weight*	*Remarks*
	315 inches or $26\frac{1}{4}$ feet					
I		26 inches	none	130 lbs through 18 inches	($\sqrt{1.5} \times 8.01 \times 130$ lbs =) 1274 lbs	No injury.
II			none	130 lbs through 33 inches	1726 lbs	One of the wood-bars torn in two.
III		26 inches	52 lbs	130 lbs through 34 inches	1735 lbs	No injury done.
IV		22 inches	52 lbs	130 lbs through 30 inches	1645 lbs	One of the wood-bars torn in two.
V		21 inches	52 lbs	130 lbs through 34 inches	1735 lbs	No injury.
VI		26 inches	4 persons, about 580 lbs	none		While the weight remained steady, nothing gave way. On swaying up and down, one of the wires broke. The fastenings remained sound.
VII		26 inches	Same, with additional 52 lbs = 632 lbs			While at rest, the bridge bore the weight. One person jumping broke one of the iron wires.
VIII		23 inches	none	130 lbs through 32 inches	1697 lbs	One of the wood-bars was torn in two pieces. The fastenings remained sound.

In the course of these trials, the fastenings of the bars, No. 1. drew considerably, but did not give way. Those of the bars No. 2., in which the inclination of the ends was more rapid, did not show any symptom of derangement.

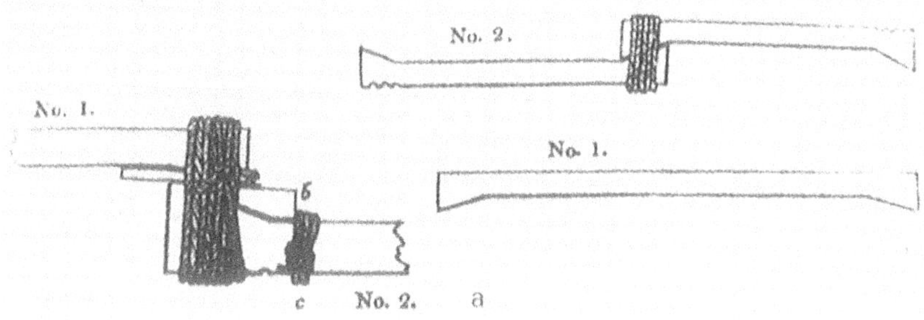

Figure 2.
From A *Memoir on Suspension Bridges*, Charles Stewart Drewry, London, 1832, p. 156.

Nothing could illustrate more clearly the type of dilemma leading engineers to use a model. Opinions differed as to whether calculation could provide the required data, and even Provis (a practical man) who felt that it could, nevertheless felt it 'more simple and satisfactory' to use a model. The expense of a model was justified on the grounds of the reassurance, or peace of mind, which it bought.

WOOD AS A MATERIAL FOR SUSPENSION CHAINS

An episode worthy of brief mention was the model test made by Charles Drewry in order to investigate an alternative material for the chains of suspension bridges. In his well-known *Memoir* of 1832 he made out a case for using wood in place of wrought iron for chain links. He admitted it would be desirable to seek a substitute for iron, but for bridges 'erected in remote places, particularly abroad, where iron and mechanics are not easily or cheaply procured',[35] and for lightly loaded structures, wood might be a useful material.

> Wood, used as a material for making the chains of a suspension bridge, is in some respects superior, in others inferior, to iron. The advantages are, that, weight for weight, its absolute strength is greater than that of iron, and that it is worked more readily and with fewer tools.[36]

He proposed joining the wooden links by means of binding with rope the tapered ends (see Fig. 2a) and it was to test the efficiency of this method that he made his experiment. As he put it:

> The author has tried this mode of fastening in a model 26¼ feet span... The bars of wood were 14 inches long, out and out. Of two kinds, viz. No. 1 of good pine, ¾-inch square in the middle, with solid wedge ends 1⅛ inches deep, and 1½ long. Some of the bars No. 1 were bound together with seven turns of common cord, of barely 3/16 of an inch diameter; others with iron wire: some had notches cut in the back, to receive the cord; others were left quite smooth. The bars (No. 2) were of very common fir, 1 inch deep by ¾ in the middle, with solid wedge ends 2 inches deep and 2 inches long.[37]

The middle 8 or 10 feet of the span was made of the wooden bars, the remainder being iron wires (see Fig. 2b). The results of his experiments are given in Table 1. The model was loaded with both static and dynamic loads and it is obvious from experiments 1, 2, and 8 that he ignored the self-weight of the model in his 'Dead Weight'. He calculated the momentum of the dropped weight correctly although he expressed the result in 'lbs'.

For example: Momentum $= m \times v = m\sqrt{2gh}$

For experiment 1: 18 inches = 1.5 ft $\sqrt{2g} = 8.01$ (gives $g = 32.08$ ft/sec^2)

therefore, as he gives: momentum $= 1.5 \times 8.01 \times 130 = 1{,}274$ *lbs*. Drewry made no attempt to calculate an equivalent static load and combine it with the dead load but merely uses the relative momentum in each case for the purposes of comparison. In experiments 6 and 7 he introduces four people as loading who obligingly acted as both dead and live load. There is no evidence that Drewry's work on wooden chains had much impact even on military engineers who might have been expected to use such bridges for temporary structures.

It is interesting to note that at the same time that Drewry was discussing wood as a substitute for wrought iron, a speaker at the summer meeting of the British Association in Oxford said:

> I am very desirous of calling the attention of the scientific world, and particularly of Civil engineers, to the serious consideration of the question, — How far ought iron to be hereafter used for suspension-bridges, since it is ascertained that a steel bridge can be built of equal strength, and superior durability, with one third or one fourth of the weight of an iron one, and at a much less expense, . . . The only doubt of this being practicable on the large scale, arises from the circumstance, that in this country iron is made with mineral coal, but in Styria with charcoal of wood.[38]

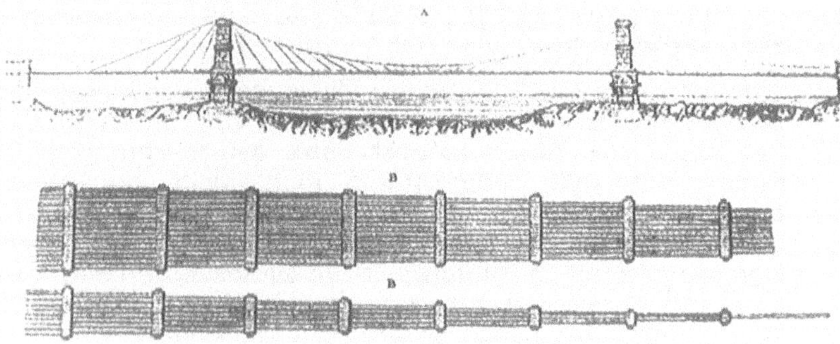

Figure 3. Typical taper chain bridge of James Dredge. From *The Surveyor, Engineer, and Architect*, June 1840, p. 100.

This echoes Telford's expressed preference for charcoal-smelted iron and the speaker added; 'I have calculated that upwards of 1000 feet span may with confidence be depended on.' This was Telford's planned span for the avant-garde Runcorn bridge but at that time the largest span built in Britain was 579 feet in the Menai bridge.

JAMES DREDGE AND THE TAPER CHAIN CONTROVERSY

In the 1830s a proposal to span the Avon at Bath with a suspension bridge led to an interesting discussion on chain design. The common chain form at that time was that of Telford and Brown as discussed above. They comprised an equal number of links between the pins giving the chain a uniform cross-sectional area throughout the span. In James Dredge's designs the area of the chain decreased progressively from the points of suspension towards the centre of the span. Although it can be argued that the idea of matching the area of the chain to the tension was not Dredge's,[39] his name became inextricably identified with the idea. A typical Dredge bridge is shown in Fig. 3.

The Victoria Bridge Company of Bath invited designs for a bridge in the centre of the city and two local men were the principal contestants — James Dredge[40] and Thomas Motley.[41] Dredge's design was chosen and the Victoria Bridge was built in 1836 on his *Taper chain* principle and is indeed still standing. The span was 150 feet, versed sine 25 feet, weight of chains 5 tons, with a platform width of 18 feet. Motley, extremely bitter about the rejection of his scheme, entered on a long exchange with Dredge in the *Bath Journal*. Dredge claimed a superior strength/weight ratio for his system and made experimental models to demonstrate its performance. He was a great propagandist and the discussion extended for nearly ten years in the technical press and elsewhere. How this west-country brewer became interested in bridges is not clear. We know that he submitted a design in the Clifton Bridge competition of 1829–30[42] and had erected a bridge over the Leven at Balloch Ferry in 1832.[43]

In 1836 Dredge took out a patent for 'Certain improvements in the construction of suspension chains for bridges, viaducts, aqueducts, and other purposes'. His improvements comprised two distinct parts, namely; 'the construction of a chain . . . which shall diminish in breadth and weight as it recedes from the point or points of suspension; also the construction of a bridge or other suspended erection by means of such chains in connection with inclined suspension rods supporting the platform'.[44] Inclined suspender rods had previously been proposed by Poyet in France and the combination of these with the taper chain in Dredge's model tests helps confuse his argument. When the patent was enrolled the Victoria bridge was under construction and the loyal local press considered the structure 'well worthy the attention of the scientific' and 'equally suitable for a large as a narrow span and the certainty of strength insured to the satisfaction of any one who may be able to know that two and two make four'.[45] Not everyone had such faith in Dredge bridges. The first reference to models in connection with the design appears in a letter from the disappointed Motley to the bridge Proprietors in July 1836:

> You are most of you aware that the project of the bridge originated entirely with me, and that I have been more or less engaged for more than 12 months endeavouring to get the object effected, and that I have incurred considerable expense in journies to Bristol, Marlboro', &c. &c. to produce what was deemed needful information, and also that a heavy expense has been incurred for models, drawings and experiments, with which all parties appeared satisfied except James Dredge, brewer, and Isaac Wilson, dentist, to whose conduct I unquestionably attribute the rejection of my plan.[46]

This gives an insight into the way in which bridge design competitions were managed at that time. No further information regarding Motley's design work has materialized, but that he was thinking in terms of a load test on a model we know from a letter written a week later urging the Proprietors to;

> direct some proper *disinterested* party to make a model upon J.D.'s first plan, beginning with eight bars and ending with two, or his second plan of beginning with ten bars and ending with three, and another model on my plan, to contain exactly the same quantity of material, each party to have the superintendence of his own plan, then appoint a time for trial, and I will engage that my model shall sustain at least double the weight which J.D.'s will do.[47]

The confident Motley added a postscript saying; 'If you object to the expence (two or three pounds) I will propose that J.D.'s model may be loaded till it breaks, and I will then arrange the same materials and make it sustain double the weight.'[48] In the subsequent argument it was never suggested that a model should represent, to scale, a particular bridge so the question of *similarity* does not arise. He merely proposed using two models to adjudicate the rival claims of the two protagonists. But he was not even comparing like with like. In a Motley bridge the roadway was supported by a series of inclined parallel rods connecting the deck to the towers without a catenary. They were, in fact, not suspension bridges but highly statically indeterminate stayed-girder bridges. He built a bridge of this type over the Avon at Twerton of 120-foot span which was described as being 'Superior to the common suspension bridge, in that it is more firm, and experiences much less friction, owing to the absence of vibration'.[49] After this, Motley the 'Grocer, Tea Dealer, and Bridge Designer'[50] withdraws and we must return to the subsequent work of James Dredge.

No evidence suggests that Dredge used a model test to justify his designs for either the Clifton or Victoria bridges, and it was not until the beginning of 1838 that he began his series of much-publicized tests. Considerable discussion followed the communication of his results to the press — invariably in a letter from Dredge. His experiments compared the strength/weight ratio (unlike Motley he did not specifically consider *stiffness*) of parallel and taper chain models of the same span/versed sine ratio, each containing 9 oz of iron wire (exclusive of hanging rods). The first trial was made in Bath on 2 January 1838 with two models of 4 feet 6 inches span, 6 inch deflection, and a platform 2 feet wide. His amusing results were given as follows — the parallel chain model 'broke down when loaded with six sacks of horse beans' and the taper (or *mathematical*) chain model 'Bore six sacks of horse beans, 7 sacks of malt, 2 cwt. cast iron, and 11 men, all this did not break it down'.[51] Considering the loading material it is to be assumed that the experiment was carried out in Dredge's brewery using materials to hand — but how, in addition to the inanimate load, he got eleven men on to the platform of so small a model is difficult to imagine. This test was witnessed by 'Messrs. Worsams from London, Ball, Cambridge, and others of Bath'.[52] These unscientific results were presumably sufficiently dramatic to demonstrate something in favour of Dredge's scheme but this must surely be discounted as a serious, well-conducted, experiment.

In the following fortnight three further tests were made on models of the same size in Bristol, 'in the presence of several eminent, commercial, scientific, and literary men, with attendant

TABLE 2

The Model Tests of James Dredge in Bristol and London

				Model Information						
				Chain Data						
				PARALLEL CHAIN			TAPER CHAIN			
DATE	LOCATION	SPAN	VERSED SINE	Weight Iron	Susp. Rods	Load Sustained	Weight Iron	Susp. Rods	Load Sustained	*Notes*
						cwt. Q. lb.			cwt. Q. lb.	
Jan. 6 1838 (A)	Bristol	4′ 6″	6″			13 3 25			34 1 25	'both broke on adding the subsequent 1 cwt.' (Ref. 2)
Jan. 13 1838 (B)	Bristol	4′ 6″	6″			13 0 0			33 0 0	'each broke on adding the subsequent ½ cwt.' (Ref. 2)
Jan. 13 1838 (C)	Bristol	4′ 6″	6″			23 2 0 *			61 0 17 **	'each broke with another ½ cwt.' (Ref. 2) 'Models made by Mr. Cross of Bristol.' (Ref. 2)
May 19 1840 (D)	Royal Gallery of Science	5′ 8½″	8½″	6½ oz.	Vert.	7 persons	6 oz.	Obl.	11 persons + dynamic load	
June 13 1840 (E)	College of Civil Engineers					9 persons		Obl.	18 persons	Steelyard used to measure load. Cords used for suspender rods.

REFERENCES:
1. *Bath Herald*, 27 January 1838, p. 3
2. *Mechanic's Magazine*, 24 March 1838, p. 423
3. *Mechanic's Magazine*, October 1838, p. 54
4. *Mechanic's Magazine*, July 1840, p. 122
5. *Civil Engineer and Architect's Journal*, June 1840, p. 194

*Given as 23 cwt. in Ref. 3
**Given as 61 cwt. in Ref. 2

engineers to superintend the uniform loading of the model bridges'.[53] This suggests the exercise of greater care in these tests, the result of which are given in Table 2. No additional information on the loading technique has been found but perhaps the consistently large difference in favour of the taper-chain models (approximately 2½:1) renders a detailed study unnecessary. For the experiment of 6 January (A) he commented that 'both broke on adding the subsequent 1 cwt.' which does not suggest small increments in the 'uniform loading', although the load carried is given in quarters and pounds. This trial was witnessed by 'Messrs. Protheroe, Guppy, and others'.[54] In the experiments of 13 January (B and C) the models were both broken on adding ½ cwt. revealing inconsistency in the loading method. The models for experiment C were 'constructed by Mr. Cross of Bristol, unknown to Mr. Dredge, in order to prove that all was fair in the former trials'.[55] These performed considerably better than those made by Dredge[56] to the identical specification and we may therefore assume that Cross was the better craftsman. These tests were witnessed by 'Messrs. Acraman, Daniels, Hillhouse, and many others of the first merchants of Bristol, Dr. Waldren and many others of Bath'.[57] These model tests were made at a time when research was not undertaken behind the closed doors of a laboratory and Dredge's meticulous lists of those present lend the events an atmosphere of theatre. He made an important statement which may be deemed as renouncing any claim to have considered *similarity* between model and full-sized structure: 'It now remains for scientific men to demonstrate why the same result should not take place in the largest chain piers and bridges.'[58]

In the early part of 1840 Dredge came under the influential patronage of Lord Western. On Thursday 30 April he said 'On Saturday last I introduced Mr. Dredge, the inventor of the improvement in bridge-building, to the Marquis of Northampton, at his evening assembly; He brought his models and drawings with him and they attracted great attention and admiration.'[59] This was, presumably, a mere *divertissement* on a social occasion but, more importantly;

> On the Monday morning following a trial of strength of the old system of building relatively to Mr. Dredge's, with two models of bridges, each formed out of the same weight of iron wire, namely six ounces, was made in the Marquis's garden, the Marquis and Lord Compton and others present. The model upon the old system was first loaded. It bore 18 half-hundred weights, and with the next half broke down. The model formed upon Mr. Dredge's principle sustained 34 half-hundreds, when upon the addition of another half-hundred, the wooden structure on which the chains were hung

> (answering to the towers of masonry, which form the fulcrum of suspension bridges) gave way altogether, and the bridge came down, but the chains were unbroken.[60]

No further details have been found, but comparing the weight of the chain with other Dredge models the span must have been between 5 and 6 feet. Western enthusiastically embraced Dredge's principle and publicly claimed that 'Mr. Dredge can build bridges on his plan at one-third the expense of the present method; he requires only one-third the quantity of iron, and his bridge will be stronger and freer from vibration and pendulous motion.'[61] On behalf of Dredge he wrote a long letter to the Prime Minister, Lord Melbourne, which was published in *The Times*, and subsequently widely discussed.[62] He described the model tests at Bristol (giving the results a very wide audience) and mentioned the 'prudent caution' of clients:

> Mr. Dredge's principle of suspension bridge building completely overthrows the theory and practice of a Telford, a Brunel, whose experience and talents we are bound highly to respect . . . can we then be surprised that the public should evince some fear, and some reluctance, hastily to adopt Mr. Dredge's novel principle or theory, in substitution of that which has been so long acted upon?[63]

He went on to say 'No human was ever exempt from error, and Messrs. Telford, Brunel, and others, must not be considered to be infallible.' But his main plea as Dredge's advocate was that he should be allowed to repeat his model experiments before the Prime Minister, the Menai Bridge Commissioners, or 'any of the most eminent engineers'. In addition he claimed that Dredge would repair the storm-damaged Menai bridge, at his own expense, and that the *superfluous* iron of the bridge would pay him and give a balance in favour of the Government. This dramatic claim, appearing in *The Times* on the last day of April 1840, must have initiated the experiments made in London during May and June. What information is available on these is contained in Table 2 (experiments D and E). That of 19 May was made at the Royal Gallery of Science, Adelaide Street, with two models somewhat larger than those used at Bath and Bristol:

> The first experiment was with a model constructed with two chains on the ordinary principle, each consisting of 3 wires laid parallel to each other . . . to which by the aid of vertical wires . . . a platform of wood was suspended, this platform was loaded with 7 full grown persons, and upon the eighth getting on, it broke down. The wire chains were fractured at the point of suspension. The weight of the wire in this model was 6½

> ounces. The next experiment was with two wire chains consisting of six wires at the point of suspension, and diminishing off to one in the centre . . . these chains supported by suspension wires placed obliquely . . . the platform which was loaded with 11 persons, without producing any fracture, until one or two of the party stamped on the platform, when it broke down, the fracture taking place at the junction of the oblique wires with the chain of suspension. The weight of the wire in this model was only 6 ounces.[64]

Here, as at Bath, there was no precision in loading and this was obviously undertaken as a dramatic public-relations exercise. The main point of interest is that not only are the parallel and taper chains compared, but the effect of vertical and inclined suspender rods. In a letter of August 1840 Dredge claimed his experiments had been made 'before the President of the Royal Society, and other noble lords'[65] and this must have been at the Royal Gallery of Science. In the following month he made his last experiments at the College of Civil Engineers, Gordon House, Kentish Town, London.[66] He again resorted to a human load and the parallel chain model 'gave way when only nine persons were standing still upon it, while eighteen persons standing on the tapered chain, had to jump before they could produce a fracture, the momentum of which must have been equivalent to a very considerable addition of weight'.[67] A more interesting experiment was made to demonstrate the relative performance of vertical and inclined suspender rods and was shown:

> in a model constructed for that purpose. This model consisted of the chains, with a spring-beam, or steel-yard at the centre, a flexible piece of timber for the roadway, and cords for the suspension rods, which could be arranged either vertically, as in chain bridges of the old construction — or diverging from the centre as in Mr. Dredge's improvement. When the cords were arranged vertically and the roadway pressed down, the strain . . . was shown by the action of the spring, which, when Mr. Dredge pressed down the roadway with both his hands, indicated a weight of 30 or 40 lbs. straining on the centre. When the suspension cords were arranged obliquely . . . the whole chain was called into action; and although the roadway was pressed down with equal or even greater force than before, the spring steel-yard remained quiescent.[68]

The steel-yard is the first reference to Dredge using a measuring instrument in conjunction with a model but the loading method was again hopelessly imprecise. No details of span, versed sine, or platform width were recorded for this experiment but it would seem to have been only a small demonstration model. The year

1840 saw the most extensive discussion of Dredge's ideas, no doubt initiated by Lord Western's letter to *The Times*.

During 1841 the debate about Dredge's system continued and five modest footbridges were built to his design in Regent's Park. In July a journal published an article by *Common Sense* on suspension bridges which extolled the virtue of Dredge bridges but had reservations about data which could only be obtained by experiment:

> experiments in such matters are not always to be trusted. They are usually made by means of models vastly smaller than the structures that are to be erected. Independently of this, it is difficult to get them done by the proper parties, or in the proper manner. The experiments of engineers are but too frequently made, not for the purpose of obtaining such results might turn out, but in order, if possible, to obtain some result previously desired by the experimenter; and in such a case, but small reliance can be placed of them.[69]

Common sense indeed. The mathematical reasoning associated with the taper-chain principle is not strictly within the scope of this paper, but it is important to note that it did receive attention. In 1841 William Turnbull published a treatise which attempted 'To trace the principle which inclined Mr. Dredge to adopt the tapering chain and the oblique suspending rods, and to prove mathematically that the principle thus adopted is strictly in accordance with the maxims of accurate mechanics'. [70] During 1843 Dredge himself published his mathematical analysis and graphic statics.[71] His mathematics was quite sophisticated and having made fundamental assumptions he showed himself adept at handling differential equations in developing his argument.

THE DREDGE AND CLIVE CORRESPONDENCE

During 1843 yet another resident of Bath entered into the controversy with Dredge and used models to justify his claims. This was the relatively brief exchange between Dredge and a Mr J. H. Clive. Like Motley, Clive was not discussing a suspension bridge but a stayed-girder bridge. His suspender rods were grouped in clusters and radiated from several points on the masonry towers (see Fig. 4). He is at least quite clear about this distinction:

> On considering my models and their indications, it appears to me, that the main chain or bearing, though one of the first and most naturally occurring modes of making a suspension bridge, is by no means the best . . . My models prove that the rods alone are equal to carrying the platform and load . . . therefore all the weight in the main chains which exceeds the

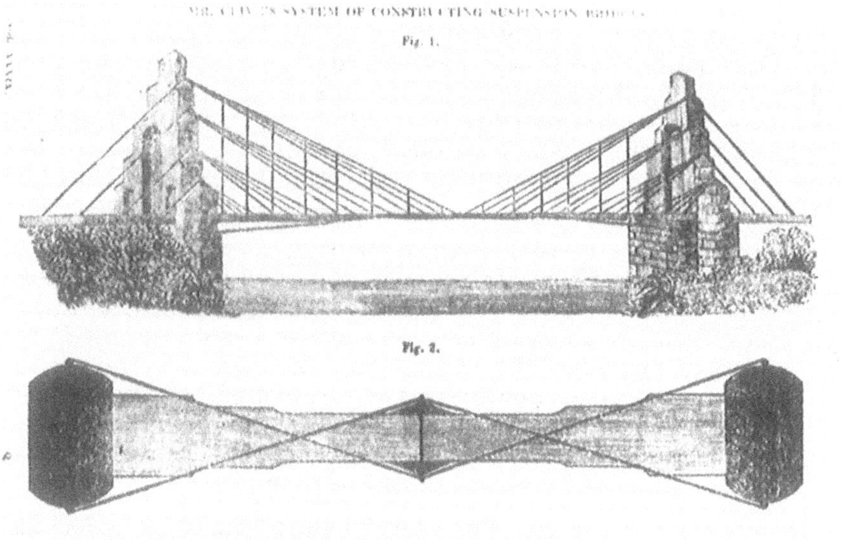

Figure 4. J. H. Clive's suspension bridge.
From *Mechanic's Magazine*, 23 September 1843.

> difference between inclined and perpendicular rods is worse then useless.[72]

Clive emerged in May 1843 when a letter of his was published in the *Mechanic's Magazine*[73] where he stated that he had seen both Dredge's and Motley's bridges over the Avon at Bath. He favoured Motley's system 'on account of its being free from any swinging motion, which chains can never be', and to test his ideas he 'constructed a rough model'. His drawing of it is shown in Fig. 5a and indicates an increasing tension in the parallel inclined suspender rods from the top to the bottom of the tower. He ended his letter with a challenge: 'If there be any fallacy in this reasoning of mine, for there is none in the model, I shall be most glad to have it pointed out.'[74] Dredge obliged in a letter a fortnight later; 'If a bracket were constructed in the manner he has shown . . . with the platform curved . . . the strain in the several tension cords would not vary as indicated by the weights.'[75] This drew a reply from Clive some six weeks later:

> Feeling assured that Mr. Dredge . . . had assumed an erroneous value for the relative forces brought into action, and thereby arrived at wrong conclusions . . . I reconstructed a model of the half bridge with four loose blocks or platform compartments, just ledged on each other (which I shall be

> happy to show him any day) in order to furnish such a near approach as a model would give.[76]

Clive's rebuilt model is shown in Fig. 5b. The subject, as presented by Clive, is confused by his individual blocks, 'just ledged on each other'. These introduce discontinuities into the deck making it incapable of sustaining bending at the junctions, rendering the system statically determinate and capable of analysis by simple mathematics. However, if such a calculation is made by taking moments about the junctions only a poor correlation with the experimental results is obtained. It is further confused by Clive's statement that the blocks were 'not of uniform texture and thickness, so as to be of equal weights in all parts'.[77] However, he admitted that in a real bridge he had no intention of letting the load 'pass over a parcel of *loose* blocks, merely kept on balance, but presupposed that they were to be suitably secured'.[78]

Clive admitted that his model experiments were 'only intended loosely to show one mode of making a projection in imitation of the horizontal bough of a tree, which is, in my opinion, an excellent example for study to all suspension bridge builders'.[79] No evidence suggests that Clive ever built a suspension bridge and we may assume that he entered the discussion in a philosophical manner, and it must be admitted that he did not shed much light on the subject.

THE BALLEE KHAL BRIDGE — CALCUTTA

Dredge's fame, and indeed the controversy, spread to the outposts of the British Empire when his principle was adopted by the Royal Corps of Engineers in India. This was its first use by the military and discussion began when a Dredge bridge in Calcutta collapsed during construction in 1845.[80] The erection of the Ballee Khal bridge was noted in the British technical press in October 1844[81] in an article quoting a letter dated July 1844 to James Dredge from Captain Goodwyn, R.E. of Calcutta. Goodwyn superintended the design and erection of the ill-fated bridge which had a span of 250 feet, a versed sine of 26 feet, and a deck 18 feet wide. He claimed to have carefully checked the preliminary design:

> With the assistance of a very able and first rate mathematician here, I have studied the theory of these bridges most thoroughly; and the model I have made 22 feet long, and 4 feet width of platform, is on so large a scale, that I have been able to test it in every possible way, and it has withstood the utmost efforts to derange its parts. The Governor-General, and all the scientific people here, have perfectly satisfied themselves of the efficiency of the system, and all these proofs, with my models, assure me that the theory is correct.[82]

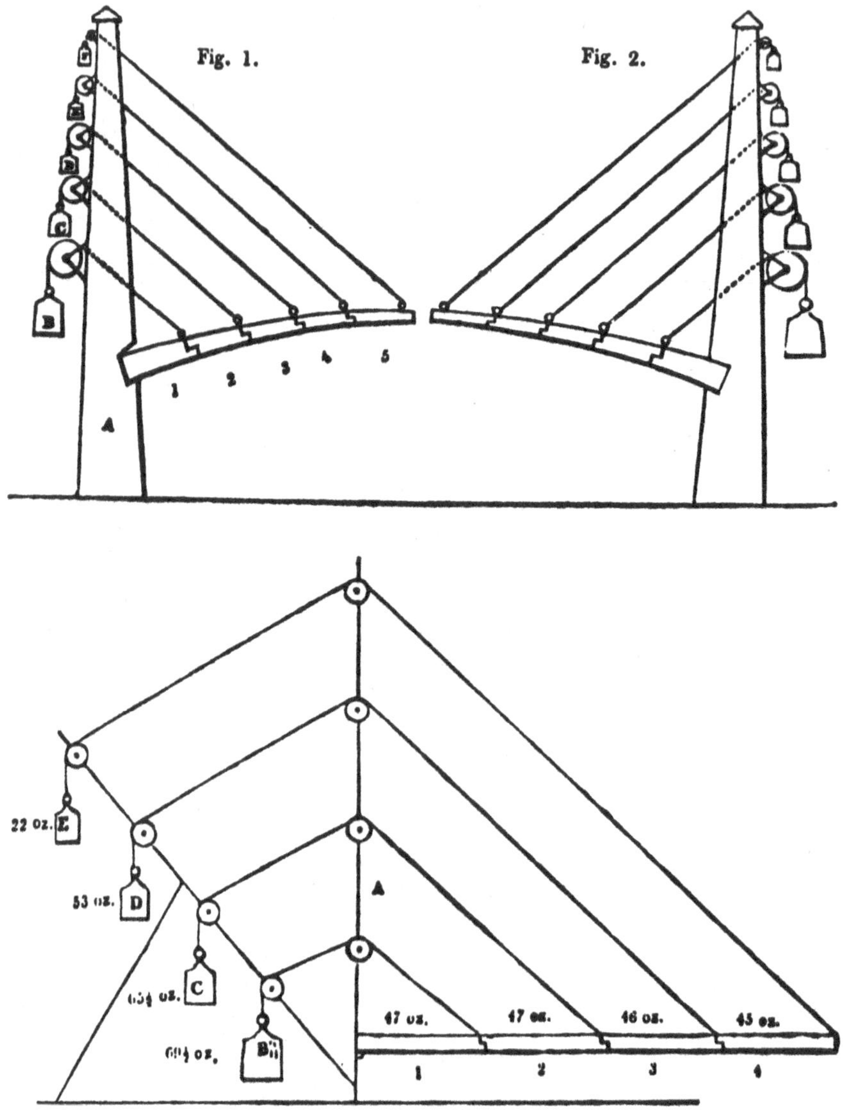

b

Figure 5. J. H. Clive's model suspension bridges — 1843.
(a) From *Mechanic's Magazine*, vol. 38, 20 May 1843.
(b) From *Mechanic's Magazine*, vol. 39, 22 July 1843.

We know nothing of either the mathematician or the theory studied but we may be sure that the model mentioned was not a scale representation of the bridge. After the collapse, a three-man committee of military engineers was directed to 'investigate the circumstances of the failure . . . and to report on the expediency of reconstructing it on the same principles as before'.[83] Goodwyn said the cause of failure was not one of principle but lack of care during construction.[84] Just before completion of the decking the overseer in charge, during Goodwyn's absence, placed a concentrated load of 40 tons at one end 'in innocent ignorance of the mischief he was doing' causing the horizontal iron edge-beams to buckle. The Committee accepted Goodwyn's analysis and, after making further experiments, decided that with modifications the bridge could safely be rebuilt. In describing the experiments the Committee made an interesting comment on the role of model tests in general:

> Useful, invaluable as models are, for rendering a particular mode of construction intelligible, all practical Engineers, as all mathematicians who are practically as well as theoretically acquainted with mechanics, know that *mere models* often lead to the most fallacious conclusions; and this from its being generally assumed (often without a shadow of satisfactory proof) that the dimensions of particular parts of the fabric only require to be increased in direct proportion to one of their linear dimensions, as for instance, in the case of bridges, directly in proportion to the spans; whereas extended or *full-scaled* experiments may be found to prove that the dimensions of some of the parts ought to be increased in some higher power than either the squares or cubes of the spans.[85]

This acknowledges the importance of the visual model and the caution expressed almost paraphrases Smeaton's words of 1759. It also reveals an awareness of the *similarity* problem, albeit rather vaguely. Experiments were made in India between October 1845 and August 1846 with the limited objective of testing the 'theory of a system based on the "resolution of forces"'.[86] Although this was a general analytical model, scaling laws were involved. The first experiment was 'on full scale as regards heights and distances, but formed of material 1/220 of the strength of the real bridge'.[87] It is not clear just what is meant by *strength* in this case; whether a material with a lower tensile breaking stress was used, or whether, in addition a scaling law related the cross-sectional dimensions of model and prototype. However, in the third experiment 'the points of suspension, the lengths of the rods and beam, heights and distances, being to a full scale . . . the sectional area of the iron was 1/196 part of reality'.[88] This obviously indicates a linear scale factor for the cross-sectional dimensions of $1/\sqrt{196} = 1/14$

making the self-weight of the model members also 1/196 of that of the prototype. The model was allegedly made to investigate *forces* but how these were measured is not described and the reason for reducing the cross-sectional area is not clear. Even though Dredge regarded the quantity of material in the failed bridge to be 'greater than was indispensibly requisite'[89] the new bridge was rebuilt, as a result of these experiments, with more material. The whole investigation was confused and involved both model tests and mathematical analysis. The Committee remarked that 'the eminent mathematician who attempted to investigate the principles of Mr. Dredge's mode of bridge construction appears only *partially* to have succeeded'.[90] One mathematician showed that suspender-rod tension increased uniformly towards the centre whilst Turnbull[91] arrived at exactly the reverse result. Indeed, Turnbull added that 'the successful application of the principle to practice must in a great measure depend on the sagacity and skill of the Engineer by whom the fabric is raised'.[92] The Committee expressed 'regret that the profound science which has been applied to its investigation has led to such an indefinite conclusion'.[94]

Other bridges were built, and more proposed, by the Royal Engineers in India at this time. Major Abbott of the Bengal Engineers visited a Dredge bridge of 120 feet span near Meerut (about 800 miles from Calcutta) and in a letter to Calcutta said:

> on being called upon to report on a project for bridging the Jumna, with spans of 500 feet, constructed on Mr. Dredge's system I no longer delayed to inquire more fully into the soundness of the principle... If I am troubling you with views that have been already set forth, you must attribute it to the distance which separates me, as well as many others of my brother officers from the scientific world. I am utterly destitute of the means of experimenting in support of my theory. I have referred the subject to Calcutta, where there are models, and on a large scale, but as yet I have received no reply.[94]

News of Goodwyn's model tests had obviously reached remote parts and the whole story is an interesting example of the way in which field military engineers undertook research at this time. The army's use of Dredge bridges in India ended with a catastrophe. Writing in October 1846, Goodwyn (now Major) said, 'A bridge on Mr. Dredge's principle, which was put up a short time since across the Kubudduck River, near Jessore, has fallen.'[95] On testing, the bridge was found to be weak even though '*guaranteed* by Mr. Dredge to stand any traffic' and before its use could be restricted, 'some grand festival took place on the water, to witness which, a crowd of natives rushed on the bridge, nearly filling the platform, and after a few minutes it gave way, and drowned nearly one hundred and fifty of them'.[96]

Although the Ballee Khal bridge committee, writing of Dredge's system, felt that 'the Civil Engineers of Europe have shown a marked disinclination to having any concern with its mathematical merits' one Englishman at least expressed strong views. After the news of the Ballee Khal bridge failure reached Britain an irate letter was published in the *Civil Engineer and Architect's Journal*:

> Any new plan for a suspension bridge unaccompanied by its correct mathematical theory is, in my opinion, unworthy of confidence. I do not dispute the fact of Mr. Dredge's having constructed bridges with a smaller quantity of material . . . but I do question *their safety* . . . We see here a laudable desire to take every precaution to ensure success — the theory is studied, a model is made, 1/12 of the size of the bridge, yet somehow theory, common sense, and experience all fail, and the blame is laid on the contractor; in my opinion there was another weighty reason.[97]

The writer accused Dredge of supporting his theory 'in face of its palpable absurdities' and charged him to 'state distinctly what is given and what is to be found by theory, and also what theory it was that the Indian engineer studied'.[98]

Dredge, and his use of models in bridge design have been dealt with here at some length, not so much because he extended model testing techniques (that he certainly did not do) but largely because of his considerable influence on the *discussion* of suspension bridge design. A study of the literature of the period confirms this. It is true that Dredge began his rather dramatic model experiments at a time when a young, rapidly growing, technical press must have been looking for *copy*, and he almost certainly received a greater share of attention than he deserved. Nevertheless many Dredge bridges were built. The first of 1832 in Scotland was followed by many others in England, Ireland, India, and some, during 1851, in Jamaica. It was during 1851 that the design and manufacture of the bridges was put on a formal footing, although not many were built after that date. Messrs Charles D. Young & Co., with extensive iron works in Edinburgh, published a pamphlet stating that they had 'completed arrangements with Messrs. Dredge and Stephenson C.E., London, whereby they become the Manufacturers and Contractors for the Patent Taper Suspension Bridge invented and patented by Mr. Dredge' but that 'the entire engineering, and the immediate superintendence of the erection of all such structures devolve upon Messrs. Dredge and Stephenson'.[99]

The real justification for Dredge's place in this account of structural model testing is his role in stimulating discussion on suspension bridge design — *the* subject for experiment in the decade 1830–40. The engineering world divided itself into two camps — either *for* or *against* the taper chain principle. Dredge's greatest contribution to our topic was to provoke argument and keep alive the *idea* that a model (however crude) could be part of an engineer's analytical equipment. This he did superbly well in the years before the tubular girder problems were to arise in the mid-1840s. In later large-span suspension bridge design the development of the spun wire cable, essentially of *uniform* cross section, rendered the discussion obsolete. The next experiments on model suspension bridges were in the second half of the century in connection with railways.

SUSPENSION BRIDGES AND RAILWAYS

That the rapid growth of railways in Britain made enormous demands on structural designers is probably nowhere better demonstrated than in the attempt to adapt the suspension bridge for railway traffic. The massive increase in loads in conjunction with the light tension structure posed problems of *stiffness* which were the subject of much discussion, analysis, and experiment in the middle of the century. The extreme caution felt by British engineers is reflected in the evidence to the Commissioners for the Application of Iron to Railway Structures in 1848. Only two engineers contributed opinions with respect to suspension bridges — Robert Stephenson and Isambard Brunel. Asked whether such bridges were applicable to railway structures, Stephenson replied: 'I do not think, with the prospects of our weights increasing upon railways, that you can run a locomotive engine over any chain bridge in existence.'[100] The Chairman then asked if that would apply to Dredge's bridges and he replied: 'Not so strongly perhaps.' It is worthy of note that Dredge was still considered sufficiently significant to warrant a separate question and that Stephenson regarded his bridges with somewhat more favour than the normal. But basically Stephenson's attitude is summed up in the following:

> I do not think that a railway bridge could be made on suspension principles; we have one at Stockton, which I replaced by one of these iron girder bridges, and it was fearful when the engine went on to it . . . when the engine and train went over the first time . . . there was a wave before the engine of something like 2 feet, just like a carpet.[101]

The dramatic failure of this Tees bridge, designed by Samuel Brown, undoubtedly set back work on British railway suspension bridges by many years. Brunel, however, was rather more sanguine and considered that they were 'applicable in the strict sense of the term, and that circumstances might arise in which they would be the best construction'.[102] However, he felt that Dredge's bridges were 'less applicable' than the normal type.

The problem of achieving adequate *stiffness* received considerable attention during 1857. In London during May a paper was read to the Institution of Civil Engineers 'On the Disturbances of Suspension Bridges, and the modes of counteracting them'[103] in which the joint authors summed up the contemporary situation:

> Suspension bridges, requiring considerably less material for their construction than girder bridges, and being practicable for far greater spans, than have ever been attempted with structures of a more rigid nature, would probably be frequently employed for railway traffic, as well as for other purposes, were they not so liable to disturbances of various kinds.[104]

They attempted to classify the 'various kinds' of disturbance and suggested remedies in a purely qualitative manner. During the discussion only one reference to a model was made. On the question of increased stability arising from the use of inclined suspender rods Mr Lukin said:

> This was one of the advantages of the inclined rods. When he made experiments upon a model, it was evident, that it was altogether so much stiffer, in proportion, than would be the case with a real bridge, that he merely noticed general results; but if the load caused a considerable deflection at any one point, without causing a corresponding elevation at another point, so far it might be argued, that there would be no material undulation. The model was for a span of 240 feet, at a scale of a quarter of an inch to a foot. When a truck loaded with 7 lbs. or 8 lbs, was sent across the loaded bridge, at various speeds he found that a deflection at one point did not produce an elevation at any other point; there was, in short, no undulation.[105]

This small model, of five-foot span, could obviously only provide the most general, unquantifiable, data, and to be fair Lukin admitted that he 'did not desire to draw any positive conclusions from such experiments'.[106] Later in the discussion P. W. Barlow rose and expressed faith in a girder-stiffened suspension bridge. As Engineer to the Londonderry and Eniskillen and Londonderry and Coleraine Railway Companies he had recently received instructions to design a bridge crossing the river Foyle at Londonderry, forming a junction between the two lines and effecting a road improvement. He proposed a two-platform bridge: 'the upper one to carry the ordinary traffic, and the lower one to support the railway carriages drawn by horses'.[107] The limitation to horse traction was significant as it removed the *hammer-blow* effect associated with the steam locomotive. Barlow suggested that the upper deck should be 'connected with the lower platform by lattice work'[108] resembling Roebling's Niagara bridge system. It could well be argued that the lattice stiffening system was anticipated thirty years earlier by Davies Gilbert when he said; 'to counteract and restrain undulating motion, the ballustrades may be carried to any desired height, and rendered inflexible by diagonal braces'.[109] However, Barlow did not refer to any model tests at this stage. Stephenson, as chairman of the meeting, said the paper 'touched upon an important point, which had deeply engaged the attention of engineers — the comparison of suspension bridges with those of girder construction, for use on railways'.[110] In summing up he remarked that the 'discussion went to show, that difference of opinion existed on this subject . . . and he trusted . . . they would be able to reconcile these differences, and to reduce the question to a simple form'.[111]

P. W. BARLOW'S EXPERIMENTS — 1857

The subject arose again at the Dublin meeting of the British Association later in the year. Charles Vignoles introduced the topic 'as appearing to possess sufficient interest for discussion, from the circumstance of difference of opinion amongst civil engineers having thrown doubt upon the feasibility of applying the principle of suspension to the purposes of railway transit'.[112] He referred to the success in America and considered it 'a striking set-off against the failure in this country, which occurred upwards of five and twenty years ago, under circumstances which have militated against any attempt to repeat the experiment'.[113] Like many other British engineers Vignoles was still very conscious of the Stockton bridge failure. He referred to the Civils meeting in May: 'from which many engineers were absent', and said the Dublin meeting would provide a further opportunity to discuss 'the intended application of a suspension bridge to carry a railway across a navigable river in the North of Ireland'.[114] This referred to the Londonderry bridge and Barlow, engineer to two Irish railway companies, read a lengthy paper to the meeting. He claimed great advantages in using a suspended girder and began by comparing the Britannia and Niagara bridges: 'the two largest railway openings yet constructed' revealing a very adverse span/weight ratio for the British bridge.[115] Although aware of the problems of live loading he said 'This is not, however, the subject I now submit for discussion; the first step in the enquiry is the simple mechanical problem of the strength and deflection with stationary loads, on which no doubt ought to exist.'[116] Although he used the word *strength* here his experiments really sought to establish the *stiffness* of the system, and for the first time the *deflection* of a model suspension bridge was measured. He claimed that his model tests 'distinctly prove that a suspended girder, as designed for the Londonderry Bridge, is rendered equally rigid with less than 1/25 of the metal required in the girder alone'.[117] His experiments are of particular interest in that they were the first to attempt the consideration of *similarity* between a model suspension bridge and the full-scale structure. Barlow 'had the model made . . . on a scale of 1/33 part of the actual span, the length being 13′ 6″ between the bearings'.[118] The principal object was 'to ascertain the deflection of the wave of a girder attached to a chain, as compared with the deflection of the same girder detached'.[119] Barlow then made an interesting statement about the scale of the model:

> My first intention was to make the experiments with a girder which was a correct model of the actual bridge, which would have indicated 1/33 of the actual deflection, but I found the deflection of the wave to be so small that it was difficult to measure it with sufficient accuracy, and I therefore had a wooden box made of the correct depth, with the sides as thin as it would stand, viz. 1/4 deal plank, in order to obtain greater deflection of the wave, with the correct depth of the girder, and with the chain attached to it as in the proposed bridge.
>
> I could no longer obtain the actual deflection of the Londonderry Bridge by multiplying the experimental deflections by 33, but knowing that the deflection of a model on the correct scale would be 1/33 of the Londonderry girder, and knowing by experiment how much the model was deflected when unattached, the actual deflection of the Londonderry girder is obtained by reducing the observed experimental deflection in the ratio of the rigidity of the actual model to a true model, and then multiplying by 33.[120]

The crucial question here is: What did he mean by a *correct* model of the actual bridge? If he scaled down all dimensions by a common geometric scale factor of 1/33 it would not, of course, have indicated 1/33 of the actual deflection. The wooden box girder, meant to simulate the two decks of the bridge, was described as being of the 'correct depth' but no dimensions were given. The problem is further complicated by his reference to a 'high tower' and a 'low tower'. It is not clear whether he was referring to a difference in height of the tops of the towers. This would produce an unsymmetrical cable profile, and was assumed to be the case by a contemporary writer,[121] but is rather unlikely. Even if the topography of the river Foyle provided one tower *footing* higher than the other, that was no reason (and certainly no

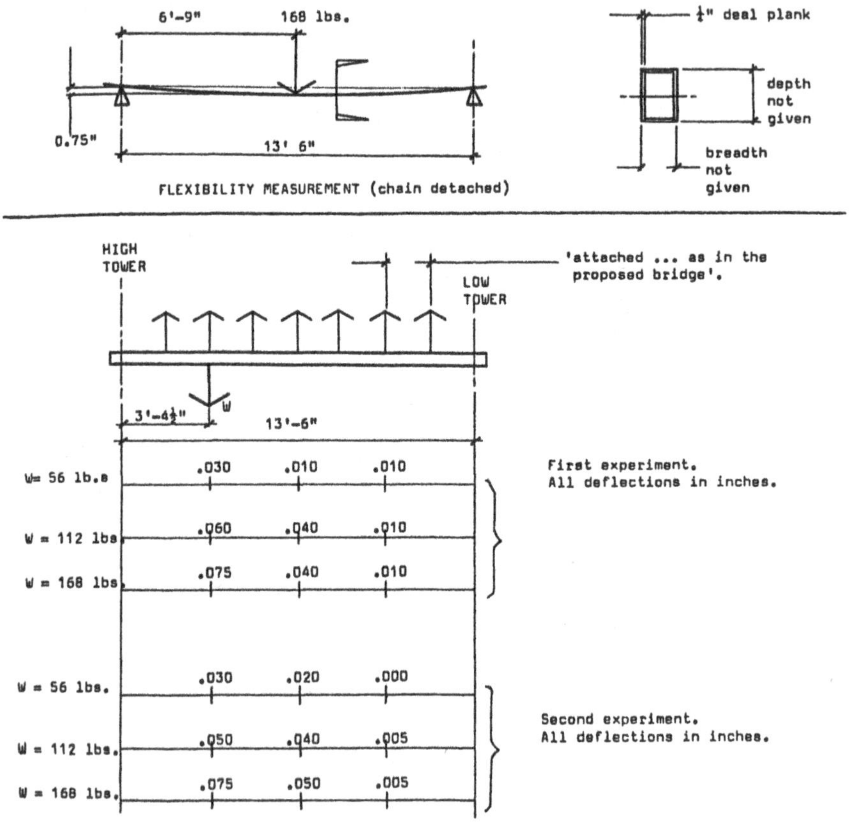

Figure 6. 1857 — Barlow's suspended wooden box girder.
From *British Association Report*, Dublin Meeting 1857, p. 240.

engineering advantage would follow) for having the *tops* of the towers at different levels. But here again Barlow is not explicit.

He made a series of experiments on four model girders of progressively reduced stiffness — a wooden box girder, an angle iron girder, a wooden plank 7½ inches wide by ¾ inches thick, and a plank 6 inches wide by ¾ inches thick. The suspended girder in each case was *two-hinged* and the system was therefore statically indeterminate. In each case deflection ordinates were recorded under different systems of loading and the deflected profiles are plotted in Fig. 6 to 9. He did not say how deflection was measured but as the smallest value recorded was 0.005 inches it must have involved either a vernier or micrometer. In analysing his results he was only concerned with the *maximum* deflection and compared that on the simply-supported unattached girder with the

Denis Smith 199

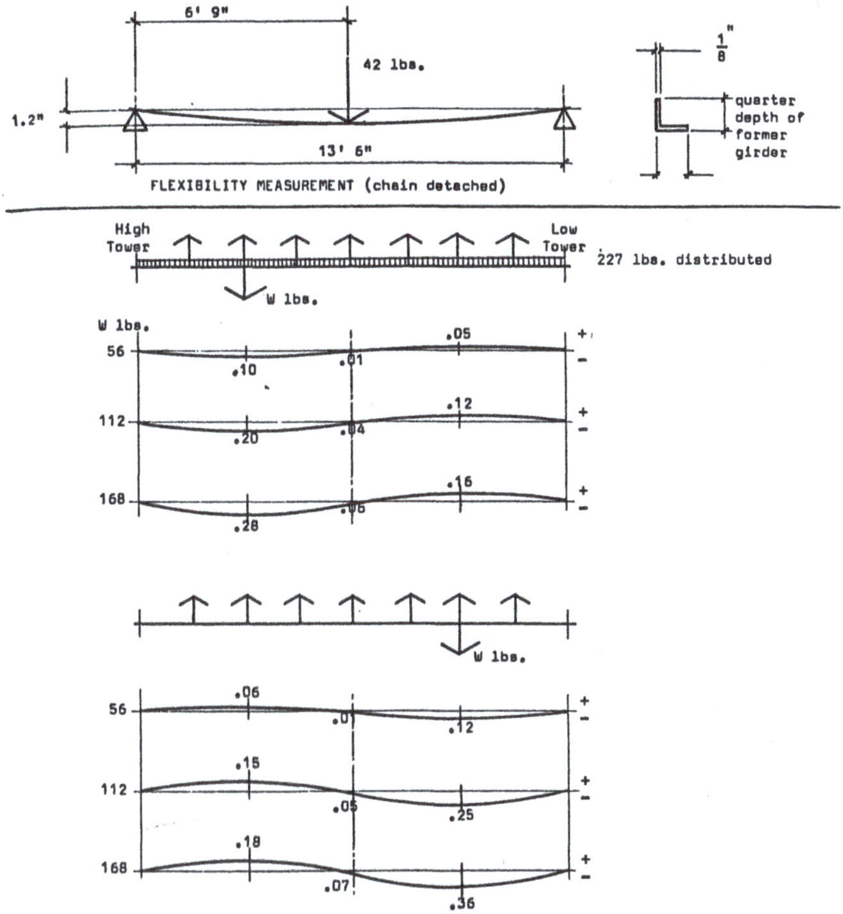

Figure 7. 1857 — Barlow's suspended angle girder.
From *British Association Report*, Dublin Meeting 1857, p. 240.

maximum on the suspended girder — even though these maxima did not occur at the same point on the span. The results with the wooden box girder are shown in Fig. 6. This box girder was so stiff that there was no upward deflection at any point and to refine the experiment Barlow 'decided in order to magnify the wave and make its amount more distinct, to have a girder made of angle-iron 1/8 inch thick and a quarter the depth of the former girder'.[122] This is shown in Fig. 7. He did not describe the method of suspending the girder beyond saying thay it was 'simply suspended from and not attached to the chain'.[123] This does not help and no drawing was published. Had he hung it with a leg vertical ('L') the maximum deflection would not have been in a vertical direction —

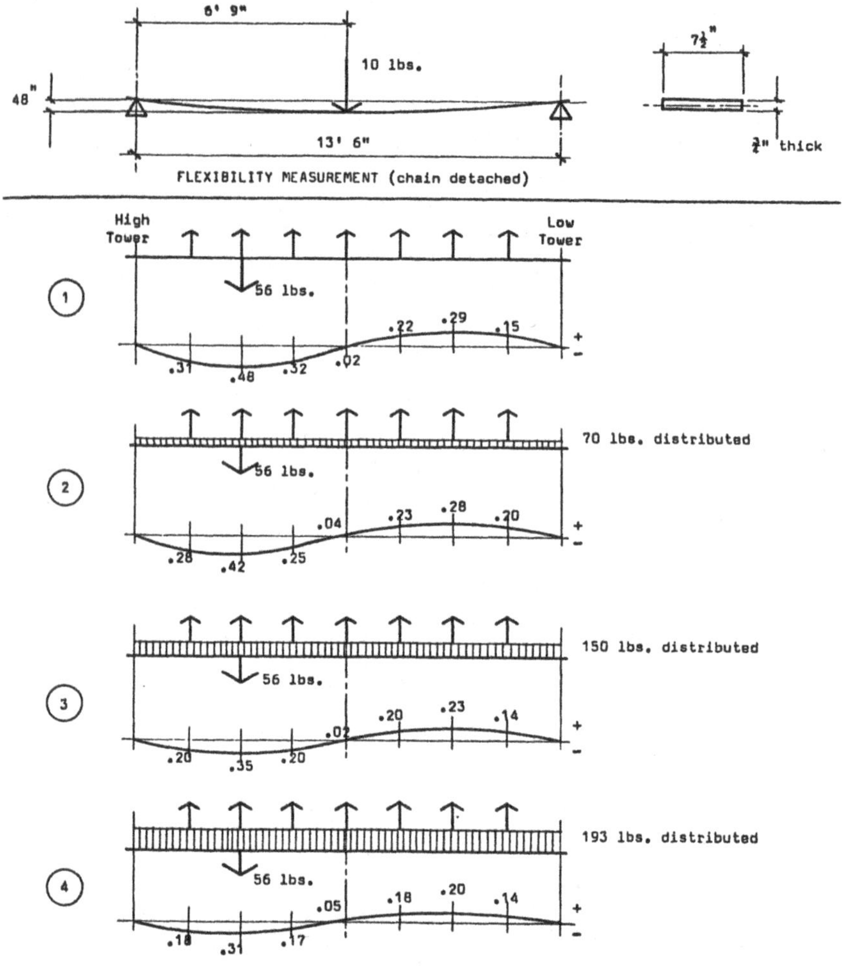

Figure 8. 1857 — Barlow's first plank girder.
From *British Association Report*, Dublin Meeting 1857, p. 241.

this could only be achieved by supporting it symmetrically disposed in the form of an inverted 'V'. However, comparing maximum deflections he said; 'The deflections here averaged .32 inch with 168 lbs., equal to 0.08 inch with 42 lbs., or 1/15 the deflection of the girder without the chain.'[124] He must have arrived at this as follows: $(0.28 + 0.36)/2 = 0.32$ ins which is $(0.32 \times 42)/168 = 0.08$ ins with 42 lbs, therefore $1.2/0.08 = 15$. Barlow incorrectly assumed linear load/deflection characteristics for the system. In conclusion he said:

> It was still obvious from the deflection at the centre and little rise exhibited in the wave, that the stretching of the chain to

> bring the metal surfaces to bear, still sensibly influenced the result, and I had another wooden girder made, consisting of a plank 7½ inches in width and ¾ of an inch thick, in order still more to magnify the wave, and to diminish the error from the stretching of the chain.

The results from this first plank girder are shown in Fig. 8[125] and indicate its reduced stiffness. In addition to the concentrated load at the quarter-span point he added an increasing uniformly distributed load in experiments 2, 3, and 4. Although the total load on the system increased, he showed that the damping effect of the u.d.l. was to decrease the maximum deflection. Barlow did not attempt to relate the model deflections to those of the Londonderry bridge by means of mathematical similarity. He did, however, attempt this by means of calculations involving the model deflections and *actual* deflections measured on three completed railway girder-bridges (without chains): the Boyne Viaduct, Newark Bridge, and the Britannia Bridge.[126] Deflection data existed for these bridges and he reduced the *average* maximum deflection, when carrying a central load of 100 tons, to 33 inches. Speaking of experiment 4 (Fig. 8) he said:

> The deflection here indicated with the model loaded with a weight representing 96 tons on the bridge (which experiment was several times repeated), was .31 with 56 lbs. = .055 with 10 lbs., or 1/26 of the deflection of the girder without the chain; 33/26 = 1.27 is therefore the deflection of the wave indicated by the experiment of the Londonderry Bridge, with a load of 100 tons at 1/4 from the tower.[127]

This is confusing. What factor related the load on the model to the '96 tons on the bridge'? He again compared maximum deflections occurring at different parts of the span and assumed linear load/deflection characteristics for the suspended girder. He added:

> To obtain the comparative rigidity of the experimental girder we have here as —
> 206 lbs. : 10 lbs. :: 1 in. :.0485, the deflection of a true model with 10 lbs.; 1.48/.0485 or 1/30.5 represents the rigidity of the experimental girder; .31/30.5 × 33 = .335, the deflection by a weight on the bridge of 56 × 33^2 = 27 tons. 27 : 100 :: .335 : 1.27, the deflection as previously calculated.[128]

Here again there are difficulties. Why did he use 206 lbs, the total load in experiment 3 (i.e. 150 + 56), whereas the load producing the maximum deflection of 0.31 inches in experiment 4 was 193 + 56 = 249 lbs? Similarly, why did he multiply the model load by the square of the geometric scale factor in relating it to the full-sized structure? Barlow made one further experiment using a

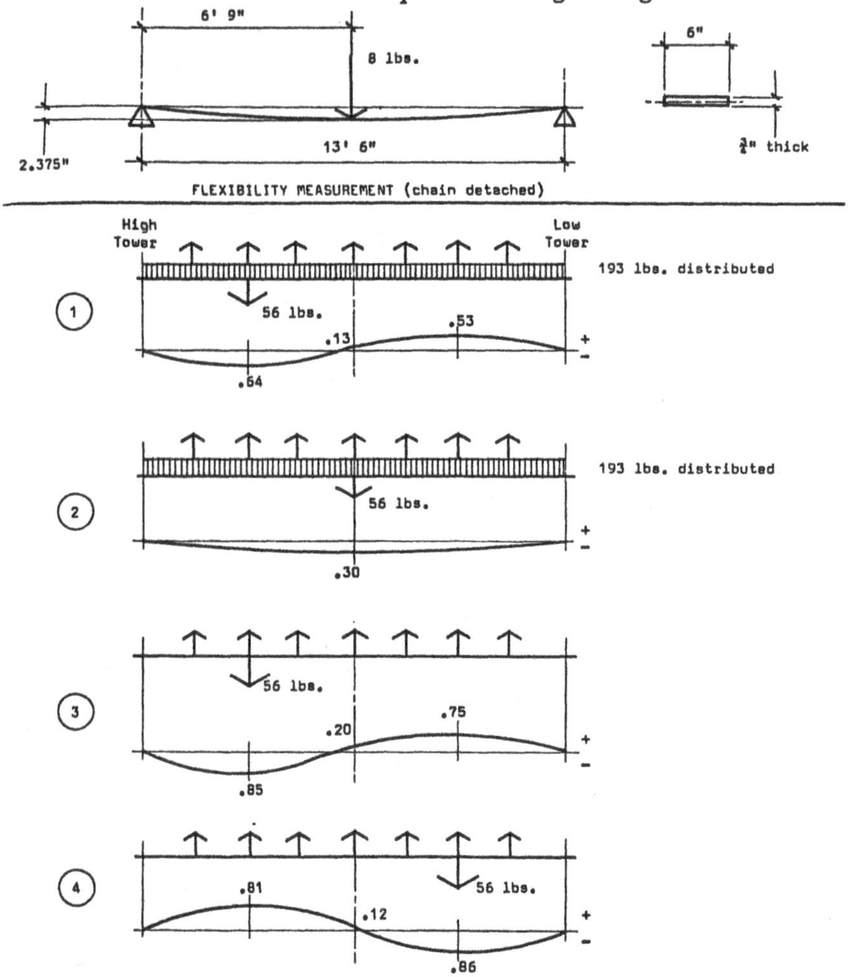

Figure 9. 1857 — Barlow's second plank girder.
From *British Association Report*, Dublin Meeting 1857, pp. 241–2.

plank 6 inches wide and $\frac{3}{4}$ inch thick with the results shown in Fig. 9, and from these he concluded that:

> the deflection is decreased by loading the bridge to 1/20 of that of the girder unattached, and if the chain were without weight it would be still further reduced; in practice, however, the weight on the bridge will much exceed that on a model, and 1/25 will be the least amount that will arise, a result so at variance with the preconceived notions of many engineers, that it is to be expected in some instances it will be received with incredulity.[129]

Although his account is rather muddled, Barlow's experiments did draw attention to the fact that a relatively light girder might effectively spread the load over the chain providing a stiffer system. This is quite different from the conclusion that Stephenson arrived at in connection with the Britannia bridge. Barlow's concluding observations were:

> 1st. That in suspension bridges it is essential that the platform should be stiffened with a girder to prevent vertical undulation.
> 2nd. That the deflection of the wave of a girder attached to a chain . . . will not exceed 1/25 of the deflection of the same girder not attached to the chain.
> 3rd. That theoretically the saving of metal to give equal strength in a suspension bridge is only one-half of that of a girder . . .[130]

It might well be thought that this dramatic claim would have convinced the Railway Company but, despite the success of Roebling's Niagara bridge, they still had reservations about railway suspension bridges. So they referred the question to John Hawkshaw who recommended a normal girder bridge.

In the summer of 1860 P. W. Barlow repeated his paper, with additions, to the Oxford meeting of the British Association. The paper was re-published in the August issue of *The Civil Engineers and Architect's Journal*[131] which led to an even wider discussion of the problem. At this time the Londonderry girder bridge was under construction and Barlow, rather bitter at Hawkshaw's decision, said:

> It is no doubt the safest course for an engineer to recommend what has been done before, and to avoid experiments; but this, it is submitted, does not render it less desirable that the subject . . . should be understood as a mechanical question.[132]

He reproduced the results of all his experiments except that of his first wooden box girder. He recalled that when he read his paper at the Dublin meeting in 1857, 'there were several girder engineers present, but no observation was made during the discussion upon the especial object of my experiments, except by Prof. Rankine, who . . . investigated the subject on this point, and arrived at nearly the same result'.[133] In fact, one of the most important results of Barlow's model experiments was to induce Rankine to develop and publish his approximate analytical technique which has been so widely used since that time. This assumed that, for a three-hinged or two-hinged stiffening girder, a concentrated load is spread *uniformly* over the whole span to the cable which retained a parabolic profile. This was the origin of the *elastic theory* — the first to attempt a rational account of the interaction between cable and girder. During the discussion of Barlow's 1860 paper his results were much contested, but he was supported by the President of the British Association, Professor Rankine. Although Barlow constantly referred to Roebling's structure 'it appeared that very little was known to the speakers of the success of the application of the principle in the chief example referred to — the Niagara Bridge'.[134] Barlow concluded with a plea for further work:

> It may be true that public feeling is against suspension bridges, from the repeated failures arising from want of metal, and from having no means of curing the wave; which arises, as my model has shown, by the smallest weight at any one point; but this is only a reason why men of science should look at the question, and correct an error which is a check upon useful enterprise and upon the progress of public improvement.[135]

The publication of Barlow's paper in August 1860 led to a two-part analytical article in November and December entitled, 'Suspension Bridges for Railway Traffic'. The anonymous author summed up the contemporary situation:

> the suitability of suspension bridges with rigid platforms to the requirements of railway traffic, is as yet to be regarded as a matter of opinion . . . if we consult the opinions even of the men most competent to pronounce on such a question, we are led to very opposite conclusions; and until the experiment is actually made on a large scale in some railway bridge, our only remaining resource lies in experiments on models, or in theoretic deductions . . . Great allowances are necessary in drawing any conclusions as to a proposed work of considerable magnitude from experiments on a model, however carefully conducted . . . It is therefore now proposed to inquire, what solution theory affords to the problem of the suspended girder; and while doing this it may be useful to note how far the results arrived at accord with Mr. Barlow's experiments.[136]

There followed some incomplete mathematics making it impossible to follow the development of his argument. He attempted to allow for the stretching of the chains and 'the inequality in the height of the towers, which is such as to increase this wave by 1/73 part when the load is placed one-fourth of span from the high tower, or to reduce it by the same amount when the load is placed one-fourth of span from the low tower'.[137] None of the accounts give the relative heights of the towers so it is impossible to check this statement. The un-named writer left the subject with a parting challenge:

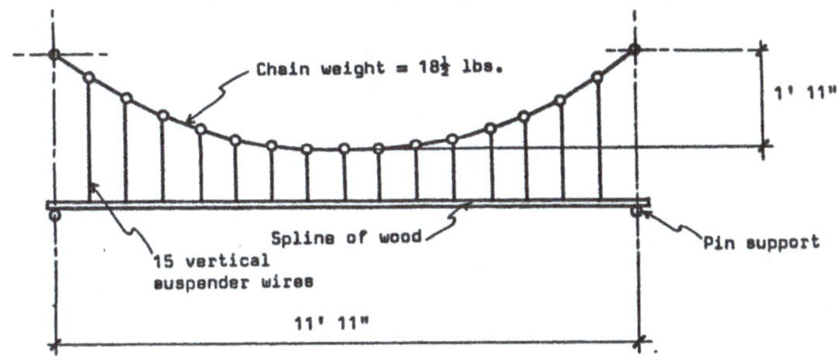

Figure 10. 1867 — Airey's experimental model.
'The upper surface of the spline was furnished with fifteen screw-eyes, at nearly equal intervals.'
From *Proc I.C.E.* vol 26, p. 260.

> Experiment and theory have now done their part. Each involve some known, and possibly also some unsuspected, sources of error. It remains for the practical engineer, by the bold test of some actual work, at once to correct and verify the conclusions of abstract science, and remove the question forever from the province of discussion.[138]

However, no such bridges were built and the subject was not re-opened experimentally until the Astronomer Royal investigated the subject some seven years later.

GEORGE BIDDELL AIRY'S EXPERIMENTS — 1867

A decade after Barlow's model experiments, Airy made a similar series of tests on an experimental rig of 11 feet 11 inches span (see Fig. 10). Airy never used the word *model* to describe this apparatus and so the question of *similarity* was not considered. This was a general analytical model. His results were made public in a paper read at the Institution of Civil Engineers in February 1867.[139] His attention had first been drawn to the question of suspended girders some seven years earlier in discussions with Stephenson about the Britannia bridge. At that time he had thought there were better methods available for wide spans than simple tubular girders. After Britannia 'No necessity for perfecting the theory arose for many years'.[140] But in 1863/4 he had again been involved in a plan for a large-span railway bridge and had proposed a suspended girder. It is interesting to note that by this time his mathematical analysis of the system presented 'no positive difficulty' but 'the complexity of symbols became so great' that he decided to 'refer to theory for considerations of a general class only, and to rely on experiment for the numerical determinations'.[141] If the Astronomer Royal could publicly admit to being overwhelmed by a 'complexity of symbols' it is not surprising to find engineers reluctant to adopt the analytical approach. At no time did he refer to Barlow's experiments but he came to very similar conclusions,

206 *Models in British Suspension Bridge Design*

In all experiments the unit of vertical deflection = 1/32 inch

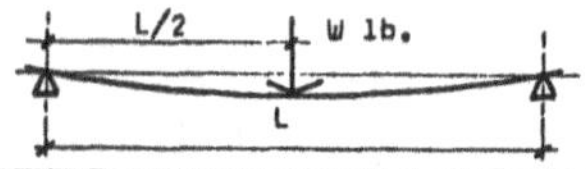

Flexibility measurement
THICK SPLINE

	Experiment Number	Span (L) ft. in.	Central deflection (W = 10 lbs.)	Central deflection (W = 20 lbs.)
1	1 & 2	3 0	1	2
	3 & 4	3 6	1.6	3.2
	5 & 6	4 0	2	4
	7 & 8	4 6	5	10
10	9 & 10	5 0	7	14

11–14

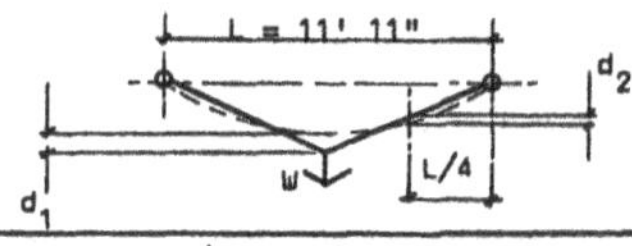

W (lb.)	d_1 (↓)	d_2 (↑)
10	78	42
20	93	64

15–20

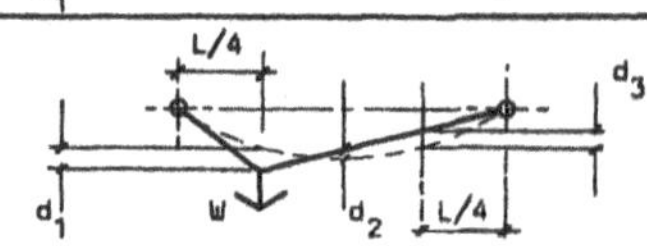

W (lb.)	d_1 (↓)	d_2 (↑)	d_3 (↑)
10	108	62	100
20	140	109	160

21–22

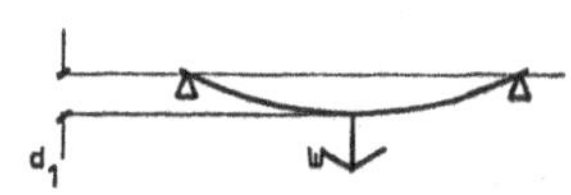

W (lb.)	d_1 (↓)
10	44
20	82

23–24

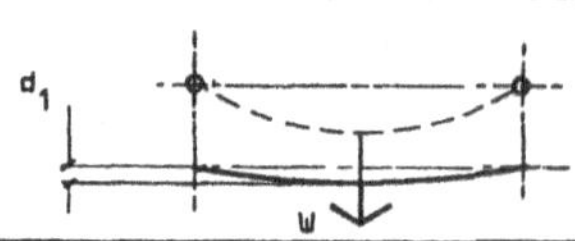

W (lb.)	d_1 (↓)
10	31
20	54

25–30

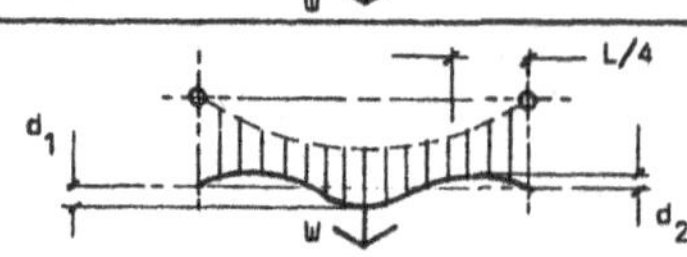

W (lb.)	d_1 (↓)	d_2 (↑)
10	2	1
20	5	1

31–36

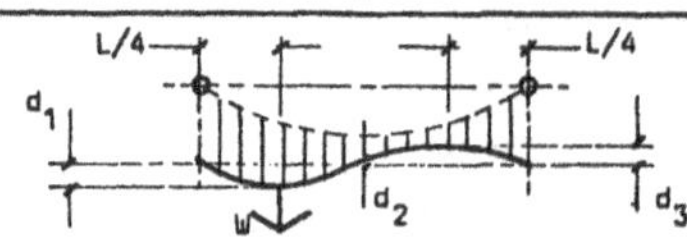

W (lb.)	d_1 (↓)	d_2 (↑)	d_3 (↑)
10	4	1	2
20	8	1	4

Figure 11. George Biddell Airey 1867.
From *Proc. I.C.E.* vol 26, p. 261.

namely, that the deflection of a suspended girder would be 1/25 to 1/30 of that when simply supported. He did not give the mathematical reasoning leading to this result, but for his experiments:

> A suspension-chain was provided, somewhat exceeding 12 feet in length, weighing 18½ lbs., attached at its ends to two pins at the distance of 11 feet 11 inches; the central dip of the

> chain being about 1 foot 11 inches. A spline of wood exceeding 12 feet in length (intended to represent the stiffened roadway) was lodged at its ends on pins, vertically below attachment-pins of the suspension-chain. The upper surface of the spline was furnished with fifteen screw-eyes, at nearly equal intervals, and these were connected with the suspension-chain by fifteen suspending wires, adjusted to support the spline in a horizontal position, and to be all (as nearly as possible) in bearing at the same time.[142]

No drawing was published but Fig. 10 is based on this description. The test rig was somewhat smaller than Barlow's but his objectives were identical. The experiments were made in two groups — the first with the stiffening girder represented by a thick plank of wood (no sizes were given), and the second with a thinner plank. The results are given in Figs. 11 and 12. He first measured midspan deflection of the 'thick spline' when simply-supported over spans varying from 3 to 5 feet (experiments 1 to 10). These showed, as was to be expected, a linear load/deflection characteristic. In experiments 11 to 20, the load was suspended directly from the catenary chain and the load/deflection characteristic is *non-linear*. This exposes the inherent analytical complexity of the stiffened suspension bridge problem — when a girder system with linear characteristics is attached to a non-linear chain the problem becomes statically indeterminate and, in engineering terms, only an approximate mathematical solution is possible. Experiments 21 and 22 measured the central deflection of the thick plank over the span of 11 feet 11 inches when carrying central loads of 10 and 20 lbs, giving deflections of 44/32 and 82/32 of an inch respectively. If these are compared with the central deflection in experiments 25 to 30 (with the plank attached to the chain by fifteen suspender rods) we see that they are reduced to 2/32 and 5/32 respectively, giving reduction factors of 44/2 = 22 and 82/5 = 16.4. Or, as Airy said;

> The comparison of the deflections in 25, 26, 51, 52, with those in 5, 6, 41, 42 shows that the central depression in the composite bridge equal to that of a tubular or girder bridge of the same section as that in the composite bridge but of one-third of the length, or less.[143]

Unlike Barlow, Airy's conclusions led him to propose a practical *design* method;

> For planning the tube or girder of such a bridge, the rule founded on known constructions is very simple. The Engineer will decide mentally on the amount of deflection that he will tolerate in the bridge as the effect of a given load. He will then, by the rules of engineering, compute the section of the tube or girder which will permit that deflection in a bridge

208 *Models in British Suspension Bridge Design*

In all experiments the unit of vertical deflection = 1/32 inch

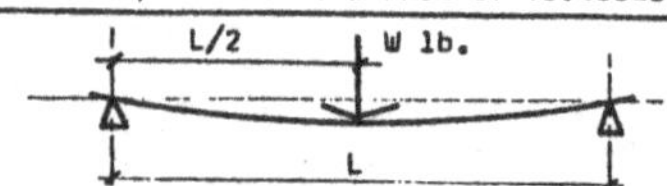

Flexibility measurement
THINNER SPLINE

Experiment Number	Span (L) ft. in.	Central deflection (W = 10 lbs.)	Central deflection (W = 20 lbs.)
37 & 38	3 0	2.5	5.5
39 & 40	3 6	4	8
41 & 42	4 0	5	10
43 & 44	4 6	8	16
45 & 46	5 0	10.5	21

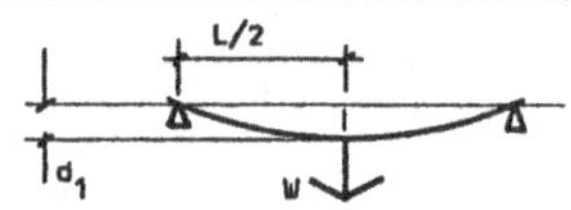

Experiment	W (lb.)	d_1 (↓)
47	10	138
48	20	240

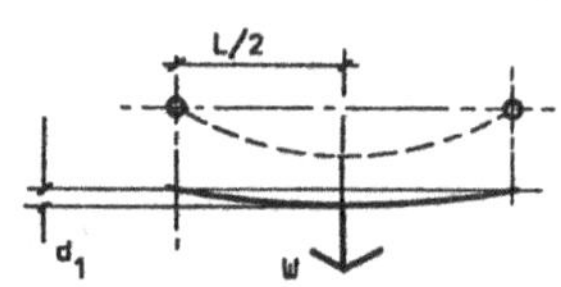

Experiment	W (lb.)	d_1 (↓)
49	10	40
50	20	60

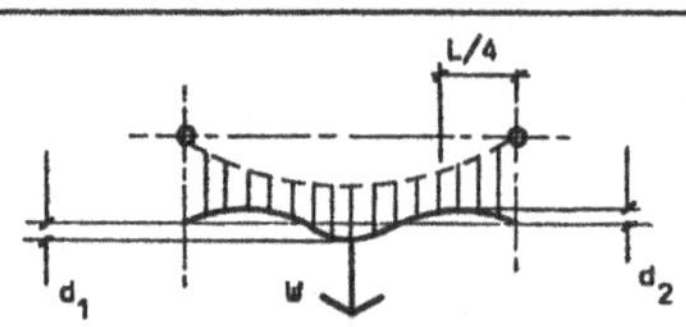

Experiment	W (lb.)	d_1 (↓)	d_2 (↑)
51	10	6	1
56	20	10	2

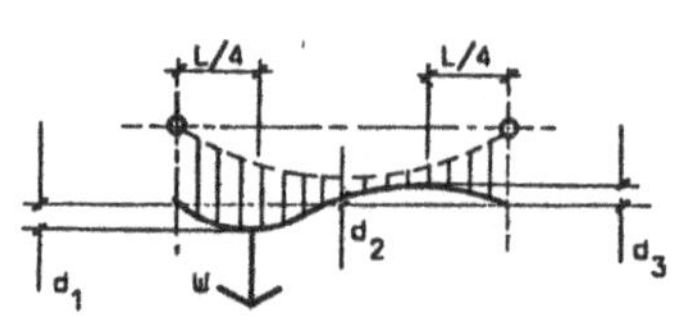

Experiment	W (lb.)	d_1 (↓)	d_2 (↑)	d_3 (↑)
57	10	9	1	3
62	20	17	1	6

Figure 12. George Biddell Airey 1867.
From *Proc. I.C.E.* vol. 26, pp. 261–2.

of 1/3 of the length in question or less; and this will be the section to be used for his long bridge.[144]

Here we have a model used, in the classic circumstances of analytical complexity, to provide the basis for an approximate design method to achieve a pre-determined degree of stiffness. It was the last model used in nineteenth-century suspension bridge design in Britain. Why was this? In answering this question it must be remembered that suspension bridge building largely lapsed in the last quarter of the century; in fact no purpose-built railway suspension bridge was ever constructed in Britain. The leaders in analysis and design of suspension bridges were now on the Continent and in America. Another factor reducing the need for model tests was the development of the more accurate *deflection theory*. Although it was known in the first half of the century that the displacement of a chain, without a stiffening girder, under the action of a concentrated load was non-linear, the first mathematical statement of this non-linearity was published in the British technical press of 1862.[145] Although Airy's experiments demonstrated the effect, the first non-linear deflection theory of suspension bridges was not evolved until 1888 by J. Melan.[146]

What role then did models play in the attempts to adapt the suspension bridge for railway traffic? In the decade 1857–67, connecting the experiments of Barlow and Airy, the plea was: 'Must the railway engineer . . . forego the advantages peculiar to the suspension bridge? . . . If it is in the power either of experiment or of analysis to throw any fresh light on this vexed question, the study cannot be deemed unimportant or uninteresting.'[147] British engineers were still haunted by the inadequacy of the Stockton and Darlington bridge, but as we have seen, two experimental attempts to grasp this analytical nettle were made. Barlow and Airy's small-scale tests did at least examine the system in which the future of large-span suspension bridges lay. Barlow's work is, on the whole, more interesting not only in that it preceded Airy's by ten years, but also as it sought to provided data for the design of a particular bridge — even if not subsequently built. But even Barlow had second thoughts about his model experiments. He was present at Airy's paper and said that he had

> listened with attention to the Astronomer Royal's observations and experiments, the fact being that similar experiments were made by himself in the year 1857, when he . . . arrived at similar results . . . However, the mere result of experiments of that kind he did not think sufficient to determine the best construction of a suspension bridge. Many practical points arose which could not be determined by experiments on a small scale.[148]

However, they drew the attention of others to the problem of the interaction of chain and stiffening girder, and this was perhaps their major contribution to the discussion.

Notes

1. *Book of Reference relative to the bridge at or near Runcorn Gap.* Cheshire Record Office, Chester, Ref: QDP 29, MSS dated 30 September 1813. Signed: Wm. Nicholson.

2. *Salopian Journal*, Vol. 21, No. 1077, Wednesday 21 September 1814.

3. *Mr. Telford's Reports*, Estimates, and plans for improving the Road from London to Liverpool, 3 April 1829, p. 7, (Warrington Reference Library).

4. Institution of Civil Engineers, London, Notebook No. 2, Ref. 52A.

5. *Parl. Pap.*, 1817, 'Report of the Select Committee, appointed to consider the most practicable and Expedient mode of effecting the proposed BRIDGE AT RUNCORN, submitted to the General Committee on the 8th. April 1817'; p. 11, 'Mr. Telford's Report'. (I.C.E. Lib. ref: Tract 4to, Vol. 13).

6. *Encyclopaedia Metropolitana*, Vol. 15, London, 1845, p. 362.

7. *Life of Thomas Telford* (autobiography), ed. Rickman. (1838), p. 542.

8. Ibid., p. 555.

9. *Proc. I.C.E.*, Vol. XVI, 1856–7, p. 133. Memoir of James Meadows Rendel (1799–1856). If Rendel joined Telford in 1814 he would only have been fifteen years of age.

10. As note 5, p. 13.

11. W. A. Provis, *An historical and descriptive account of the Suspension bridge constructed over the Menai Strait*, London, 1828, p. 16.

12. *Proc. I.C.E.* Vol. CLXV, 6 March 1906, 'The Widnes and Runcorn Transporter-Bridge' by John James Webster. A footnote on p. 88, speaking of the earlier suspension bridge scheme says:

> Telford made a model of this bridge to a scale of 20 feet to 1 inch, and this interesting memento is still in existence in a building belonging to the Shropshire Union Canals at Ellesmere, Shropshire.

13. Provis, *op. cit.*, p. 16.

14. As note 5, p. 13.

15. Charles Drewry, *A Memoir on Suspension Bridges*, London, 1832, p. 14. Details also given in Telford's autobiography (1838), p. 550.

16. As note 5, p. 4.

17. As note 5, p. 13.

18. *Patent No. 4137*, 10 July 1817: 'Invention or Improvement in the Construction of a Bridge, by the Formation and Uniting of its Component Parts in a Manner not hitherto Practised.'

19. Ibid., p. 4.

20. Ibid., p. 4.

21. Ibid., p. 5.

22. As note 5, p. 13.

23. *Tilloch's Philosophical Magazine & Journal*, Vol. 54, July 1819, p. 17: 'Third Report from the Select Committee on the Road from London to Holyhead. 29 April 1819'.

24. As note 12, p. 87.

25. John Nicholson, *The Operative Mechanic*, 4th. edition, London 1853, p. 805.

26. As note 12, p. 87.

27. *Edinburgh Philosophical Journal*, Vol. 5, 1821, pp. 140–56 and 267–81.

28. Ibid., p. 268.

29. Ibid., p. 268.

30. *Parl. Pap. 1819 (256)*. Vol. 5, p. 138: 'Third Report from the Select Committee on the Road from London to Holyhead'.

31. Provis, *op. cit.*, p. 48.

32. Provis, *op. cit.*, p. 89: Appendix No. 5.

33. Provis, *op. cit.*, p. 48.

34. Charles Drewry, *A Memoir on Suspension Bridges*, London, 1832, p. 160.

35. Charles Stewart Drewry, *A Memoir on Suspension Bridges*, London, 1832, pp. 154–8: 'Of the Use of Wood for the Supporting Chains of Suspension Bridges'.

36. Ibid., p. 154.

37. Ibid., p. 156.

38. *Report of the British Association for the Advancement of Science*, Oxford meeting, 1832, pp. 608–10: 'On the Steel Suspension Bridge recently built over an arm of the Danube at Vienna; and on the mode by which the exceeding tough steel employed was manufactured in Styria, at a small advance upon the cost of Iron', by James J. Hawkins.

39. See *Phil. Trans.*, 1826, part III, p. 202: 'On the Mathematical Theory of Suspension Bridges, with tables for facilitating their construction', by Davies Gilbert, in which he demonstrates the varying tension in a catenary. Several subsequent authors acknowledge this paper.

40. James Dredge (1794–1863) of Walcot is described as a brewer in his patent specification of 1836.

41. Thomas Motley, a grocer, gives his address as 'Mount Pleasant Cottage, Beechen Cliff' in his correspondence with the *Bath Journal.*

42. See *Surveyor, Engineer, and Architect*, 1 September 1840, p. 174: letter from J.D. dated, Bath, 11 August 1840, gives details of his Clifton bridge design. See also *Mechanic's Magazine* Vol. 30, No. 794, October 1838, p. 54.

43. The bridge had one span of 200 feet, platform width of 20 feet, and was erected for Sir James Colquhoun.

44. Patent No. 7120 dated 17 June 1836.

45. *Bath Journal*, Monday 11 July 1836, p. 2.

46. *Bath Journal*, Monday 18 July 1836, p. 2: open letter from Motley.

47. *Bath Journal*, Monday 25 July 1836, p. 2: open letter from Motley.

48. Ibid.

49. *British Association Report*, Newcastle Meeting, 1838, p. 157. A great deal of confusion exists over this bridge. It is invariably described in the technical press as 'Tiverton' bridge. The confusion is possibly explained in Collinson's *History of Somerset* (Taunton, 1898) where the Bath place-name is given as 'Twiverton or Twerton'.

50. He was so addressed in a letter from William Turner (the Bath engineer who made the Victoria Bridge chains) published in the *Bath Journal* for 15 August 1836. This letter appears to end the acrimonious exchanges in this newspaper.

51. *Mechanic's Magazine*, Vol. 30, No. 794, October 1838, p. 54: letter from J.D. dated Bath, October 19 1838.

52. *The Times*, Thursday 30 April 1840, p. 6: letter from Lord Western.

53. *Mechanic's Magazine*, Vol. 28, 24 March 1838, p. 422: letter to Editor from J.D. dated Bath, 14 March 1838.

54. *The Times*, Thursday 30 April 1840, p. 6. The 'Guppy' mentioned must be Thomas Richard Guppy, the Bristol merchant who was a principal promoter of the Great Western Railway.

55. Ibid.

56. Both Cross's models performed better than Dredge's and if the load (in quarters) which they carried is compared, a performance quotient can be obtained:

Parallel chain model: $$\frac{\text{Cross}}{\text{Dredge}} = \frac{94}{52} = 1.8$$

Taper chain model: $$\frac{\text{Cross}}{\text{Dredge}} = \frac{244}{132} = 1.85$$

57. *Surveyor, Engineer, and Architect*, No. V, 1 June 1840, p. 101.

58. *Mechanics Magazine*, Vol. 28, 24 March, p. 433.

59. *The Times*, Thursday 30 April 1840, p. 6: preamble to Western's letter to Melbourne.

60. Ibid.

61. Ibid.

62. *The Times*, Thursday 30 April 1840, p. 6: Lord Western's letter to Lord Melbourne. This letter was also published in full, but not in its original sequence, with illustrations and editorial appendix in: *Civil Engineer and Architect's Journal*, Vol. 3, No. 33, 1 June 1840, pp. 193–4. An abridged version of the letter was also published in the *Mechanic's Magazine*, Vol. 32.

63. *The Times*, Ibid.

64. *Civil Engineer and Architect's Journal*, Vol. 3, No. 33, June 1840, p. 194, (Editor's appendix to Western's letter).

65. *Surveyor, Engineer, and Architect*, 1 September 1840, p. 174: letter from James Dredge dated 'Bath, August 11 1840'.

66. This recently-formed College had only appointed its staff in the spring of 1840 and received considerable adverse criticism in the press. See the *Surveyor, Engineer, and Architect* for September 1840.

67. *Mechanic's Magazine*, Vol. 33, No. 884, 18 July 1840, p. 122.

68. Ibid.

69. *Surveyor, Engineer, and Architect*, Vol. 2, 1 July 1841, p. 131: Article, '"Common Sense" on Suspension Bridges'.

70. William Turnbull C.E., *The Mathematical Principle of Dredge's Suspension Bridge*, London, John Weale, 1841.

71. See *Mechanic's Magazine*, Vol. 38, 18 March 1843, pp. 213–15: 'Mathematical demonstration of the Principles of Dredge's Patent iron bridges, by the Inventor'. (Article dated: 'Bath Feb. 15 1843'.) Article completed in the issue of 28 March 1843, pp. 234–9.

72. *Mechanic's Magazine*, Vol. 39, 19 August 1843, pp. 133–4.

73. *Mechanic's Magazine*, Vol. 38, 20 May 1843, pp. 406–7.

74. Ibid.

75. *Mechanic's Magazine*, Vol. 38, 3 June 1843, p. 463.

76. *Mechanic's Magazine*, Vol. 39, 22 July 1843, pp. 57–8.

77. Ibid., p. 57.

78. Ibid., p. 58.

79. Ibid., p. 58.

80. See *Professional Papers of the Royal Corps of Engineers*, Vol. IX, 1847, pp. 83–144: 'The Taper Chain Tension Bridge at Ballee Khal, near Calcutta, in its renewed form after the Failure in June 1845', by Captain Goodwyn, E.I.C. Engineers.

81. *Mechanic's Magazine*, No. 1106, 19 October 1844, pp. 258–61: 'Description of a Suspension Bridge on Mr. Dredge's principle, erected over the Ballee Khal, for the Indian Government, from the designs, and under the superintendence of Captain Goodwyn, R.B.E.8. (A woodcut of the bridge appears on the front cover of this issue.)

82. Ibid., p. 261.

83. The committee comprised Lieut-Colonel E. Garstin, Superintending Engineer, L.Ps.; Lieut-Colonel W. N. Forbes, Engineers, Mint Master; and Lieut-Colonel A. Irvine, C.B., Engineers, and was convened by an order dated: 'Fort William, 18th. June, 1845. Their report was dated: 'Calcutta 6 September 1845' but was not published in Britain until 1847 (see *Professional Papers* above).

84. As note 80, pp. 126–130. Appendix to report signed (rather surprisingly) 'H. Goodwyn, Captain, Civil Architect'.

85. As note 80, Report, p. 121.

86. As note 80, pp. 130–6: 'Letter from Captain Goodwyn to Major Greene, Secretary to the Military Board'. Dated: 'Iron Bridge Department, Fort William, 28 August 1846'.

87. Ibid., p. 132.

88. Ibid., p. 134.

89. Ibid., p. 115.

90. Ibid., p. 114.

91. William Turnbull, C.E., *Mathematical Principle of Dredge's Suspension Bridge*, John Weale, London, 1841.

92. Turnbull, *op. cit.*, p. 35.

93. As note 80, p. 115.

94. *Professional Papers of the Royal Corps of Engineers*, Vol. I, 2nd ed., 1844, pp. 24–6: 'An Inquiry into the Principle of Mr. Dredge's Tension Bridge', by Major F. Abbott, Bengal Engineers; p. 24: Letter from Abbott, to Captain Denison, R.E.

95. As note 80, p. 138: Letter dated, 'Calcutta, October 1846'.

96. Ibid.

97. *Civil Engineer and Architect's Journal*, October 1845, p. 313: 'on Dredge's Suspension Bridges' by F. Bashforth, B.A., Fellow of St. John's College Cambridge and of the Cambridge Philosophical Soc.

98. Ibid.

99. *I.C.E. Tract*, Vol. 69, Pamphlet, (dated 1851): 'Description of Suspension Bridges on Dredge's Patent taper principle', published by Charles D. Young & Co., London and Edinburgh, p. 5.

100. *Parl. Pap.* 1849 (1123) XXIX. *Report of the Commissioners appointed to inquire into the Application of Iron to Railway Structures.* p. 357: Robert Stephenson giving evidence on 16 March 1848.

101. Ibid.

102. *Parl. Pap.*, 1849, (1123), XXIX, *op. cit.* p. 376: I. K. Brunel giving evidence on 13 April 1848.

103. *Proc. I.C.E.*, Volume XVI, 19 May 1857, pp. 458–78: paper by A. S. Lukin and C. E. Conder.

104. Ibid., p. 458.

105. Ibid., p. 474.

106. Ibid.

107. Ibid., p. 476.

108. Ibid.

109. *Phil. Trans.*, 1826, Part III, p. 211.

110. *Proc. I.C.E.*, Volume XVI, *op. cit.*, p. 473.

111. Ibid., p. 478.

112. *British Association Report*, Dublin 1857, pp. 154–8: 'On the Adaptation of Suspension Bridges to Sustain the Passage of Railway Trains', by C. Vignoles, C.E., F.R.S., p. 154.

113. Ibid.

114. Ibid.

115. *British Association Report*, Dublin, 1857, pp. 238–48: 'On the Mechanical Effect of combining Girders and Suspension Chains, and a comparison of the weight of metal in Ordinary and Suspension Girders, to produce equal deflections with a given load', by Peter W. Barlow, F.R.S; p. 239: Barlow made the comparison:

	Span	*Weight of Material*
Niagara Bridge	*820 feet*	*1,000 tons*
Britannia Bridge	*460 feet*	*3,000 tons*

116. Ibid., p. 246.

117. Ibid., p. 239.

118. Ibid.

119. Ibid.

120. Ibid., p. 240.

121. See *The Civil Engineer and Architect's Journal*, December 1860, p. 352, article by anonymous author, 'Suspension Girder Bridges for Railway Traffic'.

122. *British Association Report*, 1857, *op. cit.*, p. 240.

123. Ibid.

124. Ibid.

125. Ibid., p. 241.

126. Ibid., p. 246: Appendix A.

127. Ibid., p. 242.

128. Ibid.

129. Ibid., p. 242.

130. Ibid., pp. 244–5.

131. *The Civil Engineer and Architect's Journal*, August 1860, pp. 225–30.

132. Ibid., p. 226.

133. Ibid., p. 229.

134. Ibid.

135. Ibid.

136. *The Civil Engineer and Architect's Journal*, November 1860, pp. 317–18.

137. *The Civil Engineer and Architect's Journal*, December 1860, p. 352.

138. Ibid., p. 356.

139. *Proc. I.C.E.*, Vol. 26, 19 February 1867, pp. 258–64: 'On the Use of the Suspension Bridge with Stiffened Roadway, for Railway and other Bridges of Great Span', by George Biddell Airy.

140. Ibid.

141. Ibid.
142. Ibid.
143. Ibid., p. 263.
144. Ibid.
145. *The Civil Engineer and Architect's Journal*, August 1862, pp. 236–7: 'The Statics of Bridges — the Suspension Chain'.
146. *Handbuch der Ingenieurwissenschaften*, 2nd edition, Leipzig, 1888, 'Theorie der eisernen Bogenbrucken und der Hangebrucken', by J. Melan.
147. *The Civil Engineer and Architect's Journal*, November 1860, p. 318 Article: 'Suspension Girder Bridges for Railway Traffic'. Anon.
148. As note 139, p. 278.

5
Brunel and the design of the Clifton Suspension Bridge

R.F.D. Porter Goff

Introduction

The River Avon flows through Bristol to meet the Severn estuary 9 miles away at Avonmouth. About 1 mile downstream from the Bristol docks, the river enters a gorge with cliffs rising precipitously to some 300 ft on the Gloucestershire side at Clifton, but with slopes ascending less steeply through Leigh Woods from the Somerset bank. The river itself is tidal in these reaches and is about 300 ft wide from bank to bank (Figs 1 and 2).

2. The project to span the Avon at Clifton lasted more than a century. In 1753 Alderman William Vick of Bristol left £1000 in trust for the building of a stone bridge.[1] By 1829 the Trustees had in hand a sum approaching that thought by Vick to be sufficient for the work. The accumulation of the trust fund, however, had not increased as rapidly as the estimated cost of a masonry structure. The Trustees therefore advertised for designs of an 'iron suspension bridge',[2] and in March 1831 appointed Isambard Kingdom Brunel (1806–1859) as their Engineer. Site preparation began three months later but the disturbed state of the country, as evidenced by the Bristol Riots of October 1831, prevented the raising of sufficient loan capital to allow the work to proceed. It was not until August 1836 that the foundation stone was laid (Fig. 3). Construction of the abutments and suspension towers then advanced until 1843 when the Trustees' funds were exhausted. The project was formally abandoned in 1853 and the ironwork sold to meet the debts of the Trustees. Brunel died in 1859. The following year certain members of the Institution of Civil Engineers formed a company to finish the bridge as a monument to their 'late friend and colleague' and in December 1864 the work was finally complete.

3. The Clifton bridge was Brunel's first professional commission and this he obtained in unusual circumstances.[3] Telford, acting as assessor of the 1829 competition, had rejected all the designs, including Brunel's, submitted

PORTER GOFF

Fig. 1. Aerial view of Avon Gorge at Clifton (looking north)
[*Cambridge University Collection: copyright reserved*]

by the ten entrants.[4] The Trustees had then invited him to make alternative proposals and had issued a prospectus based on his design—a three-span bridge suspended from immense Gothic towers founded on each bank of the river. The initial response of the public had been encouraging but lack of sufficient financial support for Telford's scheme had constrained the Trustees, late in 1830, to seek further competitive designs. Telford, now a competitor, re-submitted his earlier design; Brunel put forward fresh proposals, and from among the remaining submissions the Trustees selected the designs of three other entrants for further consideration. On this occasion the judge was Davies Gilbert, immediate past-president of the Royal Society, assisted by John Seaward, partner in a foundry. As a matter of policy the Trustees chose men who were not professional engineers, although both had published papers on suspension bridges.[5,6] In their joint report Gilbert and Seaward set Telford's plan aside as too costly; Brunel's proposal was placed only second in order of merit on the grounds of deficiency in detail design. The Trustees, however, referred the report back for more information. Gilbert, in the absence of Seaward, sought further explanations from Brunel and then advised the Trustees that he now judged Brunel's design, with certain mutually agreed modifications, superior to all others. The Trustees accepted this decision and Brunel became their Civil Engineer.

4. Brunel's calculation books record his design work at Clifton in three distinct periods. In the first, 1829–30, Brunel prepared his design proposals. He offered four alternative schemes in the first competition, and conceived

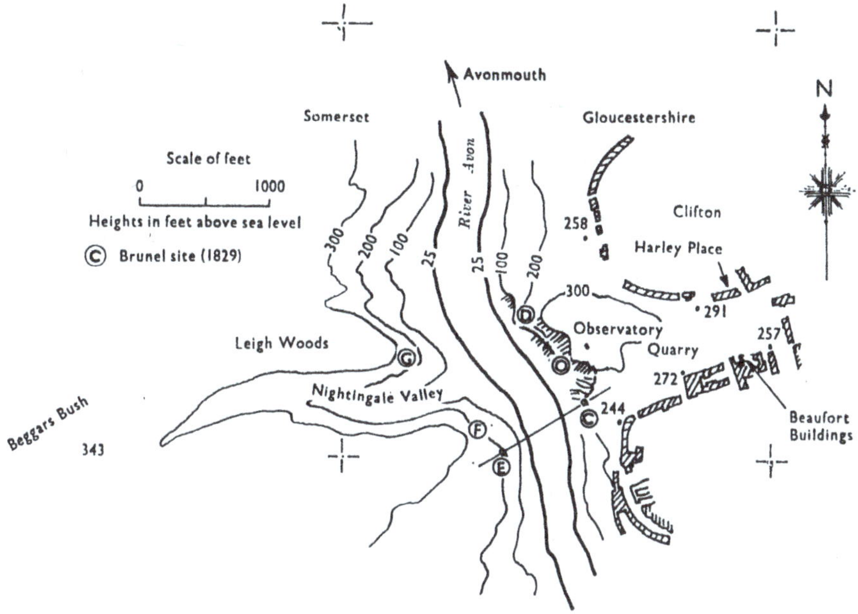

Fig. 2. Topography of Avon Gorge at Clifton

[*Topographical data, Crown Copyright reserved*]

two further designs in the light of earlier criticism for submission in 1830. In the second phase, from October 1838, he undertook detailed design of the bridge ironwork, following the Trustees' decision to go out to tender. Tower calculations were done in the summer of 1839 as the abutments approached completion. Finally, late in 1850, with the possibility of new finance being available, Brunel explored designs for a deck that would enable him to complete the bridge with a minimum of ironwork additional to that already in hand.

5. The papers of the Trustees complement those of Brunel. Technical matters are recorded as well as financial. Brunel's 1835 scheme (not adopted) to build the bridge in two stages, with initially a restricted capacity, is noted. The possibility of using wire suspension cables, as at Fribourg, is raised. His opinion of Dredge's Patent is sought. But the Trustees are silent about schemes for completing the bridge after 1843. Most notably Serrell, designer of the 1040 ft Lewiston Bridge, proposed in 1857 to use at Clifton a wire cable suspension. Financial support, however, was again lacking and the plan was dropped.

6. Brunel designed only one other suspension bridge, the Hungerford footbridge across the Thames. Of 676 ft main span, it was completed in 1845 and incorporated many design features also intended for Clifton.[7] It was particularly appropriate, therefore, that when it was dismantled in the early 1860s to make way for the Charing Cross railway bridge, its chains should have been acquired for the completion of Clifton.[8]

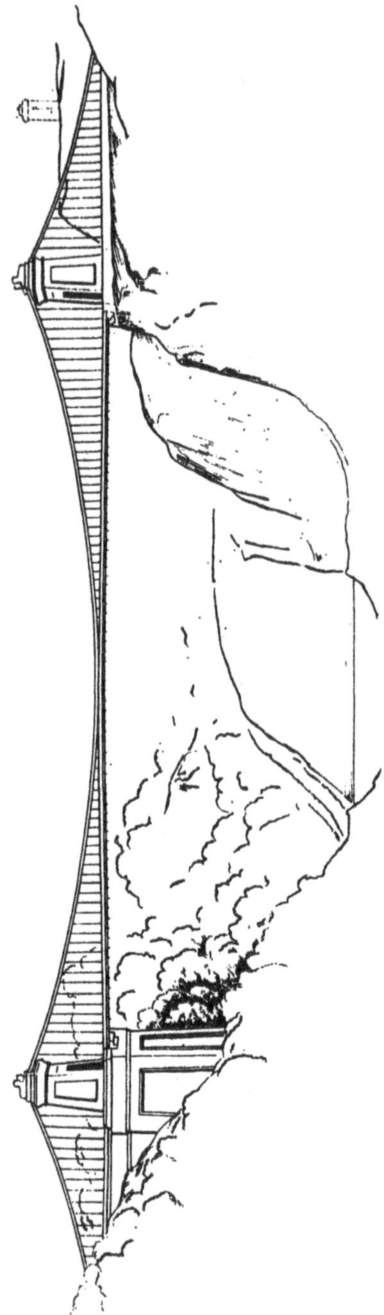

Fig. 3. Brunel's 1836 design[7]: view downstream

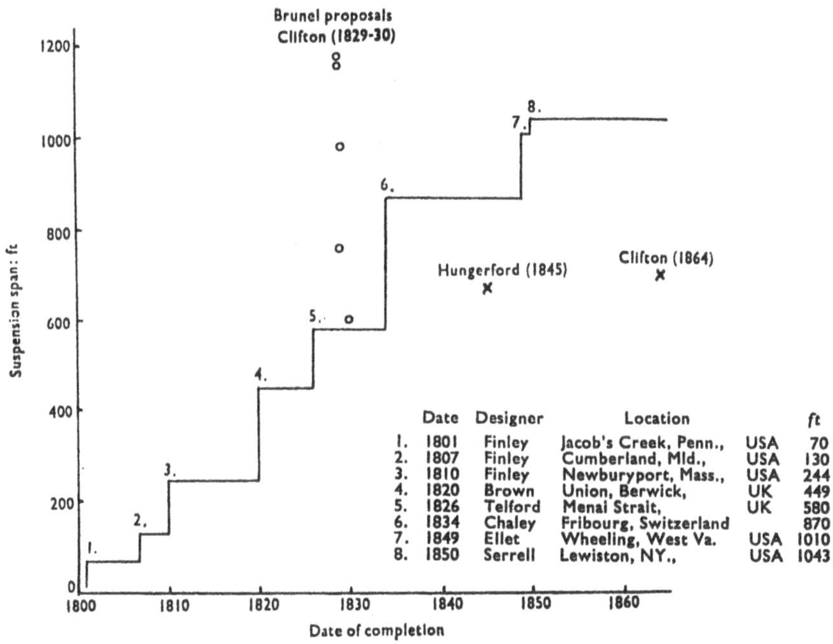

Fig. 4. Suspension bridges of greatest span, 1800–65[25]

The conception

7. Brunel's design proposals of November 1829 and December 1830 afford a notable insight into the methods he adopted to evolve alternative designs. His imagination was stirred when he first heard of the Trustees' call for plans for a suspension bridge.

> 'I was struck with the grandeur of the idea and the splendid effect which might be produced by a Bridge of this construction were it made to harmonize with scenery so peculiarly suited to such a work of Art.
>
> 'Fully appreciating the opportunity thus offered of executing a work which must reflect such Credit upon the Engineer, I determined to spare neither time nor expense in studying well the locality and profiting from the opinions and ideas of those who had probably for a length of time turned their attention to the subject.
>
> 'I immediately came to Clifton and made a correct survey of the Avon. . . .'

He took cross-sections of the gorge at every possible site for a bridge and deepened his familiarity with the area by making numerous sketches. He became convinced that such 'imposing scenery' demanded a work that was bold, simple and unobtrusive.

8. The objective of the bridge was to provide a direct route from the

vicinity of Harley Place and Beaufort Buildings in Clifton to the Beggars' Bush, a mile away across the river to the west (Fig. 2). Brunel took the heights of these points as about 250 ft and 350 ft above sea level respectively, and from the topography at Clifton concluded that the elevation of the bridge deck could hardly be less than 240 ft. This would give a navigation clearance of 215 ft above the river high water mark, well in excess of the 150 ft minimum required. The line of the bridge should be chosen to give as short a span as possible.

9. The distance across the gorge at the required elevation was greater than the clear span of any bridge up to that time. The 200 ft contours on either side of the river approach within about 650 ft of each other at the southern end of the Observatory rocks. At the 300 ft level it is 1000 ft across the gorge from below the Observatory to the southern spur of Nightingale Valley. These distances may be compared with the span of 580 ft achieved by Telford on the Menai Bridge, completed in 1826 (Fig. 4). The question therefore arose as to whether the bridge over the Avon could be built with a single span or whether intermediate piers would be necessary.

10. Brunel argued cogently for a single span. The distance across the river was, he believed, well within the limit of span considered possible for a suspension bridge. He may have had in mind Telford's Runcorn project of 1814, with a proposed central span of 1000 ft. He therefore thought that the use of intermediate piers constituted a timid solution, reflecting unfavourably on the skill of the profession. Furthermore, piers would be inconsistent with the concept of simplicity. A tower raised from the floor of the valley on the Clifton side would be sited so close to near-vertical rocks as to appear a costly irrelevance, while the tower on the other bank would obtrude into the centre of view and detract from the natural grandeur of the gorge. Finally, Brunel estimated that the cost of piers of sufficient height would outweigh the economies from reducing the span.

11. The topography suggested four possible sites. The first was the most economical and the line of the bridge (CE, Fig. 2) was close to that finally adopted. It was also the shortest span, with a bridge deck 720 ft long and with suspension points 760 ft apart. Brunel, however, pointed out certain disadvantages of this scheme. Towers would be necessary to carry the chains and these would tend to obstruct the view. The proximity of the houses on the Clifton side would restrict the choice of architectural style. The location at the end of the gorge might raise the question in an observer's mind as to why it was necessary to erect the bridge at such an elevation when lower ground was close at hand.

12. The second site was, in Brunel's opinion, a more elegant solution. He proposed to attach the chains directly to the rock on either side of the gorge, with the suspension points 980 ft apart (OF, Fig. 2). From Clifton the bridge would be approached by a tunnel, entering from the quarry to the east of the Observatory and emerging on to the 890 ft bridge deck at the cliff face.

> 'I thought that the effect of thus bursting upon this splendid scene from the Giant's Hole, arriving on the opposite side amongst the woods and winding along the valley by a road which might be compared to parts of the Simplon Road, the elegance of the festoon formed by the chain . . .

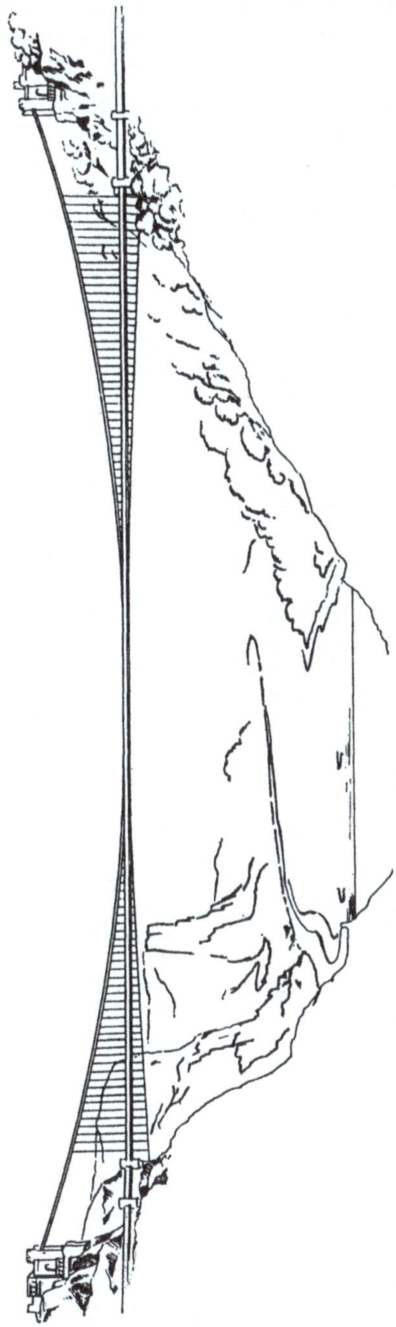

Fig. 5. Brunel's 1829 proposal[7] : suspension span 1160 ft view upstream ;

> the great height of the bridge above the water . . . and the simplicity of the whole . . . would altogether have formed a work perfectly unique and the grandeur of which would have been consistent with the situation.'

This scheme would be about 4% more expensive than the first. The increased cost of ironwork and of the longer floor, together with the cost of the approach tunnel, would offset the saving in tower masonry. However, the design would be simplified without having to provide for chain movement due to expansion over the towers. The construction of the approach road on the west would also be helped by a natural rock ledge on the south side of Nightingale Valley.

13. The bridge lines of the remaining two schemes both ran from the northern end of the Observatory rocks (DF and DG Fig. 2). The tunnel approach was avoided, but the chains were again attached directly to the rocks. The length of the bridge deck, however, exceeded 910 ft in both cases and the estimated costs were 14% and 18% respectively above that of the first scheme. Nevertheless Brunel evidently considered the less costly of these two schemes to be an impressive alternative (Fig. 5) to the preferred Giant's Hole design.

14. In presenting his proposals to the Trustees, Brunel admitted that he had been unable to make detailed estimates of cost. He believed, however, that he had fairly represented the relative costs of the alternative sites and he pointed out that economies could be achieved by omitting ornamental masonry and by accepting less 'excess of strength' in the chains than he had provided.

15. Brunel's designs of 1829 were criticized on two grounds—the magnitude of the span and the lack of a 'striking outline'. These criticisms ran counter to his own technical and aesthetic judgement and his proposals of 1830 must therefore be seen as an accommodation to the objections of Telford and to the opinions of the Trustees. He suggested building an abutment on the western slopes 'as symmetrical as possible with the bold rock which forms a natural pier on the east' thus giving a bridge deck only 590 ft long, comparable with the span at Menai. Towers of 'almost any style of architecture according to the taste which may prevail' would then carry the chains to suspension points about 60 ft above the roadway. Alternatively, should intermediate piers be considered indispensable—a decision he would 'sincerely regret'—then he would propose 'a sort of Egyptian column', simple and plain up to the level of the roadway. Telford's Gothic towers he judged 'inadmissable'. Traditional architecture was intended to be seen from below: at Clifton the majority of visitors would see the bridge from a height.

16. While he might defer to the wishes of his clients in the siting of the bridge and in its general form, Brunel believed he could still effect a number of improvements on what had previously been achieved in suspension bridge design. His principle was to adopt what had proved to be sound and to remedy the 'several evils' which were apparent in contemporary designs.

Design influences

17. It is worth reviewing the influences which guided Brunel in his design. First, there was his own direct experience. In 1823 his father, Marc Brunel, designed two small suspension bridges of about 130 ft span for use on the Isle de Bourbon in the Indian Ocean. These structures were manufactured in Sheffield and erected temporarily under the younger Brunel's supervision

before being sent abroad. The success in this design of inverted chains in preventing bridge deck movement undoubtedly coloured Brunel's later proposals.

18. Second, there was Brunel's observation of contemporary English bridges. At the time of the Bourbon bridge design he visited 'all the works of this class' in Britain. The most notable of these was the 449 ft span Union Bridge, near Berwick. Designed by Samuel Brown, it was completed in 1820 and is still in use today. In view of Brunel's 1829 Clifton schemes, it may be noted that on the English side of the Tweed the chains of the Union Bridge are suspended directly from an abutment set into the overlooking rock.

19. By 1829 both the Menai Bridge (1826) and Tierney Clarke's Hammersmith Bridge (1827) had been completed. In his first proposal Brunel specifically refers to the latter as a precedent for bringing the chains down almost to deck level and for 'McAdamizing' the roadway. It was, however, to the Menai Bridge that he looked for particular guidance. He spent several days at the bridge on two occasions and also studied Provis' account of the work.[9] Although he was critical of the design of the chains, he knew that he could profit from Telford's experience and indeed he proposed in a number of details to 'follow implicitly what has succeeded so well at Menai'.

20. The third influence on Brunel was the experience of foreign bridge builders. It appears likely, however, that this may have been by way more of warning than of guidance. Finley's pioneer designs in America[10] during the first decade of the 1800s established the suspension bridge as a viable structural type, but Brunel acknowledges no practical ideas from this source. The anchorage failure of Navier's Pont des Invalides in 1826, just before the bridge was completed, may have had a more profound effect. Navier had inspected the Bourbon bridges in England[11] and Brunel must have met him on that occasion. Brunel also knew personally the Seguin brothers who first introduced iron wire cable bridges in France. He had himself 'some schemes for remedying the practical difficulties . . . which render foreign bridges [of this type] anything but examples for us to follow: but I . . . abandoned them'.[12]

21. Fourth, Brunel had the benefit of sound advice and informed encouragement. Before submitting his 1829 proposals he referred them to his father. Marc Brunel had his own professional reputation to maintain but he nevertheless expressed his approval. Brunel also showed his plans to Davies Gilbert—a man well known for his insight into the principles of mechanics and for his willingness to be consulted. He confirmed that the chains were strong enough and that the means of restraining deck movement was 'well adapted to its purpose'. It would be a slur on the evident integrity of Gilbert to suggest that his opinion in 1829 affected his judgement in 1831. Finally, with regard to the manufacture of bridge components, Brunel sought the advice of Joshua Field, Vice President of the Society of Engineers. Field was partner in Maudslay and Field, a firm which had recently built a suspension bridge in Ceylon and with which Marc Brunel had long been associated. He too considered the proposals quite feasible.

22. At the age of 24 Brunel may have been thought too inexperienced to be responsible for the design and construction of a major engineering project. There is no evidence to suggest, however, that he did not approach the task well able to capitalize on the experience of others.

Design conditions

23. By the late 1820s there was sufficient experience of suspension bridges to establish a basis for structural design. The suspension chains were required to carry, as Brunel put it, 'the greatest possible load that can come upon the bridge . . . besides the weight of the floor and of the chain itself.' Apart from this fundamental static loading of the bridge there were also dynamic effects to be provided for. Oscillations could be excited in several early bridges with the passage of pedestrians or a carriage:[13] the failure induced in Broughton bridge in 1831 by a column of marching soldiers merely highlighted a problem that was already appreciated. The wind, too, was recognized as a source of excitation, although the nature of its action was a matter of opinion. Although hardly relevant to the situation at Clifton, it had also been noted that suspension structures such as Samuel Brown's Brighton Chain Pier (1823) might be subjected to buffeting from the sea.

24. The loading of bridges by dynamic phenomena was largely observed qualitatively. The detailed design of the chains and the deck, however, often reflected the engineer's physical appreciation of the problem. Static loading, on the other hand, was more tractable. The weight of the structure could be calculated and the 'adventitious weight' (as Davies Gilbert described the maximum static live load) could be defined. The wind, apart from its dynamic effect, was not thought to be significant as a design load.[14]

25. In England there was some variation in assumptions about the adventitious weight. Peter Barlow, in his assessment of Telford's Runcorn project,[15] assumed an arbitrary 100 ton load distributed over the 1000 ft span. The Parliamentary Committee of 1818 investigating the Menai Bridge proposals obtained an estimate of 330 tons as the weight of a drove of cattle on the Runcorn span. Taking the bridge deck as being 30 ft wide, these values correspond to uniformly distributed loads of 7½ lbf/sq. ft and 24½ lbf/sq. ft respectively. At the opening ceremony of the Union Bridge in 1820 a crowd of about 700 people broke through the barricades and forced their way onto the bridge. The bridge survived this unlooked-for proof loading, but discussion of the appropriate design loading for suspension bridges appears subsequently to have been in terms of the weight of a crowd.[13] Even so there were variations in the assumed weight and girth of the standard Englishman—typically a 10 stone man occupying 2 sq. ft. In the Clifton design competition Rendel[16] took values corresponding to a uniform loading of about 45 lbf/sq. ft, while Davies Gilbert compared all the reserved designs on the basis of 84 lbf/sq. ft. Since Brunel had designed his chains to carry an adventitious weight of 100 lbf/sq. ft, it is perhaps not surprising that his proposal alone was considered strong enough.

26. In France there were Government requirements to be met by all suspension bridges.[17] Successful proof loading to 200 kgf/m^2 (41 lbf/sq. ft), held for 24 h, without permanent deformation of the structure, was a necessary condition for approval of the bridge as a public thoroughfare. As a concession, an initial 'demi-épreuve' was permitted with the full proof test applied after six months of restricted bridge use, thus allowing the mortar in the piers to develop its full strength. Furthermore an annual inspection of the structure was required with an additional proof test at the discretion of the Préfet if the bridge had suffered any untoward occurrence during the year. However, the

relatively low adventitious weight specified in the French regulations may well have been a contributory factor to Brunel's disapproval of foreign bridges.

27. Brunel's estimate of the load to be carried at Clifton was therefore conservative by contemporary standards. The economies he sought to achieve were not to be made at the expense of the safety of the structure.

The suspension chains

28. The ironwork of the main chains and of the suspension rods contributed about 45% of the total estimated cost of Brunel's 1829 proposals. A saving in material and labour in this part of the design was therefore significant in reducing total costs, and Brunel achieved this by adopting a new method of linking the bars in the chains. There were, however, a number of other factors to be considered in the design.

Dip : span ratio

29. Brunel's contemporaries knew that the shape of the curve of a suspension chain affected the forces in it for a given distributed load.[5] It was recognized that if the curve was too shallow, the forces would be unnecessarily high; but if the dip was too great, the chain would then be liable to oscillation. Davies Gilbert, in his report on the 1830 Clifton competition, accepted Brunel's dip:span ratio of 1 in 10 but recommended that Brown's ratio of 1 in 12 should be increased. It may be noted however that the ratio for Brunel's Giant's Hole design, determined largely by the topography, was about 1 in 12·2 while the ratio at Menai was 1 in 13·5.

30. Figure 6 shows how the tensions in a chain supporting a uniformly distributed load vary with dip:span ratio. Since the length of a chain with

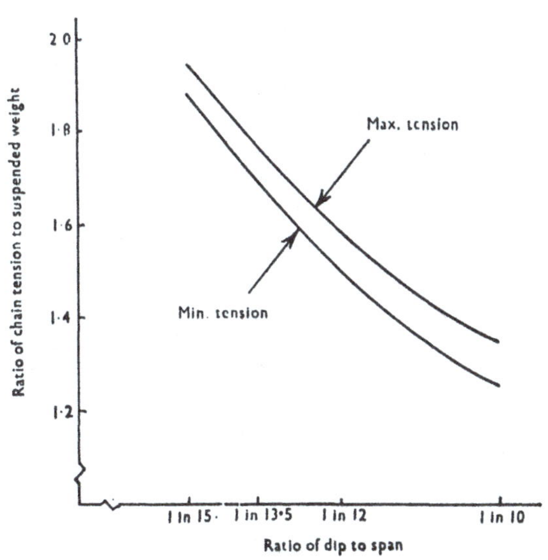

Fig. 6. Variation of chain tension with dip : span ratio

ratio of 1 in 10 is only about 3% greater than the basic span, the weight saved by decreasing the dip is more than offset by the increase in material needed to resist the greater tensions. For a given suspended load, a chain with ratio of 1 in 12 is 16% heavier than a chain with ratio 1 in 10. The figure also shows that, with a ratio of 1 in 10, the minimum tension (at the centre of the span) is about 7% less than the maximum tension (at the suspension points), and Brunel in his 1838 calculations took advantage of this by reducing the chain cross-sectional area accordingly.

Maximum working stress

31. Early 19th century experiments on the strength of wrought iron, carried out notably by Telford and Brown, showed that an ultimate tensile stress (UTS) of at least 27 ton/sq. in. was normally obtainable: the excessive deformation due to yielding occurred at about half this stress level.[15] There was, however, a wide variation in opinion as to the appropriate working stress to be adopted for design purposes. At the Menai inquiry, John Rennie favoured a safety factor of 4 or 5 on the UTS, Chapman suggested 3 and Barlow (an applied mathematician, not an engineer) recommended 2. In the event Telford followed Rennie: the chains were designed to a stress of 5½ ton/sq. in., and individual bars were proof tested to twice this stress before assembly in the chain.

32. At Clifton Gilbert also adopted 5½ ton/sq. in. as the desirable maximum working stress. It was a cautious decision, although he believed that he was not requiring 'greater strength in the materials than prudence would naturally dictate'. Brunel became more conservative in his working stress as time went on. In 1829 he considered that 'iron should not be loaded with more than 8 ton/sq. in.'; in 1830 he took 6½ ton/sq. in., 'a degree of strength not usually given to suspension bridges but which . . . I consider advisable to adopt here'; in 1838 he designed the chains to 5 ton/sq. in. He also specified that the chain bars should be proof tested before assembly.

33. As an indication of practice elsewhere it may be noted that Lamé,[18] writing from Russia in 1824, quoted 8 ton/sq. in. for the design of chains with a proof test to 16 ton/sq. in. without yield: while Drewry,[19] reviewing the subject in 1832, believed that 6 to 9 ton/sq. in. (according to circumstances) was appropriate. Drewry's qualification indicates the variation of opinion as to how safety factors should be applied, whether to the adventitious weight or to the material strength.

Disposition of the chains

34. Brunel's ideas on the disposition of the chains appear to have been formed from a critical appraisal of the Menai design. At Menai there were four lines of chains, each line comprising four chains set one above the other (Fig. 7(a)). The bridge deck was supported on a set of transverse beams, each beam itself being suspended by four rods, attached one to each line. Successive rods in a given line were bolted on to alternate pairs of chains, as shown.

35. These features Brunel considered unsatisfactory: the load from the deck would not of necessity be equally shared between the chains. If, however, each beam was supported at only two points, each rod would carry the same load, irrespective of the adjustment of the length of the rods or the motion of the deck. This implied only two lines of chains—as with the Union Bridge.

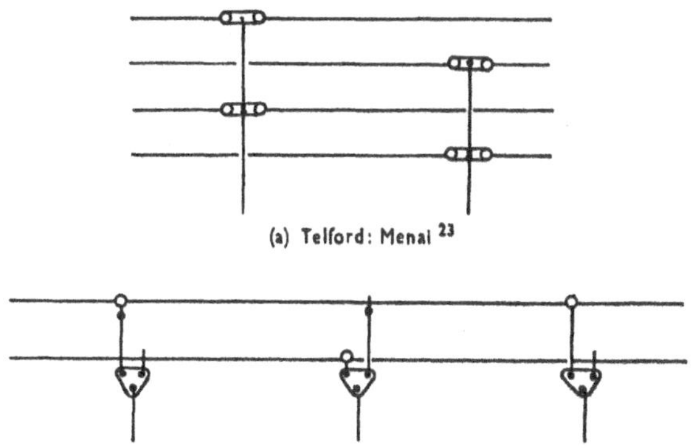

(a) Telford: Menai [23]

(b) Brunel: Hungerford [18]

Fig. 7. Attachment of suspension rods to chains : schematic arrangement

An incidental effect would be to reduce the 'forest of suspension rods' and 'increase the attraction and consequently the revenue of the bridge'. Furthermore, Brunel proposed to have just two chains in each line. Each suspension rod would then be connected to both chains by a link, spreading the load equally between them. (Brunel appears not to have been concerned about the bending stresses induced by the transverse loading at mid-link, for he adopted this scheme at Hungerford, Fig. 7(b)). The system would be statically determinate and would not be affected by the relative movement of the chains due to temperature or wind effects.

36. When the Menai chains were first erected it was observed that the windward chains would swing across the line of the bridge, raising the deck below and setting up a wave motion, which then induced vibration in the other chains. The problem was somewhat ameliorated by linking the four lines of chains together at intervals with spacing beams. Brunel proposed that the chains should be brought down at mid-span to deck level, thus using the deck itself as a restraint against transverse chain deflexion.

37. These proposals were criticized by Gilbert and Seaward. They believed that there should be three or four 'clusters' of chains, possibly on the grounds that otherwise the failure of one cluster would endanger the stability of the bridge. Brunel agreed with Gilbert to divide the chains into four but gained the concession that their position was 'of no consequence'. No doubt he then used this clause to argue that his four chains were effectively four clusters: in any event he did nothing about it. He also agreed to keep the deck below the curve of the chains 'so as to preserve the catenary entire'—a point of aesthetic rather than technical significance.

Joints

38. Brunel fully appreciated that a joint between two bars in a chain introduces material in excess of that required for a continuous uniform bar (Fig. 8(a)). The manufacture of the lug ends and the assembly of the joint involves additional labour. He therefore considered uneconomical the contemporary practice of joining chain bars with short intermediate links, as

exemplified by Telford's Menai Bridge design (Fig. 8(b)). He proposed to dispense with the short links and adopt a scheme recently introduced by Messrs Maudslay and Field. Some redesign of the bar lug end was necessary (Fig. 8(c)). A machined shoulder was added to carry temporary ties when replacing an adjacent chain bar in situ: in Telford's design the ties were attached to the extreme ends of the short links. Additionally, the connexion of the suspension rods to the joints required separate provision.

39. Gilbert and Seaward again criticized the design. It was 'not judicious', it was 'inferior as regards symmetry and neatness' and was 'not convenient for attaching the suspension rods'. One senses that this was Seaward's judgement. When Brunel had the opportunity of explaining the reasons for his modifications to Gilbert alone, the latter was satisfied that Brunel's joint design was sound and acceptable.

40. The detailed design of the lug ends and of the joint pins depended on experiment rather than structural theory. Doubtless Brunel was initially guided by the proportions of Telford's links. In the 1838 calculations he determined the excess bar material with pins of $4\frac{1}{4}$ in. and 5 in. diameter; and concluded that $4\frac{5}{8}$ in. diameter 'will then be a very good size for the pins'. Certainly he subsequently tested a series of finished bars to ensure that the lug strength was consistent with that of the bar. He was also satisfied that the welds, formed to attach the undrilled lugs to the rolled bar, were not a weak feature of the design.[20]

41. From the start he realized the advantage of increasing bar length. The pitch of the Menai chain was 10 ft but by 1829 bars 16 ft long could be 'manufactured with perfect certainty'. Brunel argued for the use of the longest bars available, on the grounds of reducing the number of joints, and thus saving labour. He further pointed out that the longer the bars, the less the initial stresses in parallel plates due to inaccuracy in locating the pin holes, a problem that Telford had experienced. By 1838 he was able to think in terms of 20 ft bars and in the early 1840s he specified for the Hungerford Bridge bars 24 ft long.

42. Brunel was aware that by reducing the 'useless weight' of the chain he could also reduce the necessary cross-sectional area, since the chain had to carry its own weight as well as that of the rest of the structure and the live load. He did not evaluate this further saving, but it amounted to about another 10% reduction in material. The thrust of his argument, however, was that Telford's chain carried over 40% excess weight due to joints. If the number of joints could be reduced by a factor of three by lengthening the bars and changing the joint design, a significant economy would be effected.

Temperature effects

43. Brunel knew that temperature changes would affect the bridge structure. With regard to the main chains, the change in their length between the towers and the anchorages would be allowed for by movement over the tower saddles. The total change in length of each chain would cause, he estimated, a change in dip of about 1 ft between extreme conditions of temperature. An increase in dip of similar magnitude, he thought, would result from the initial bedding-in and wear of the chains. He proposed to allow for both effects by ensuring that the towers were high enough to permit setting up the deck with a rise of 2 ft towards mid-span.

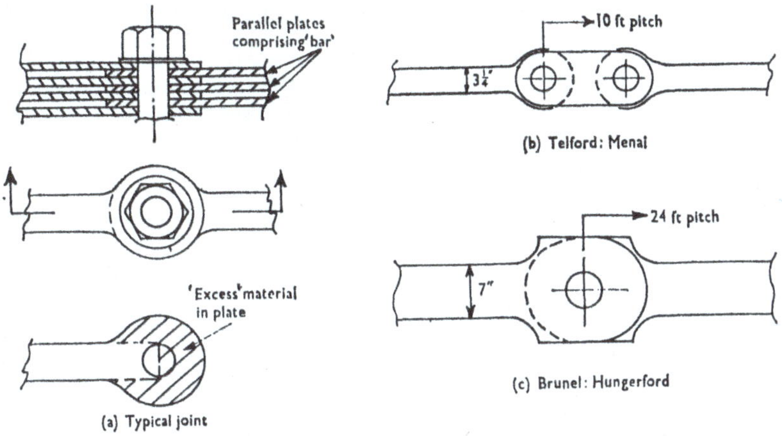

Fig. 8. Bar-chain joint design

44. He allowed for differential expansion between chains on the same side of the bridge by his method of connecting the suspension rods to the chains. There would also be, however, differential expansion between parallel plates in the same chain link, due for example to sun on one side and a cold wind on the other. At Menai this had resulted in a twisting of the chain, although the effect was contained when the cross-framing to reduce chain swing was introduced. Brunel was concerned that a temperature difference of only 15 deg F would produce twist or 'if confined, the cold part of the chain would have to bear the greatest part of the weight'. He therefore proposed in 1829 to shield the chain by adding sheet iron plates on either side. He may on reflection have considered this an unduly cautious provision. The differential stress would have been less than $1\frac{1}{2}$ ton/sq. in. for 15 deg F temperature difference and he may have felt that his chain, being over twice as deep as those at Menai, would be less liable to twist. He did not mention this design feature in his 1830 proposal nor adopt it a decade later for the Hungerford Bridge.

Tower saddles, land ties and anchorages

45. In the 1829 submission Brunel proposed to follow the Menai design in regard to 'passing the chains over points of bearing and the mode of attaching them to the rock'. There appears however to have been some development of his ideas during the following decade. By 1840 he was intending to carry the chains over the towers on saddles supported by rollers on a horizontal bed,[21] as distinct from the Menai design in which the roller bed was arched. This arrangement ensured that, apart from friction, no horizontal force could be transmitted to the tower.

46. The possibility of friction at the tower saddles implied that the land ties—that part of the chains between the towers and the anchorages—must be strong enough not only to take the tension due to the suspension loads but 'when contracted by cold to overcome all friction as well'. In the 1838 calculations he estimated the coefficient of friction as 0·05 and calculated (by

a rather dubious analysis) that a 4% increase in land tie chain area was necessary.

47. Gilbert and Seaward objected to his proposal for the chain anchorages. It presented 'many inconveniences'. The criticism suggests that in this respect also Brunel had sought to improve on the Menai design although no evidence remains to indicate his intentions.

Suspension rods

48. The length of the suspension rods depends upon the level of the bridge deck and the form of the chains. The latter was calculated at the detail design stage by interpolation from the tabulated values given by Davies Gilbert for the 'simple catenary' and the 'catenary of equal strength'.[5] At the same time the lengths of the chain links were calculated for a uniform spacing of suspension rods. The links were then grouped into sets, each set being given a mean length. The number of links in each set, however, was limited by how far out of vertical the rods could be.

49. The cross-sectional area of the rods was calculated on the basis of each being 'capable of bearing the load of two when the bridge is loaded, at a maximum stress of 5 ton/sq. in.'

Abutments and towers

50. The distance between the abutments in Brunel's scheme of 1830 was 590 ft: when the abutments were built the clear span had increased to 630 ft. One possible reason for the change was the increase in the height of the bridge deck recommended by Brunel for aesthetic reasons. He may then have decided to set the Leigh abutment further back up the slope to avoid a substantial increase in its massiveness—and hence cost. He also established the level of the Clifton abutment 3 ft higher than the other to avoid the illusion of the roadway falling towards the Observatory cliffs.

51. The design of the towers was subject to little formal analysis. Brunel evaluated the total vertical force being applied by the chains and then estimated the height of a block of masonry, placed on top of the tower, that would have the equivalent weight. This he found to be 44 ft. He made no written comment, although he presumably deduced from this value that he would be able to spread the chain load through the saddle over a sufficient area to avoid crushing the stone. His calculations on other parts of the tower were no more than an evaluation of the area required at different cross-sections to maintain a uniform compressive stress throughout. He made no formal statement about the expected strength of the masonry—unlike Rendel, who even in his competition proposal quoted the crushing strength of the stone he intended to use.

52. Brunel appears to have given considerable thought to the architectural style and the ornamentation of the towers. His criticism of Telford's Gothic piers shows that he could rationalize his aesthetic decisions. His reference, at different times, to Egyptian, Grecian, Roman, Moorish, Saxon and Gothic styles in his own schemes reveals a certain lack of prejudice, an open-mindedness in seeking an appropriate solution. His biographers[7,22] mention his intention to decorate the Egyptian towers with 'series of figure subjects, illustrating the whole work of constructing the bridge'. He noted in his calcula-

tion book possible topics for illustration and made doodles, in the margin, of Alderman Vick on his deathbed and of Bridges' and Burge's proposals for masonry bridges at Clifton. The gaunt towers that Economy finally forced on the project reveal little of the artistic imagination of their designer.

Bridge deck

53. Brunel conceived the Clifton bridge deck as a single broad carriageway, bounded by the suspension rods and flanked on either side by a footpath outside the line of the chains. In his competition proposals he stated his technical objectives: the deck must be adequately braced to prevent undulation arising from moving loads or the action of the wind; its safety must not be affected by the loss of as many as three 'contiguous' suspension rods; it must be possible to repair and even replace parts of the bridge without having to close it to traffic.

54. His ideas on how to brace the bridge deck appear to have changed with time. His original plan was to modify his father's scheme of using restraining chains curved downwards and outwards beneath the deck. At Clifton, however, there would be significant movement of the bridge due to temperature changes. The brace chains would have therefore to be held at a constant tension by a system of weights and levers rather than be rigidly fixed to the abutments, as in the Bourbon bridge design. Figure 5 shows these chains in one of the long span projects of 1829. At a discussion on wind effects at the Institution of Civil Engineers in 1841,[23] Brunel stated his intention of using brace chains at Clifton in a horizontal plane, to check lateral movement of the deck. The bridge would also have 'a complete system of trough shaped triangular bracing, which would render it quite stiff'. Rendel doubted that brace chains were effective. He had studied the problem before repairing the Montrose Bridge and had there introduced 10 ft deep trusses, running the length of the bridge.[24] It is interesting therefore to note that when in 1850 Brunel sought to revive the Clifton project, he sketched a bridge deck cross-section showing two longitudinal timber beams over 6 ft deep. The roadway, as understood by Barlow in taking over the design,[8] was to be 'formed of trussed woodwork and stiffened by an arrangement of vertical struts, with diagonal tie-bars of iron'.

55. Brunel visualized the appearance of his bridge from many viewpoints, as is shown by his sketchbooks and his competition proposals. No doubt in his imagination he looked over the side of the bridge. 'I have adopted an apparently heavy and massive balustrade not only as suited to the style of architecture but to prevent the unpleasant effect of looking down a great height. . . .' Whether he also expected to obtain increased deck stiffness from the balustrade he did not say.

56. Brunel probably saw the need to design against the failure of one or more suspension rods because of the experience at Menai. Rods too rigidly attached at deck level had there been broken by the movement of the bridge. Furthermore, differential movement of the chains had led to failure of those transverse beams which were supported by more than two suspension rods. This was an additional argument for two-point suspension but it emphasized the need for some degree of longitudinal load spreading should a failure occur. It is not clear from his proposals what were his original intentions. He

believed, however, that a timber structure would be the most economical and would allow local repairs to be made easily without closing the bridge.

Conclusion

57. There is a sense of anticlimax as one concludes a study of Brunel's papers on the Clifton bridge. The work is unfinished. The last calculations (in 1850) reflect no confidence in the flicker of renewed interest. They concern a deck of 'somewhat reduced dimensions' and Brunel explores the possibility of a smaller design live load with increased allowable chain stresses. After 20 years the vision of a graceful span has been realized in two stark unconnected towers and 700 tons of unassembled ironwork, with the Trustees' creditors seeking settlement.

58. Yet the Clifton records convey a vigorous picture of Brunel as a designer. The challenge of the task stirred his ambition. He observed as he sketched and his imagination was stimulated. Physical insight was reinforced by empathy, knowledge was refined by criticism. His assumptions were cautious and tested by experiment. His decisions were tempered by circumstance. But his intellectual independence was untouched, and one doubts if he would really have approved of the bridge that his colleagues completed.

59. In the perspective of history, Clifton and Hungerford are away from the mainstream of suspension bridge development. They stand out for boldness of span, unmatched in Britain by their 19th century successors.[25] Technically Brunel contributed to the development of bar-chains and his influence was seen in Page's Chelsea bridge of 1858.[26] It was however to the wire cable of France and America that the future belonged. The 1113 ft span of the Florianopolis chain[27] is but a reminder of Brunel's vision a century before. Yet one may speculate that if the tides of Reform had not disturbed the confidence of Bristol wealth and if the topography of Britain had called for more road bridges of considerable span, then Brunel would have seen a generation of bridges of which he would have been justly proud.

Acknowledgements

60. The Author is particularly indebted to the Librarian of Bristol University for permission to study the Brunel papers in his care, together with the records of the Trustees of the Clifton Suspension Bridge. He is grateful to the Committee for Aerial Photography, Cambridge University, for permission to publish Fig. 1; and to the Director General of the Ordnance Survey for permission to include the topographical data in Fig. 2.

References

1. Local Act of Parliament. An Act for building a Bridge over the River Avon. 11Geo4, c 69, 29 May 1830.
2. *Aris's Birmingham Gazette*, 4 October 1829.
3. *Proceedings of the Trustees of the Clifton Suspension Bridge*, 17 June 1830, *et seq.*
4. Gibb A. *The story of Telford.* Alexander Maclehose, London, 1935, p. 247.
5. Gilbert D. On the mathematical theory of suspension bridges. *Phil. Trans. R. Soc.*, 1826, March, 202–208.
6. Seaward J. On suspension chain bridges. *Mechanics Mag.*, 1824, **1**, Jan., 22, 339–40.

BRUNEL AND THE CLIFTON SUSPENSION BRIDGE

7. Brunel I. *The life of Isambard Kingdom Brunel.* Longmans, Green and Co., London, 1870.
8. Barlow W. H. Description of the Clifton Suspension Bridge. *Proc. Instn Civ. Engrs*, 1867, **264**, 243–257.
9. Provis W. A. *Historical account of the suspension bridge over the Menai Strait.* London, 1828.
10. Pope T. *A treatise on bridge architecture.* New York, 1811.
11. Navier C. L. M. *Mémoire sur les ponts suspendus.* Paris, 1823.
12. Brunel I. K. Letter to the Clifton Suspension Bridge Committee, 3 February 1835.
13. Stevenson R. Description of bridges of suspension. *Edin. Phil. J.*, 1821, **5**, Oct., 237–256.
14. Papers relating to . . . bridge over Menai Strait. *Parl. Rpt*, 1819 V, 328.
15. Barlow P. *Essay on strength and stress of timber.* London, 1817, Appendix.
16. Rendel J. M. *Particulars descriptive of the accompanying design for an intended bridge over the Avon at Clifton.* Plymouth, 1830.
17. *Ann. Ponts et Chausées, Mémoires*, 1832, 411.
18. Lamé G. Sur les ponts de chaines (de Russie). *Ann. Mines*, 1825, **10**, 311.
19. Drewry C. S. *A memoir on suspension bridges.* London, 1832.
20. Howard T. Description of a method for rolling bars for suspension bridges. *Min. Proc. Instn Civ. Engrs*, 1849, **8**, Discussion, 278 and 281.
21. Phipps G. H. Discussion of Barlow (ref. 8), 301.
22. Rolt L. T. C. *Isambard Kingdom Brunel.* Longmans, Green and Co., London, 1957.
23. Provis W. A. Observations on the effect of wind on the suspension bridge over the Menai Strait. *Min. Proc. Instn Civ. Engrs*, 1841, **1**, 74–80.
24. Rendel J. M. Memoir of the Montrose Suspension Bridge. *Min. Proc. Instn Civ. Engrs*, 1841, **1**, 122–9.
25. Jakkula A. A. A history of suspension bridges in bibliographical form. *Bull. No. 57*, Texas Engng Expt Station, 1941.
26. Humber W. *A complete treatise on cast and wrought iron bridge construction.* London, 1870.
27. Steinman D. B. *A practical treatise on suspension bridges.* Wiley, New York, 2nd ed., 1929, Appendix B.

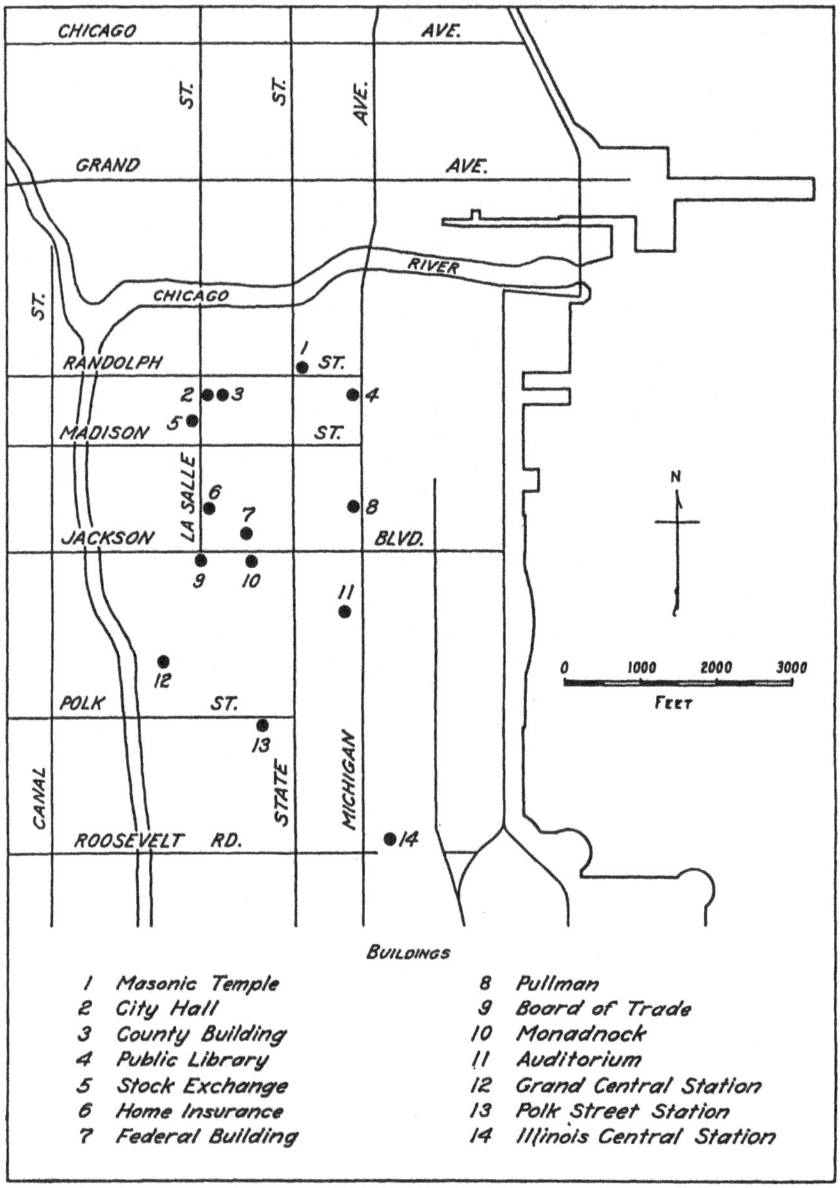

FIG. 1. MAP SHOWING LOCATION OF BUILDINGS IN AND NEAR LOOP AREA

6
History of building foundations in Chicago

Ralph B. Peck

I. Introduction

Scope of Study

The history of building foundations in Chicago between 1871, the year of the great fire, and 1915 epitomizes the development of foundation engineering throughout the world. Within less than half a century, and within the confines of the small area known as the Loop, the art of constructing building foundations grew to maturity by a process of trial, error, and correction. Seldom in any field of engineering has so much experience been concentrated in so little time and space.

Many factors worked together to produce this unique experience. The most important, undoubtedly, was the presence of a deep bed of compressible clay which underlies the Loop, and without which no foundation problems would have arisen.

Of almost equal importance, however, was the rapid growth of the city. At the time of the great fire, on October 9, 1871, the population was 300,000. By 1880 it was 500,000; by 1890, 1,100,000; and by 1900, 1,700,000.

The phenomenal growth of the city after the fire was accompanied by the continual construction of new and larger buildings. The progressive settlement of these structures on the soft clay bed became apparent, but the pace of new construction afforded little opportunity to observe the behavior of one structure before designing the next. Therefore, different architects and engineers diagnosed the difficulties differently, and many variations in foundation practice were introduced. These variations constituted full-scale experiments in the field of foundation engineering.

The final factor which made the Chicago foundation experience unique was that the period of phenomenal growth of the city was coincident with the transition from the traditional wall-bearing structure to the modern skeleton-frame building. Differential settlement became a matter of prime importance because it led to distortion of the masonry and of the structural framework and introduced supplementary stresses which were difficult to consider in the design. The quest for an unyielding foundation for the new skyscraper had an important influence on the history of Chicago foundations.

A scientific study of the records of the settlements of the structures built in Chicago during this period is a promising field for research in soil mechanics and foundation engineering. However, even a non-technical history of the developments provides the practicing engineer of today with a valuable background, and helps him develop an insight into present foundation practice. It is the purpose of this bulletin to present such a history. Although the subject-matter is likely to be of particular interest to Chicago engineers, it should be of some significance to the foundation engineer everywhere. In addition, the non-technical history presented herein will serve as a framework for a more detailed, scientific study of the records of individual buildings.

Acknowledgments

This bulletin represents part of the work of the Joint Committee on Soil Mechanics and Foundations of the Western Society of Engineers, the Illinois Section of the American Society of Civil Engineers, and the Engineering Experiment Station of the University of Illinois.

The personnel of the Committee consists of FRANK A. RANDALL, Chairman; RALPH B. PECK, Secretary and Editor; FREDERICK W. REICHERT, Associate Editor; A. E. CUMMINGS; V. O. MCCLURG; P. C. RUTLEDGE; C. P. SIESS; and K. TERZAGHI.

The assistance of many Chicago engineers in locating information pertinent to this study is gratefully acknowledged by the committee.

II. Soil Conditions

Original Topography

The general nature of the soil deposit in the Loop area at the present time is shown in Fig. 2, which represents the summary of a number of boring logs. All elevations refer to Chicago City Datum (El. 579.94 M.S.L.) which corresponds roughly to the level of Lake Michigan. The site of the city was originally swampy and low-lying ground, and the earliest buildings experienced foundation troubles due to the unstable nature of the wet, silty soil. Furthermore, because of the low elevation, the area could not be drained adequately. The early

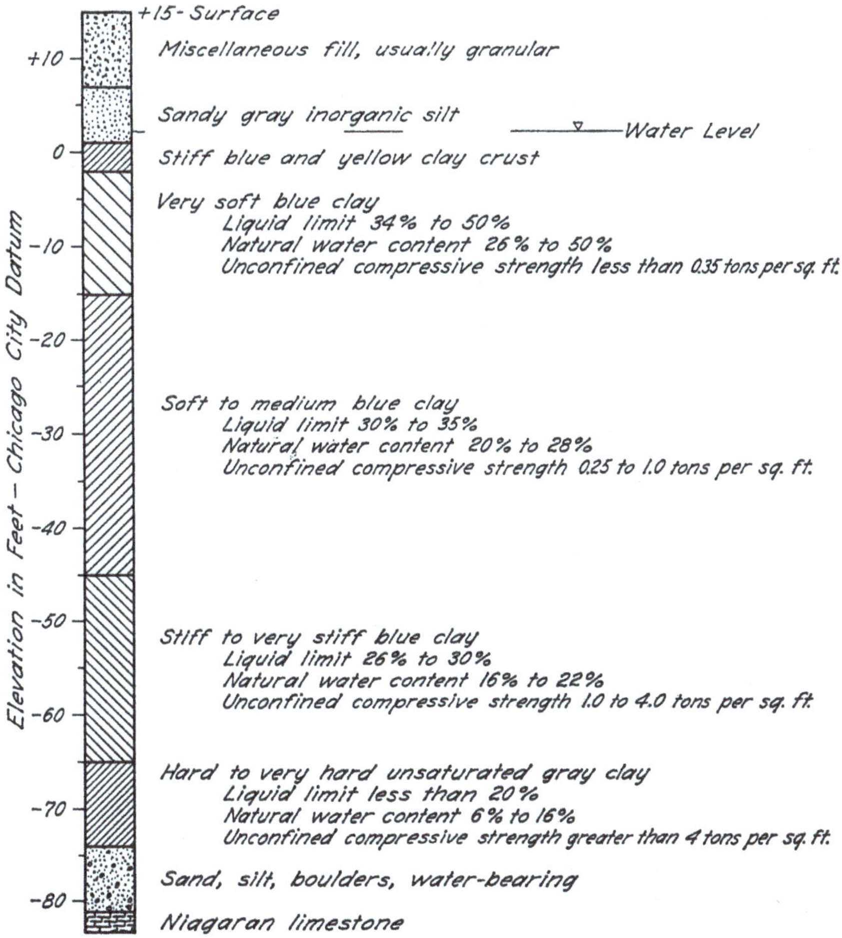

Fig. 2. Typical Soil Conditions in Loop Area

conditions were described in 1883 by John M. Van Osdel, Chicago's first architect, who arrived in the city in 1837. At that time, he states,

"From the Fort [on the south side of the river, just west of Michigan Avenue] . . . south on Michigan Avenue, the surface of the ground was . . . about nine or ten feet above the surface of the lake. [At this time, the elevation of the lake and river was probably about + 3 feet, C.C.D.]. The surface drainage was from Michigan Avenue west to the river, and from State Street west was nearly a level plain, elevated some two or three feet above the river [i.e., at elevation + 5 or + 6]. The topography of the north division was similar, the surface declining from Rush Street toward the west. The surface-water cut large gullies in the soil, known as sloughs; three of these sloughs opened into the main river. One at State Street was about sixty feet wide at the mouth, and extended in a southwesterly direction to the site of the present Tremont House [S.E. corner Dearborn and Lake Streets]. Another had its outlet between Clark and LaSalle Streets, and extended across Lake Street. The third and most formidable one was on the north side, near Franklin Street, being eighty feet wide at the river, and extending north through the Kingsbury and Newberry tracts to Chicago Avenue. In 1838 . . . the depth of water in the slough was sufficient to float . . [a] . . scow."(1)*

Raising the Grade

The progressive nature of the young city was demonstrated by the decision, in 1855, to raise the grade of the entire area. According to Van Osdel,

"The grade established by the city . . . made it necessary to build stone walls around each block to retain the filling of the streets . . . . Mr. James Brown, of Boston, an expert in raising buildings, contracted with J. D. Jennings to raise his store on the northeast corner of Dearborn and Randolph Streets. This was done in 1857, and was the first brick building raised to the new grade. An entire block of heavy buildings on the north side of Lake Street, extending from Clark to LaSalle Street, was raised simultaneously. This was accomplished in 1859. The raising-business extended through seven years, from 1857 to 1864."(2)

After the fire of 1871, the grades were raised once more, and debris from the fire was used for much of the filling. These final grades,

* The parenthesized numbers throughout the bulletin refer to Appendix B.

which refer to the sidewalk level at the building line, were El. + 15.0 on Michigan Ave., El. + 14.5 on Wabash, and El. + 14.0 in all other parts of the south division, making a level grade from State St. west to the river.

Subsoil Conditions

On account of the raising of the grade, the upper eight feet of soil indicated in Fig. 2 consist of miscellaneous fill. Beneath the fill is commonly found a gray, lake-bottom silt of variable quality, but usually loose and compressible. It has a thickness of about six feet. Approximately at datum is a crust of partially oxidized blue and yellow clay which acquired its stiffness by desiccation during a temporary low-water stage of the geological forerunner of Lake Michigan. Although the strength and thickness of this crust vary, the layer is one of the most important features of the Chicago subsoil, because it is the only material of high quality at shallow depth. Most of the early skyscrapers were founded upon it, and the crust undoubtedly served to spread the loads from their footings before the pressure was transmitted to the soft clay beneath.

The soft clay deposit beneath the oxidized crust was the principal seat of settlement of the early buildings. It extends to El. − 40 or − 50. Beneath it the clay becomes stiffer, and hard dry clay is found in some localities at El. − 60 to − 70. Bedrock is encountered at elevations between − 80 and − 120. The hard clay is an extremely variable deposit. In places it is missing altogether. It may be underlain or replaced by waterbearing sand, silt, or boulders, or it may extend to the bedrock.

Within the area bounded by Roosevelt Road on the south, Canal Street on the west, Division Street on the north, and Lake Michigan on the east, the subsoil conditions are represented with reasonable accuracy by Fig. 2. Actually, the subsoil contains many local variations which must be taken into account in design and construction. In this general survey, however, the preceding information is sufficient for an understanding of the principal foundation problems.

III. Foundation Practice Immediately After the Great Fire (1871-1878)

Primitive Practices

The buildings erected after the fire of 1871 were different in no important respects from those constructed during the previous two decades. The driving urge in the reconstruction period was to restore the city to its former condition. Indeed, when Van Osdel observed the approach of the flames, he buried his building plans and records, and after the conflagration many structures were rebuilt according to the exact plans of their predecessors. Most of these structures were from three to six stories in height, with bearing walls and few interior columns. The foundations consisted of dimension stone footings with their bases from three to five feet above the surface of the stiff clay crust. The width of the footings varied in a general way with the number of stories in the building, but almost without exception there was no attempt to proportion the width to obtain a uniform soil pressure under all parts of a structure.

Information regarding these foundations comes largely from the observations of men who examined them while the buildings were being demolished. For example, in 1910 E. C. Shankland wrote:

"The load per square foot on the soil under these buildings is often very great, and averages much greater than those under buildings erected later. There are buildings of this type where the pressure on the soil from the dead weight of the buildings is from 14,000 to 15,000 pounds per square foot.

"These excessive loads caused very great and very uneven settlement, but the buildings being of brick and wood could withstand all distortion produced by the great variation in settlement, — variations which would not be allowed in the more recent steel frame buildings.

"Then, too, the settlement was not noticed, because the street grades varied so much. Only a few years ago it was common to have steps in the sidewalk, caused by the different grades of adjoining buildings, and several such sets of steps would be found in the same block. Where the street varied in height several feet in the same block, the settlement of the buildings, however great it might be, would attract little attention."[3]

The Contribution of Frederick Baumann

The settlement of the buildings and the distortion and cracking of the structures did, however, attract the attention of at least one Chicago architect, Frederick Baumann. For a number of years he had observed the behavior of the buildings in the Loop area, and also experimented with possible improvements in his own practice. In 1873 he published his conclusions in a pamphlet entitled "The Art of Preparing Foundations for All Kinds of Buildings, with Particular Illustration of the 'Method of Isolated Piers' as Followed in Chicago." This pamphlet of 38 pages, written in clear and forceful language, is undoubtedly the first statement made in the United States of principles for the scientific design of foundations. It had a far-reaching influence upon the future development of Chicago foundations.

Baumann stated two principles which should govern the design of footings. First, "the areas of base must be in due proportion to the superincumbent loads"; second, "the centers of these areas of base must coincide with the axes of their loads." He added that these "principles are self-evident, well known, and often loosely mentioned, yet so seldom observed."

For the use of the architect he summarized his principles into a set of three rules:

"First Rule.—Resolve the building, upon its ground plan of the lower story, into isolated parts, and independently apportion to each its proper share of foundation.

"Second Rule.—Estimate the weights of all those (really and ideally) isolated parts, in order to apportion to each its due share of foundation.

"Third Rule.—Determine, upon the ground section, centers . . . of all . . . parts, which in upright section will be the axes . . . of these parts, and place the . . . bases so that the centers of their areas of contact will coincide with the first centers. . . . Observation of this rule is of the utmost importance, for upon it will depend the perpendicularity of all the walls and corners of the structure. . . . There will . . . be no particular need of any anchors, except for temporary use, while in the contrary case the strongest and best applied anchors will not suffice to preserve the *exact* normal position of the walls and corners."[4]

Baumann claimed no originality for these rules, which he grouped together as the "Method of Isolated Piers." Rather, it was the extent to which he carried the application of the rules that was original and that influenced the work of later architects so strongly. Since the buildings of Baumann's time were wall-bearing structures, continuous footings were the rule, and the guiding principle was to obtain all the bearing area possible. In contrast, he advocated "a principle which makes isolated piers the rule in all cases, and continuous foundations

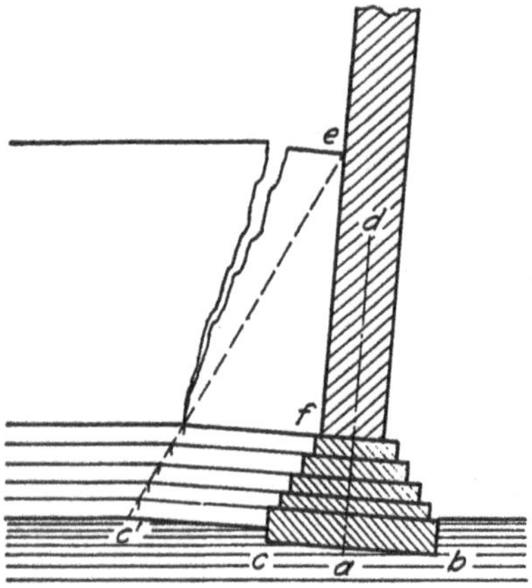

FIG. 3. APPLICATION OF METHOD OF ISOLATED PIERS TO OUTER WALL AND ABUTTING DWARF-WALL

the exception, where, for instance, piers of uniform weights are so close to each other that the bases will interconnect."

As an example of the consequences of failure to follow his principles, Baumann considered the pier of an outer wall and an abutting dwarf-wall (Fig. 3). His discussion was as follows:

"If, as the old method would suggest, in order to furnish 'all the bearing possible,' the dwarf-wall is connected with the pier at its line of intersection, *ef*, the pier will be thrust outward and the dwarf-wall crack as indicated. The cause may readily be found. Construct the axis, *da*, of the pier, and see whether it coincides with the center of area occupied by the base of the pier. Were the dwarf-wall *not connected* at *ef* and $ab = ac$, i.e., were the construction made in accordance with the 'method of isolated piers,' there would be no thrust against the pier. But the usual old random mode of 'all the bearing possible' extends the area of base insidely to *c'*, and thereby shifts the axis of the pier *off the center* toward the outer edge of the supporting base, *bc'*, which causes the ground to be pressed into an inclined surface and, consequently, the pier to be thrust outward. Were the base of the dwarf-wall made so narrow as to cause a settling of the dwarf-wall equal to that of the pier, it would at first sight appear as though then the wall might be connected. Yet this is, nevertheless, forbidden by the circumstance that the base of the dwarf-wall would receive *all* its load before the pier would, say, one-fifth part of it."[5]

He concluded that the axis of the load should "always strike a little way *inward* from the center of the area of the base, in order to make sure that it shall not be toward the *outside*," because inward inclination is resisted by the floor beams. Further, he recommended that the dwarf-wall should be built independently from the outer pier or wall, with perhaps a clear space of four inches to be walled up afterward. In case certain foundations would receive their load more quickly than others in the same structure, he suggested pre-loading the latter with pig iron so that all foundations would settle at the same rate.

Additional examples discussed in detail the common types of defects in Chicago buildings due to differential settlement, and in a convincing manner each failure was ascribed to a deviation from the strict application of the method of isolated piers. These examples, explaining for the first time the cause of well-known defects, made a considerable impression on Chicago architects. However, the full importance of Baumann's work was not felt for five or ten years after its publication. The last example presented in the pamphlet was a critical discussion of the foundation of the new Federal Building, then under construction. According to Baumann, the mat foundation of this structure violated every principle, and would prove unsatisfactory. The complete and prompt failure of the foundation greatly increased the reputation of the method of isolated piers, and influenced the design of foundations for at least two decades.

Engineers of today have an understanding of the manner in which stresses are transmitted through the subsoil from individual footings, and are aware that the compression of a deep bed of clay is not appreciably altered, as Baumann believed, by minor details in the arrangement of the footings. This understanding has been obtained largely from a study of the theory of elasticity, which was, at the time of Baumann's pamphlet, in its infancy. Indeed, the method of isolated piers antedates the theory of Boussinesq, now the basis for settlement computations, by more than a decade. Hence, Baumann's work must be regarded as an important forward step. It concentrated the attention of Chicago architects and engineers on the rational design of foundations and introduced conservative values for allowable soil pressures. In this connection, Baumann states that the most suitable bearing capacity is "twenty pounds to the square inch [2880 lb. per sq. ft.]. . . . Such proportionate load will compress the hardpan [now known as the blue and yellow clay crust] to the extent of about one inch during construction of the building, and about one-half of an inch during the next six months."(6)

Compared to the customary loads of 14,000 to 15,000 pounds per square foot, mentioned previously, Baumann's practice was indeed conservative. With the publication of his pamphlet, Chicago foundation design became recognized as an engineering problem.

IV. Era of Dimension Stone Footings (Before 1874)

Description of Stone Footings

Until about 1874 the most common construction material for footings was hard limestone quarried at Lemont, a few miles west of Chicago. This stone was even-bedded, about eight to twenty inches thick. If the material was cut into large slabs of rectangular shape, it was known as dimension stone. According to Baumann, there "can be no better material in the whole world than this dimension stone." Footings were built up in layers (see Fig. 4a), with a mortar between the joints of "two parts of roofing gravel and one of fresh cement." The offsets were generally made somewhat less than the thickness of the stone. If the footing rested on sand, it was bedded on "a thin layer of gravel or broken stone, rammed into the surface, and grouted with liquid cement-mortar."

Smaller, irregularly shaped pieces of the same stone were laid up into rubble stone piers (Fig. 4b). The offsets were generally four inches for each foot of height. Both dimension stone and rubble stone footings were used commonly in construction. The choice between the two types was generally a matter of economy. Rubble stone was usually cheaper for narrow bases, and dimension stone for wider ones.

Concrete, made of natural cements and not reinforced, was not yet considered a reliable material. It was used by some builders for constructing mat formations, but as will be discussed subsequently, these early mats met with spectacular failure.

Polk Street Station

Figure 5 shows two cross-sections through the foundations of the Chicago and Western Indiana Railroad Station, now known as the

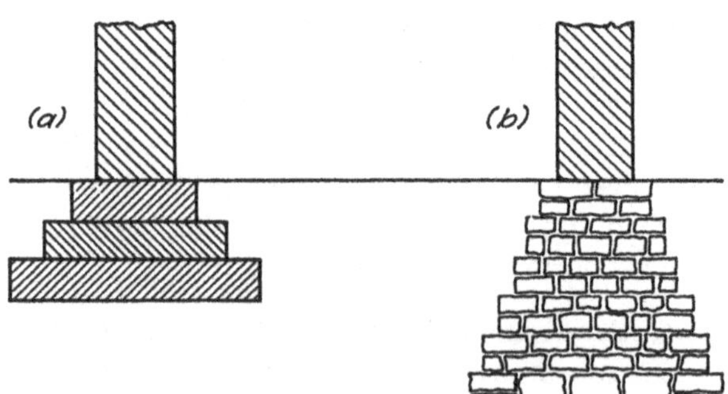

Fig. 4. (a) Typical Footing of Dimension Stone; (b) Rubble Stone Pier

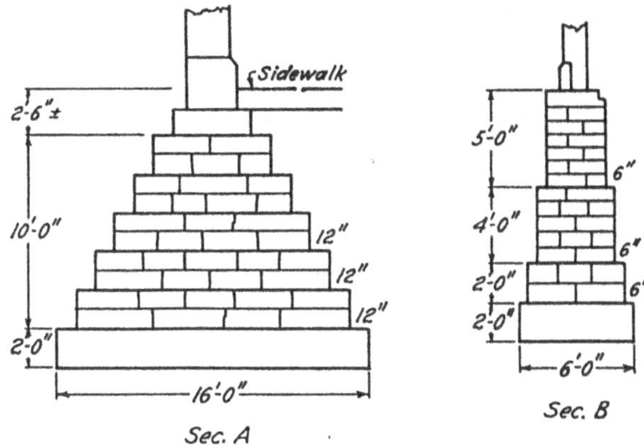

FIG. 5. SECTIONS THROUGH FOUNDATIONS OF POLK STREET STATION (1883)

Polk Street Station, on Polk St. at the foot of Dearborn St. Section A shows the foundation for the base of the tower, 170 feet high, and Section B for the three-story part of the structure. The Polk Street Station, which was completed in 1883, was one of the last important buildings now in existence in which masonry footings were used. However, in 1885 the Continental National Bank Building and the Home Insurance Building were completed, both with dimension stone footings. The former, located at 208 S. LaSalle St., was demolished in 1912, and the latter in 1931.

Home Insurance Building

The Home Insurance Building, on the northeast corner of LaSalle and Adams Streets, is now generally recognized as the first structure embodying the principle of skeleton construction. The procedure in constructing its footings was probably representative of the best practice. According to its architect, W. L. B. Jenney, the top of the blue and yellow clay crust, known at that time as the hardpan, "is used for the bottom of the masonry foundations, care being taken not to remove any of the hard pan.

"The system of foundations adopted is what is known as that of independent piers, each basement pier and each interior column having its independent foundation. . . . [Weights] on each separate foundation were carefully calculated, it being of the greatest importance that the load per square foot of footing course should be uniform over the entire building, otherwise the settlement would be unequal. . . .

"It is not easy to decide in advance as to the exact permanent load on the floors of the building. This, however, is of much importance in calculating the foundation of the interior columns which as a rule are made disproportionally large, and give trouble by not settling so much as the outside walls. In the Home Insurance Building there was added to the weight of the material actually used in the construction of the building a further weight of eighteen pounds per square foot for the average permanent load on the floors, the building being solely for office purposes. Borings were made in some twenty places over the site [to a depth of twenty feet], and the thickness of hard pan [five to six feet] was found sufficiently uniform to allow of a uniform weight of two tons per square foot to be used as the permanent load on the foundations. . . . "(7)

Figure 6 shows a section through one of the exterior piers.

Jenney's account continues:

"The alternation of rubble and dimension stone is for economy. Rubble can be laid directly on undressed dimension stone, while if large dimension stone are placed, the one directly upon the other, it is necessary to dress the surface in order to obtain an even bearing. The offsets on dimension stone can be made much greater than on rubble stone, and also the large stone serve as a bond to the pier, — so that at least each alternate course should be dimension stone, and in the instances where the weights are excessive, every course must be dimension. This must be done in order by many large offsets to obtain the necessary surface of concrete without disturbing the hard pan which is the best foundation this locality offers, every inch of whose thickness should be preserved."(8)

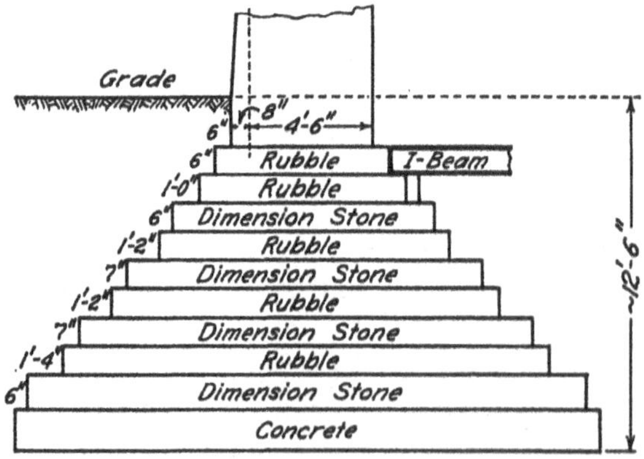

Fig. 6. Section Through Exterior Pier of Home Insurance Building

V. Grillage Footings (1878-1894)

Drawbacks of Stone Piers

Mr. Jenney's description of the foundations for the Home Insurance Building suggests the impasse which designers of office buildings seemed to be approaching. Under heavy loads, the footings had to be of dimension stone rather than rubble masonry to spread the load adequately without having to cut into the stiff clay crust, but even in the nine-story Home Insurance Building the piers, in order to attain the required width, occupied the full height of the basement. The weight of the piers themselves was at least as great as that of an additional story of the superstructure. Furthermore, as the buildings became taller the demand for basement space increased, but the greater volume of the piers reduced its availability.

Timber Grillages

The combination of the factors just mentioned appeared to set a limit to the height of structures which could be built on spread foundations. The first solutions of the problem were made by S. S. Beman in the Pullman Building, built in 1884 and still standing at 79 E. Adams St., and by W. W. Boyington in the Board of Trade, built in 1885 at 141 W. Jackson Blvd. In both these buildings the dimension stone piers rested on timber grillages probably consisting of two layers of 12-by-12-inch oak timbers. The members of the upper layer were placed at right angles to those of the lower. The offset between the lowest stone course and the edge of the timbers was made as great as two feet, whereas with dimension stone an offset was commonly restricted to one foot. Hence the timber grillage permitted use of a somewhat smaller stone pier within the basement. Use of timbers was, of course, restricted to elevations below the water table or to a distance of two or three feet above the stiff clay crust. Consequently, the timber grillages did not solve the problem completely.

Rail Grillages

A more far-reaching solution was developed by John Wellborn Root, junior partner of the young firm of Burnham and Root. In 1878 this firm was commissioned to design a 10-story office building at 64 W. Monroe St., known as the Montauk Block. The structure, which was completed in 1882, was the first tall building to be designed by the partners, and to it they brought the genius which led to their undisputed leadership during the next decade. The history of the solution of the footing problem has been related by Harriet Monroe, from whose writings the following paragraphs are quoted.

"In the planning, it was found that stone piers, of the size necessary for supporting a heavy column of fireproof vaults, would seriously obstruct, not only the basement, but also the first story. Mr. Owin F. Aldis, one of the owners, mentioned this objection to Root, and told him that he must make room somehow for the dynamos. 'A day or two later,' Mr. Aldis reports, 'he said to me: "I believe we could make the piers smaller without loss of strength by buying old steel rails, which would be cheap, and laying them row on row, with concrete around them to prevent rust." He made preliminary calculations then and there, and decided that it would be feasible. . . . '

"In the Montauk Block, the new foundation was used merely as a temporary expedient, with no thought of extending the system to other buildings. But after the Insurance Exchange had been built in the old way, the firm found that there was scarcely room enough in its stone-obstructed basement for necessary machinery. Resolving not to be caught in that way again, they studied various expedients for gaining space without digging deep into the earth, which the conditions of our soil make inexpedient. The old Montauk experiment recurred to their minds as the most available plan, and accordingly they determined to use steel-rail-and-concrete experiments throughout under the tall buildings then in the office — the Rialto, the Phoenix, and the Rookery."[9]

The Rookery Building, still standing at 209 S. LaSalle St., was completed in 1886. It was the first structure to be supported entirely on rail grillages, and during the next five years most of the spread footings were of this type. In 1891, the north half of the Monadnock Block, the heaviest office building in Chicago, was completed on such a foundation. It still stands on the southwest corner of Dearborn St. and Jackson Blvd. Figure 7 shows a cross-section through one of the footings of this building.

The rails used in the grillages were commonly of a section weighing 75 pounds per yard. At a unit stress of 16,000 pounds per square inch, each rail was capable of resisting a moment of 12,500 foot-pounds. The rails in each course were designed as if they were uniformly loaded cantilever beams with a length equal to the length of the projection beyond the overlying course. The offset was taken as three feet (sometimes less), and the number of rails per course calculated. Since the height of each course was only 4¾ inches, the new footings

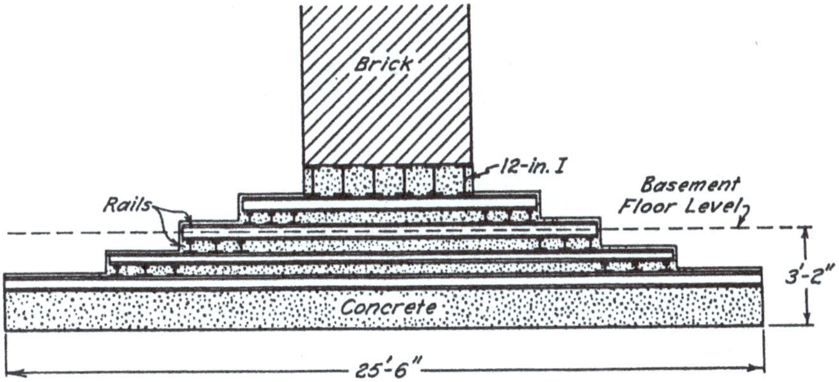

FIG. 7. SECTION THROUGH RAIL-GRILLAGE FOOTING OF MONADNOCK BLOCK

possessed a great superiority over their dimension-stone predecessors, because the basement space was very slightly obstructed. The distance from the clay surface to the base of the column on a dimension stone pier 16.5 feet square would be about 8 feet, whereas that for a rail grillage footing of the same base area would be about 3½ ft. The space between the rails was filled with cement grout to prevent rusting, and the whole grillage was bedded on a layer of concrete generally about one foot thick. The grout and concrete were not considered in the calculations as adding to the strength of the footing.

The rail-grillage footings saved both weight and basement space. They also possessed the advantage that they could be constructed more quickly than piers and, since heavy cranes were no longer required to handle stone blocks, could be installed during the winter under protection of tarpaulins. In the early 1890's, the speed of construction became a matter of considerable importance.

I-Beam Grillages

Root's invention of the rail-grillage footing made possible continued progress in the development of the tall building. By 1886, the office building had grown to 13 stories, and by 1892 to 20 stories. In 1888, the steel rails were replaced by I-beams in the grillage for the Tacoma Building, designed by the new firm of Holabird and Roche. The greater efficiency of the I-beams in resisting bending moments caused the new type of grillage to be widely adopted, and to remain the dominant type of spread footing until the introduction of reinforced concrete footings in the early 1900's.

24 ILLINOIS ENGINEERING EXPERIMENT STATION

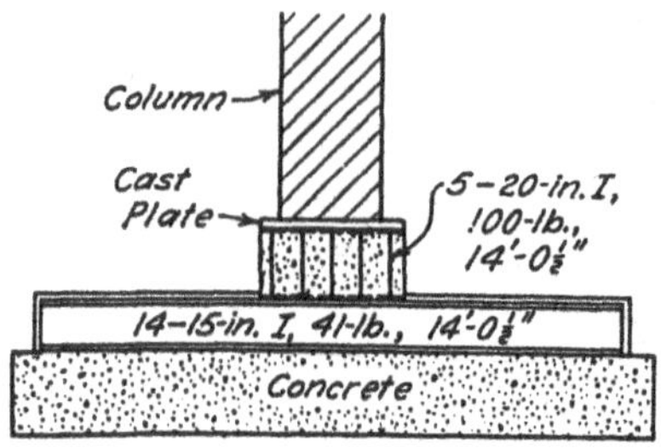

FIG. 8. SECTION THROUGH I-BEAM FOOTING OF MARQUETTE BUILDING

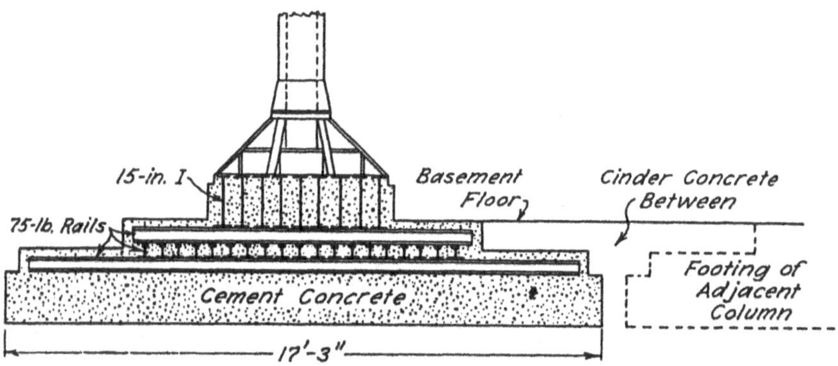

FIG. 9. SECTION THROUGH RAIL AND I-BEAM GRILLAGE OF THE FAIR STORE

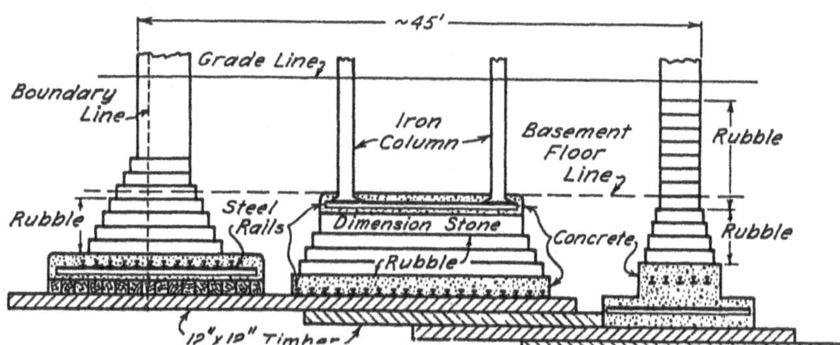

FIG. 10. SECTION THROUGH COMBINED FOOTINGS OF AUDITORIUM

A typical I-beam grillage under a single column in the Marquette Building, constructed in 1895 at 140 S. Dearborn St., is shown in Fig. 8. Although the depth of such footings was somewhat greater than that of rail grillages, the quantity of steel was considerably less.

Various combinations of the spread-footing types were used to suit the particular needs of the structures. In the Fair Store, located between State and Dearborn Streets north of Adams St., Jenney combined the rail and I-beam grillages as shown in Fig. 9. In the Auditorium, under construction in 1887, Adler and Sullivan combined rails, I-beams, dimension stone, rubble masonry, timber, and concrete. Some features of these footings are shown in Fig. 10.

Cantilever Footings

One of the principal advantages of the I-beam grillage footing is the ease with which the load from two or more columns can be transmitted to a single footing. Before the use of I-beams, the load from exterior wall columns had to be supported by footings laid out with their centers of gravity directly below the columns. As a consequence the footings projected beyond the building line. When the wall of a new building was constructed adjacent to that of an older one, it was necessary to underpin the old wall with new party-footings common to both walls. This operation was expensive, and in some instances not feasible.

In the Auditorium, a few lightly loaded columns adjacent to the Studebaker Building were supported on beams cantilevered from footings about four feet inside the building line. One of these footings is shown in Fig. 11. This appears to be one of the earliest uses of the cantilever footing. More extensive use of the principle is reported by Mr. Adler to have been made in the Chicago Edison's Building on Adams St. Construction of this building was begun at about the same time as the Auditorium, but details of the foundation are not known.

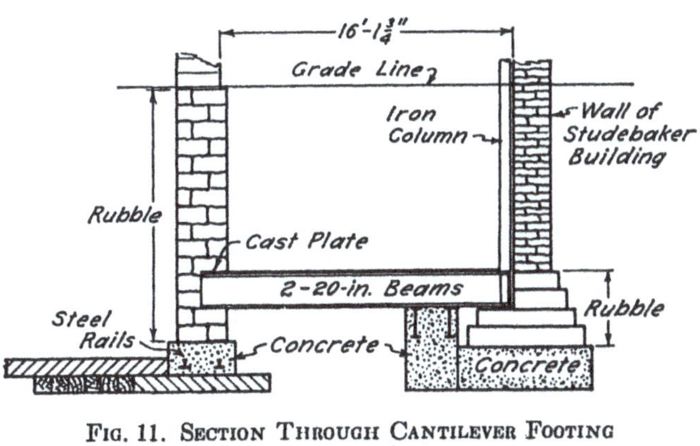

FIG. 11. SECTION THROUGH CANTILEVER FOOTING OF AUDITORIUM

The first large-scale use of the cantilever footing was made in 1890 by Burnham and Root in the Rand McNally Building located between Adams and Quincy Streets west of LaSalle St. At almost the same time, Jenney used cantilever footings to support the walls of the Manhattan Building, still standing at 431 S. Dearborn St. Figure 12 shows a cantilever footing used to support the south wall of the Old Colony Building, constructed in 1894 at 407 S. Dearborn St. This footing was underpinned a few years later because of excessive settlement (see page 58).

In the design of all of the structures on grillage footings, the architects and engineers complied strictly with the principles laid down by Frederick Baumann. Combined and cantilever footings were carefully proportioned so that the centroids of the footings were coincident with the resultants of the column loads. The amount of live load which should be taken into account when the footings were proportioned for equal settlement was a matter of some controversy. It was agreed by the foremost architects and engineers that only a small fraction of the nominal live load should be considered. For office buildings and stores, many designers considered only the dead load of the structure. A few included a floor load of five pounds per square foot, and the maximum value which seems to have been used was 18 pounds per square foot, by Jenney in the Home Insurance Building. In the Fair Store, the same architect later used the conservative value of 2880 pounds per square foot in proportioning the footing areas, with no allowance for live load.

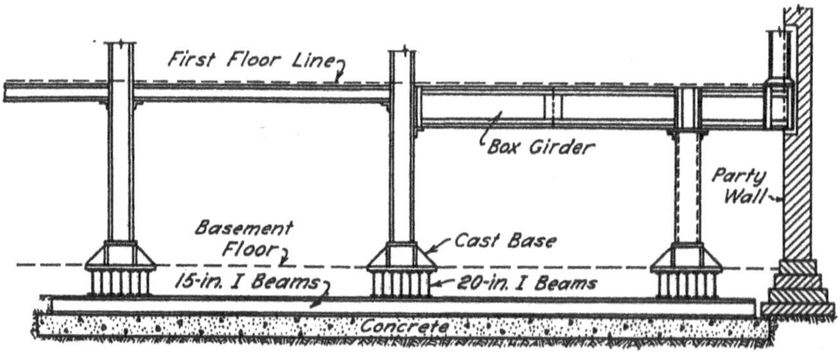

FIG. 12. CANTILEVER FOOTING SUPPORTING SOUTH WALL OF OLD COLONY BUILDING

VI. The Public Buildings, 1871-1894

Federal Building

The development and popularity of spread footing foundations in Chicago were due not only to the impact of Baumann's treatise on Chicago builders, but also in large measure to the fate of three monumental public buildings, two of which were founded on concrete mats and the third on piles.

The first of these was the Federal Building, in the block bounded by Jackson, Clark, Adams and Dearborn Streets. This structure was founded on a continuous bed of plain concrete 3½ or 4 feet in thickness, spread upon the surface of the stiff clay crust. The exterior of the structure consisted of a series of masonry piers of unequal weight, with fairly large openings between. In the interior were many lightly loaded columns, and several piers carrying loads of 380 tons or more. It was assumed that the concrete would act as a rigid base which would spread the loads uniformly to the underlying clay.

In "The Art of Preparing Foundations" Baumann commented upon this foundation, which was then under construction. His first observation was that "any masonry is essentially a yielding mass, through which the loads act, in proportion to their weights, compressing the ground beneath, forcing the masonry to accommodate itself accordingly." Thus, he recognized the fallacy of the fundamental assumption of the designer that the mat would remain rigid. In discussing the foundation of the outer wall, he presented the cross-section (Fig. 13). According to his method of isolated piers,

"The axis *da* is shifted off the center of its base; the clay beneath will consequently be compressed . . . with a tendency to thrust the wall out of its perpendicular line. This tendency need, however, not become reality, because of the very probable rupture of the bed of concrete, as indicated in [the figure]. Had this wall an independent central base, as dictated by the 'method of isolated piers,' no possible contingency could ever arise.

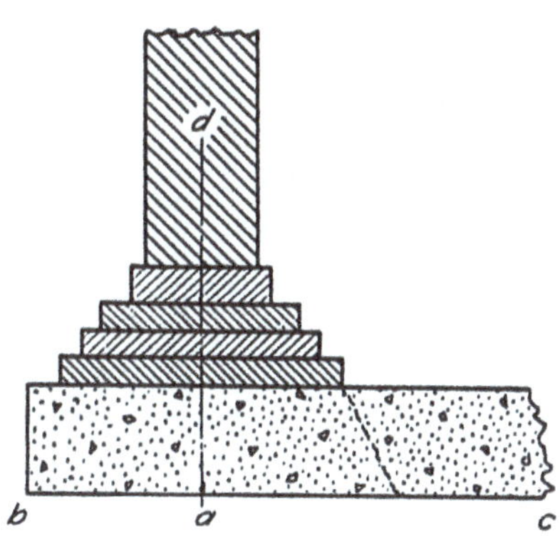

Fig. 13. Section Through Portion of Mat Foundation of Federal Building

"Considering the large inequality of the weights of the piers of the outside walls, the heavier piers will sink down, in some measure, proportionate to the weight of pier and size of base upon the clay, such as it may assume for itself, while the little piers will almost wholly retain their original levels."[10]

Baumann's judgment was more than justified. By December, 1885, six years after completion, repairs to the building due to the unequal settlement amounted to $94,000. The mat was said to have broken up when it had experienced a settlement of about 14 inches. In 1896, the building had to be demolished.

City Hall

In 1882, another mat foundation of similar construction was built to serve as the base for the new City Hall, which occupied the west half of the block bounded by LaSalle, Randolph, Clark, and Washington Streets, the site of the present City Hall. Even during the construction of the building, the settlement became noticeable. In 1891, Mr. D. C. Cregier recalled that "when we came to put in the floors, it had settled in some portions seven inches and a quarter."[11] By the time of its removal, the structure had settled differentially at least 14 inches.

Chicago engineers and architects derived considerable satisfaction from the fact that the architects of these two structures were Easterners. With Baumann's forecast to emphasize the merit of their own local methods, they dismissed mat foundations from consideration, believing them inherently unsuited to the local conditions. Even after the introduction of steel beams into foundations, so that mats of sufficient rigidity and strength became feasible, only one building of any consequence, the 14-story Owings or Bedford Building, built in 1890 at 203 S. Dearborn St., was constructed on a foundation of this type before the turn of the century.

County Building

The east half of the block bounded by LaSalle, Randolph, Clark, and Washington Streets was occupied by the County Building, completed in 1885. The superstructure of this building was the counterpart of the City Hall on the west half of the block. The foundation, however, consisted of footings supported by short piles. The character of the piling is not exactly known, but all records agree that the length of piling was 30 feet and that the tops were cut off at about city datum. One record, which may be unreliable, states that after the thirty-foot piles were driven, twenty-foot piles were driven on top of them without providing a substantial connection. It is certain that the piles did not penetrate to the bottom of the soft clay. Mr. Cregier reported that the piles were "so thick that when they drove one down where they wanted it, the next one that was driven down drove up the first one, and they had to keep driving right along; they were a good many months at it."[12]

By 1891, the County Building had experienced a differential settlement of 18 inches, and was in worse condition than the City Hall. The failure of the foundation demonstrated the inability of short piles to support heavy loads without excessive settlement. However, the statements of Chicago engineers at the time leave no doubt that the failure was ascribed to the mere fact that the structure had a pile foundation, and that no mental distinction was made between short friction piles and long point-bearing piles. Since this foundation also was designed by an Eastern architect, the local engineers were more than ever inclined to use and to refine their own specialty, the foundation of isolated spread footings.

Hence, from the time of the great fire until 1890, the spread foundation was the dominant type in Chicago, and many architects had dismissed the thought that any other type was worthy of consideration. The behavior of the public buildings in the early 1880's merely added to their conviction.

VII. THE PROBLEM OF SETTLEMENT

Introduction

In retrospect it seems obvious that one important characteristic which the three public buildings had in common was the extreme variation in the distribution of the weight of each building over the total area covered by the structure. The excessive differential settlement of the structures was attributed solely to the use of a foundation type believed to be unsuited to the local conditions. Yet, in the period between 1883 and 1887, three equally monumental buildings were constructed on spread footings, with equally poor results. These structures were the Polk Street Station (1883), the Board of Trade (1885), and the Auditorium (1889). The unsatisfactory behavior of the first two was attributed to the belief that their Eastern architects, although wisely using spread footings, did not fulfil the requirements of the method of isolated piers. The Auditorium, however, was designed by one of the ablest Chicago architectural firms, and the requirements of the method were meticulously satisfied. Its subsidence began a discussion among Chicago engineers which was not resolved for two decades. Details of the controversy are given in Chapter X (pages 56–60).

Polk Street Station

The Polk Street Station, Fig. 14, contains a brick tower, originally 170 feet high and 21 feet square at the base. Adjacent to the tower the structure was only two stories high, whereas beyond the two-story sections the height was three stories. The foundation plan of that portion of the structure facing Polk Street is shown in Fig. 15a. All the walls, including those of the tower, are supported by continuous rubble stone footings. Interior columns, which support only the floors and roof, are founded on individual spread footings. The tower footing extends to the clay surface at city datum, whereas the wall and column footings rest on silt or clay at approximately El. + 1.00.

The weight of the original tower, including the foundation, was about 1860 tons, and the area of the tower footing was 1200 square feet. Hence the unit soil pressure under the tower was 3100 pounds per square foot. The pressure under the smaller footings was approximately the same. Therefore, it appears that the structure was designed for a moderate soil pressure. It is also apparent that the tower footing and the footings for the adjacent walls were not separated from each other and that the method of isolated piers, as such, was not used.

In 1897 Mr. Artingstall stated that, to his knowledge, the tower had twice been cut away from the main building in order to allow it to settle and not wreck the building. In 1899 a set of levels was taken along the front of the building which indicated that the tower had settled about 8 inches, whereas the two-story building had settled less than one inch.

In 1943 a series of differential levels was made along the originally horizontal stone base course. The results, as shown in Fig. 15b, indicate a differential settlement of $7\frac{5}{8}$ inches between the tower and the main building; most of the settlement is concentrated in the main building within a few feet of the tower. The top of the present tower is some 45 feet lower than the original, because of the destruction of the original roof by fire.

The failure of the foundation of the Polk Street Station was well known, and was magnified by ignorance. Even Frederick Baumann, who was usually conservative and reliable in his statements, wrote that "the Polk Street Railroad Depot stands upon ground which had been softened by extinct water courses," and that [1898] the "central tower . . . has sunk more than a foot below where it ought to stand, and the whole fine building is, in view of this fact, little better than a wreck."(13)

FIG. 14. POLK STREET STATION, SHOWING THE ORIGINAL TOWER

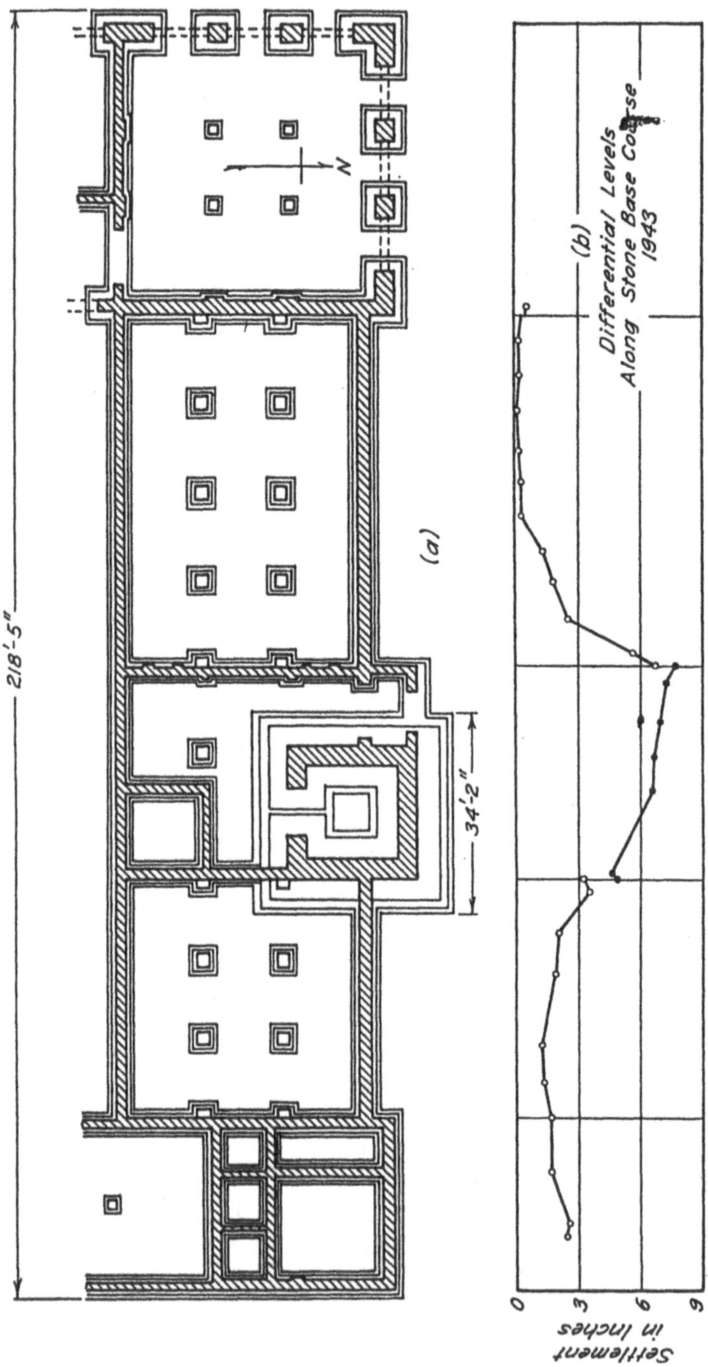

Fig. 15. Foundation Plan of Front Portion of Polk Street Station, with Differential Levels Along Stone Base Course of North Wall (1943)

Old Board of Trade

The Board of Trade Building is shown in Fig. 16. It was located on the site of the present structure of the same name, between Sherman and LaSalle Streets on the south side of Jackson Blvd. The front portion of the structure was 240 feet high. It contained the trading hall, a large room without interior columns. At the middle of the front wall, on Jackson Blvd., rose a tower 32 feet square to a height of 303 feet. The lower 225 feet of the tower were of granite masonry, whereas the upper 78 feet consisted of a cast-iron lantern. The rear of the building, 160 feet high, contained offices.

The total weight of the structure was 69,700,000 pounds, of which 53,800,000 pounds were carried by the wall footings. The foundation appears to have consisted of individual masonry piers resting on timber

Fig. 16. Board of Trade Building, Showing Original Tower

grillages at the clay surface. The average load of the structure over the entire building area, not considering the reduction in load due to the basement excavation, was only 1780 pounds per square foot However, the average soil pressure under the wall footings was 6900 pounds per square foot, and the values of pressure actually ranged from 5680 to 8800 pounds per square foot. On the other hand, the soil pressure beneath some of the interior columns was as small as 2060 pounds per square foot.

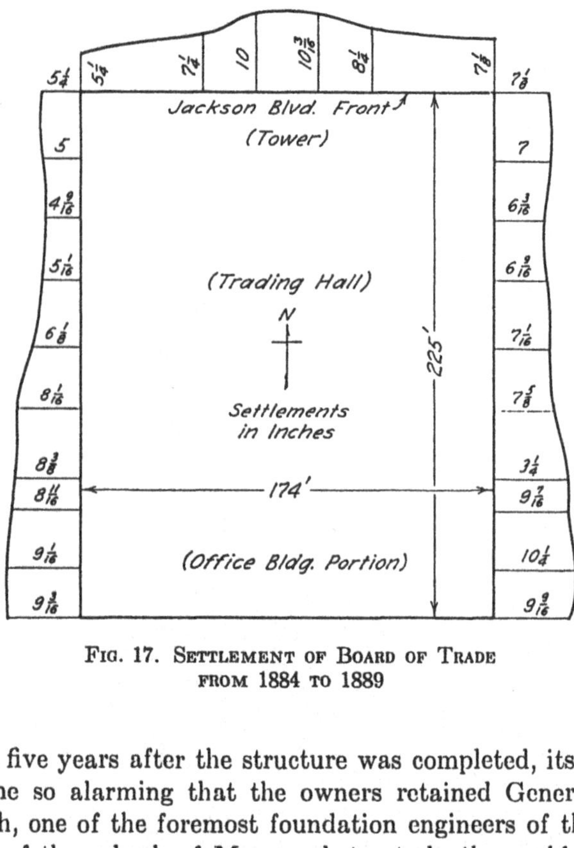

FIG. 17. SETTLEMENT OF BOARD OF TRADE FROM 1884 TO 1889

Within five years after the structure was completed, its settlement had become so alarming that the owners retained General William Sooy Smith, one of the foremost foundation engineers of the day and a resident of the suburb of Maywood, to study the problem. At this time (1889) the maximum settlement had reached 10¼ inches. Figure 17 shows the results of Sooy Smith's levels on the outer walls.

In November, 1893, Sooy Smith made a report stating that the uneven settlement demanded prompt and costly repair. He proposed that deep masonry piers should be constructed to hard foundation at a cost of $150,000. In February, 1894, a popular account stated:

"Levels taken for five years show that the building has settled at the rate of one-sixteenth of an inch a month, the settlement [of the building proper] amounting in the whole to eight inches. . . . The northwest corner has settled most and is now almost entirely separate from the rest of the structure, there being a crack ten inches wide between it and the partition wall. . . . The heavy weight of the tower has caused another depression, greater still than that of the north wall."[14]

The owners did not follow Sooy Smith's recommendation to underpin the building, but in the interest of safety the cast iron lantern was removed from the tower in 1895, and a system of steel tie rods was installed to hold the walls together.

After the removal of the lantern, the structure stood until 1929, when it was removed and the present Board of Trade Building erected. During this entire period, level readings were taken periodically and the condition of the structure was closely observed. At the time of its destruction, the maximum settlement was 21 inches.

Auditorium

The settlement of the Board of Trade was public knowledge, and was almost universally ascribed to the failure of its Eastern architect to use a constant and conservative value for the soil pressure. Of the Auditorium, however, it could hardly be said that the best Chicago practice was unobserved. Hence a history of this structure is of more than usual interest to the student of Chicago foundations.

The Auditorium (Fig. 18), upon which construction started in August, 1887, was designed to combine the facilities of a hotel, an office building, and a theater. The structure is roughly rectangular, extending 160 feet along Wabash Ave. and Michigan Ave. and 360 feet along Congress St. The portion fronting on Wabash Ave. and partly on Congress St. comprises the office building. The Michigan Ave. front and most of the rest of the Congress St. face is the hotel. The entire area enclosed by these three faces is occupied by the theater, with its stage toward Michigan Ave. and its balconies toward Wabash Ave. Above the theater is a large banquet hall. The theater entrance is located in a 19-story tower at about one quarter of the distance from Wabash Ave. to Michigan Ave. The rest of the structure is 10 stories.

The building consists of exterior and interior masonry bearing walls which support most of the floor and roof loads. Within the theater itself there are no columns except those which support the floor.

Fig. 18. Auditorium

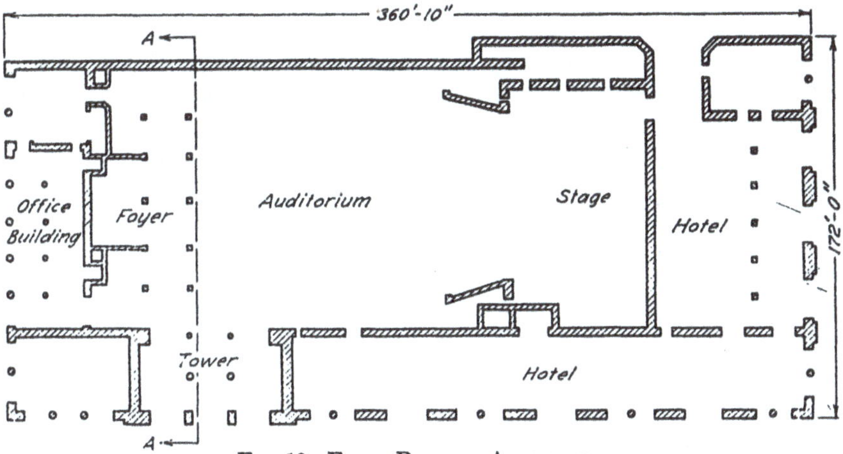

FIG. 19. FLOOR PLAN OF AUDITORIUM

The banquet hall and roof are carried by wrought iron trusses resting upon the masonry walls. The foyer at the rear of the theater contains a number of columns which support the balconies, and the interior loads of the office building, hotel, and tower are carried by cast iron columns and steel or iron floor beams. The bearing walls are continuous along the back of the building, parallel to Congress St., and for the most part in the interior. The fronts on Michigan Ave., Congress St., and Wabash Ave., however, consist of separate piers and columns in the lower stories. The general arrangement is shown in Fig. 19.

A study of Fig. 19 shows that the distribution of load over the building area is extremely non-uniform. Under the tower the concentration of load is very great. On the other hand, under the theater floor the weight of material excavated for the basement is greater than the applied building loads. Furthermore, the requirements of head room in the basement, particularly below the stage, made it necessary to establish the footings below the top of the desiccated clay crust. All these factors suggested to Adler and Sullivan, the architects, that the foundations must be designed with great care. Mr. Adler was himself an engineer of considerable skill, and he retained General Sooy Smith as consultant on foundations.

Sooy Smith made a series of borings at the site of the structure, and conducted a series of load tests. The results of these tests have not been located, but it is known that the areas of the test plates were varied and that tests were made at different elevations in the upper clay strata in order to study the effects of these factors upon the probable settlement of the footings. On the basis of the tests it was decided to locate the base of the tower footing at El. − 5 and of the others at El. − 4, and to proportion all the footings for a unit load of 4500 pounds per square foot including an allowance for live load. The intensity of dead load beneath the footings was approximately 4100 pounds per square foot.

All footings consisted of 2 layers of 12-inch timbers laid at right angles to each other in prepared beds of gravel. So great was Sooy Smith's concern lest the water level might drop below the level of the timbers that he interconnected all the gravel beds and established a free connection to the water of Lake Michigan, which at that time was only a short distance across Michigan Avenue.

Above the timbers was placed a stepped grillage of railroad rails, steel beams, or both, embedded in concrete. On top of the grillage were placed dimension stone or rubble piers. The foundation of the tower was one large footing approximately 100 by 67 feet. It is shown in cross-section in Fig. 20.

The footings were proportioned with great care for the same soil pressure. Many of the footings were combined in order to satisfy the condition of concentricity of loads. Furthermore, since it was anticipated that the 10-story part of the structure would be completed before the tower reached its full height of 19 stories, Baumann's suggestion of pre-loading was adopted. The bricks to be used for construction of the entire tower, supplemented by loads of pig iron, were placed on the tower footing at such a rate that when the entire building reached the height of 10 stories, the full weight of the tower rested upon its footing.

Thus, in every respect, the foundations represented the best practice. The amount of settlement expected is reflected in the fact that the entire building was set two inches above grade. However, Sooy Smith and Adler took additional precautions to minimize the effects of such differential settlements as might occur. In order to avoid separation of the tower from the adjoining parts of the structure, such as had already become apparent in the Polk Street Station and the Board of Trade, Sooy Smith tied the tower footing to the adjoining wall footings by a series of 15-inch I-beams, which are clearly shown in Fig. 20. Furthermore, it was feared that several lightly loaded columns which carried loads from only one story would not settle as much as the rest of the structure, and screwjacks with a run of six inches were inserted at the top of these columns to allow for adjustment in the future.

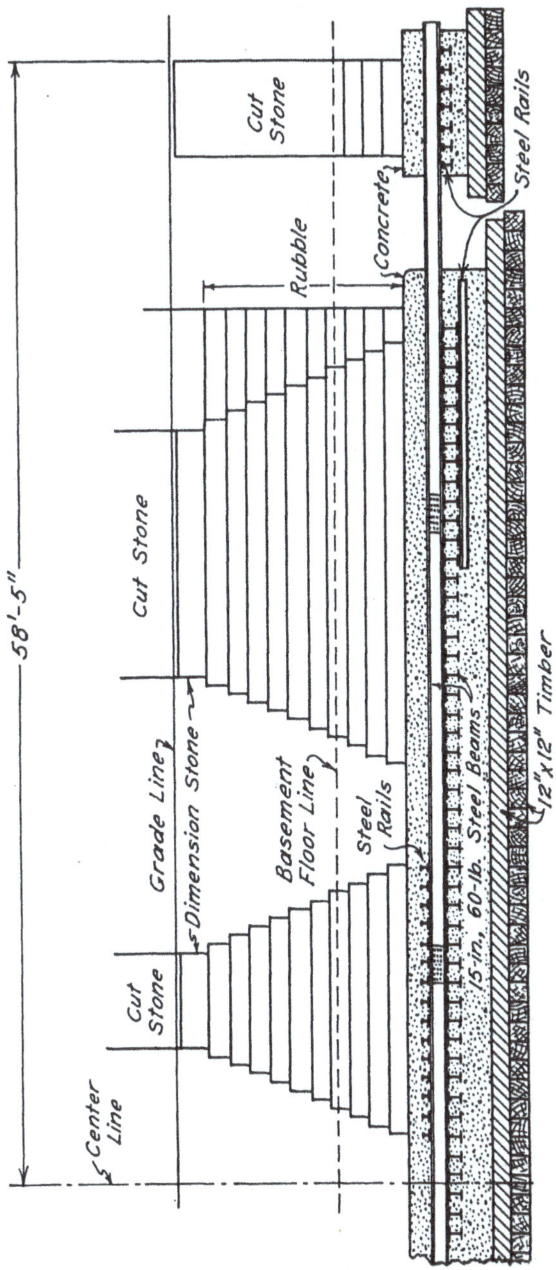

Fig. 20. Tower Foundation of Auditorium

The care with which Dankmar Adler considered the foundations for the Auditorium was representative of the design of the entire building. The artistry of Sullivan and the engineering skill of Adler produced what has generally been regarded as a masterpiece of architecture. The theater, with perfect acoustics, was the largest of its time. The decision to hold the World's Columbian Exposition of 1893 in Chicago was strongly influenced by the fact that the city possessed this remarkable building.

Yet Adler and Sooy Smith were compelled to watch the settlement of this structure grow within two years to seven inches, and in five years to eighteen. The tower did not separate from the rest of the structure, but the distortion of the frame was very great and on several occasions the safety of the building was a matter of litigation.

Fortunately, continuous settlement levels have been taken since the construction of the structure. A curve showing the progress of settlement at the point of maximum subsidence is shown in Fig. 21. Figure 22 shows a profile of the foyer floor in 1942.

Other Buildings

The full significance of the Auditorium's settlement was not felt until about 1895. In the meantime, many other large buildings were constructed on spread footings, with a considerable degree of success. Table 1 is a list which includes all downtown structures nine stories or higher, except the Auditorium and the Board of Trade, which were built on spread footings between the years 1884 and 1895.

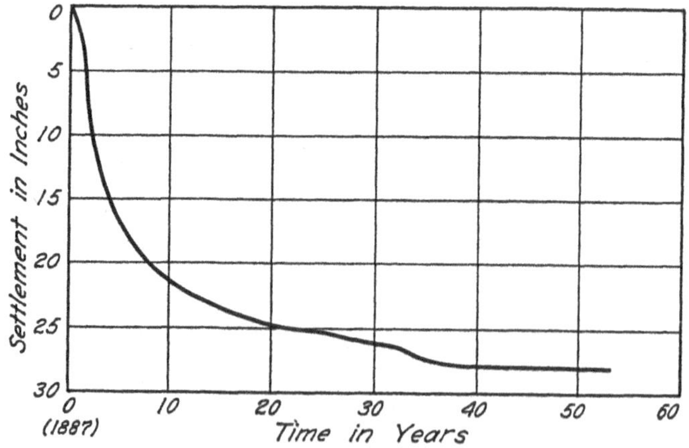

FIG. 21. TIME-SETTLEMENT CURVE FOR TOWER OF AUDITORIUM

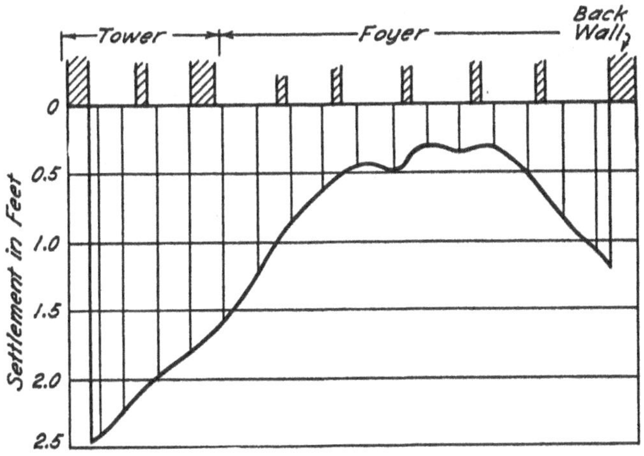

FIG. 22. PROFILE OF FOYER FLOOR IN AUDITORIUM (1942)

TABLE 1

BUILDINGS ON SPREAD FOOTINGS, NINE STORIES OR HIGHER, 1884-1895

The asterisk indicates that the structure was still in use in 1940, approximately 50 years after construction.

Date	Name	Stories	Location	Architect
1884	Calumet	9	111-17 S. LaSalle	Burnham & Root
	Mallers	12	224-26 S. LaSalle	John J. Flanders
	Pullman*	10	79 E. Adams	S. S. Beman
	Goff	9	230 S. LaSalle	S. V. Shipman
	Counselman	9	N.W. cor., S. LaSalle & W. Jackson	Burnham & Root
1885	Chicago Opera House Block	10	S.W. cor., N. Clark & W. Washington	Cobb & Frost
	Continental Nat'l Bank	9	208 S. LaSalle	Burnham & Root
	Royal Insurance	12	160 W. Jackson	W. W. Boyington
	Home Insurance	9	N.E. cor., S. LaSalle & W. Adams	W. L. B. Jenney
1886	Rialto*	9	132-48 W. VanBuren	Burnham & Root
	Union Bank	9	21-29 N. Dearborn	
	Mercantile*	9	305 W. Adams	Bauer & Hill
	Rookery*	11	209 S. LaSalle	Burnham & Root
	Austin (Western Union)*	11	111 W. Jackson	Burnham & Root
	Adams Express	11	109-19 S. Dearborn	Geo. H. Edbrooke
	Bd. of Trade Hotel (Atlantic)*	9	319-21 S. LaSalle	
1887	Como	9	443 S. Dearborn	John M. Van Osdel
1889	Tacoma	13	N.E. cor., N. LaSalle & W. Madison	Holabird & Roche
1890	Chamber of Commerce	13	S.E. cor., N. LaSalle & W. Washington	Edward Baumann, Harris Huehl
	Rand McNally	10	W. Adams west of S. LaSalle	Burnham & Root
	Manhattan*	16	431 S. Dearborn	W. L. B. Jenney
	Monon*	13	436-44 S. Dearborn	John M. Van Osdel
	Bedford (Owings) (Mat)*	14	203 S. Dearborn	H. I. Cobb & Frost
	Caxton*	12	508 S. Dearborn	Holabird & Roche
1891	Monadnock (North)*	16	53 W. Jackson	Burnham & Root
	125 S. Market St.*	10	125 S. Market	
	Virginia Hotel	10	N.W. cor., N. Rush & E. Ohio	Clinton J. Warren
	Pontiac*	14	N.W. cor., S. Dearborn & W. Harrison	Holabird & Roche

TABLE 1 (CONCLUDED)

BUILDINGS ON SPREAD FOOTINGS, NINE STORIES OR HIGHER, 1884-1895

The asterisk indicates that the structure was still in use in 1940, approximately 50 years after construction.

Date	Name	Stories	Location	Architect
1892	Venetian*	13	15 E. Washington	Holabird & Roche
	Woman's Temple	13	102-16 S. LaSalle	Burnham & Root
	Great Northern Hotel*	16	237 S. Dearborn	Burnham & Root
	Fair Store*	9	S. State, W. Adams & S. Dearborn	Jenney & Mundie
	Unity*	16	127 N. Dearborn	Clinton J. Warren
	Isabella*	11	21 E. VanBuren	Jenney & Mundie
	Chicago Title & Trust*	17	69 W. Washington	Henry I. Cobb
	Ashland Block*	16	155 N. Clark	D. H. Burnham
	City Hall Sq. Hotel (Kedzie Bldg.)*	9	87-91 W. Randolph	Edmund R. Krause
	Terminals*	14	537 S. Dearborn	J. M. Van Osdel & Co.
	Boyce*	12	30 N. Dearborn	Henry I. Cobb
	Marshall Field & Co. (So.)*	9	N. Wabash & E. Washington	D. H. Burnham
	Masonic Temple* (Capitol)	21	N.E. cor., N. State & E. Randolph	Burnham & Root
1893	Teutonic*	10	179 W. Washington	Handy & Cady
	Congress Hotel*	10	504 S. Michigan	Clinton J. Warren
	Central Y.M.C.A.*	16	19 S. LaSalle	Jenney & Mundie
	Watson	9	15-19 N. LaSalle	
	Monadnock (South)*	17	53 W. Jackson	Holabird & Roche
	Columbus Memorial*	14	31 N. State	W. W. Boyington
	Chicago Athletic Club*	10	12 S. Michigan	Henry I. Cobb
	Hartford*	14	8 S. Dearborn	Henry I. Cobb
	Lees*	12	15-23 S. Wells	James Gamble Rogers
	Security*	14	189 W. Madison	Clinton J. Warren
	Majestic Hotel*	17	29 W. Quincy	D. H. Burnham
	VanBuren*	10	210-14 W. VanBuren	
1894	Old Colony*	17	407 S. Dearborn	Holabird & Roche
	Champlain	15	N.W. cor., N. State & W. Madison	Holabird & Roche
	Chemical Bank (Iroquois)	9	21-23 N. LaSalle	Burnham & Root
	LaSalle-Monroe* (New York Life)	12	37-43 S. LaSalle	Jenney & Mundie

The fact that 38 of the 57 structures listed were still in service in 1940 is convincing evidence that the buildings served their purpose well. Furthermore, with one exception, there is no record of structural damage to any of them as a result of settlement under their own weight, such as occurred in the Polk Street Station, the Board of Trade, or the Auditorium. It is also worth noting that all these buildings exerted a relatively uniform load on the ground. None of them contained either a high tower with adjacent lower sections, or heavy load-bearing walls combined with lightly loaded interior columns. After 1885, almost every building listed was of skeleton construction and had a fairly uniform column spacing.

On the other hand, it should not be assumed that the structures did not settle or that Chicago engineers were ignorant of the settlement. In fact, it was the practice at least as early as 1884 to set the buildings above grade such a distance as would, in the judgment of the

TABLE 2

ALLOWANCES FOR SETTLEMENT

Date	Building	Allowance inches	Architect
1885	Home Insurance	2½	W. L. B. Jenney
1889	Auditorium	2½	Adler & Sullivan
1891	Monadnock (North)	8	Burnham & Root
1892	Capitol	8	Burnham & Root
1892	Great Northern Hotel	9	Burnham & Root
1892	Fair Store	4½	Jenney & Mundie
1894	New York Life	4½	Jenney & Mundie
1896	Atwood	6½	Holabird & Roche

engineers, cause the structure to be at grade after settlement. As experience disclosed the progressive settlement of the buildings, the allowance for settlement became greater. A chronological table (Table 2) showing the allowance for several structures is informative of the manner in which the importance of the settlement problem grew in the minds of the designers.

The actual settlement of a few of these structures is known in some detail, and fragmentary information exists about others. Mr. Jenney kept records of the settlements of his structures during the construction period. The maximum settlement of the Home Insurance Building at the time of completion was 2¼ inches, and the greatest differential settlement was $^{11}/_{16}$ inch. There is no record of subsequent levels, but in February, 1889, Jenney wrote: "experience has taught that foundations calculated to carry 3500 to 4000 pounds per square foot . . . will settle four to five inches."[(15)] Since the soil pressure used for designing the Home Insurance Building was 4000 pounds per square foot, his statement suggests that after five years the Home Insurance Building may have settled that amount. The Fair Store, probably designed for 16 stories with a soil pressure of 3500 pounds per square foot and actually constructed with 9 stories at a pressure of 2850 pounds per square foot, settled a maximum of 1⅜ inches during its first four years. The Central Y.M.C.A., built in 1893 by Jenney, settled $2^{11}/_{16}$ inches during its construction period.

The 16-story Monadnock Block (north half) is one of the heaviest structures in Chicago. Although built in 1891, it was of wall-bearing construction. The average footing pressure was 3750 pounds per square foot, and the average pressure over the gross area of the structure was 3300 pounds per square foot. The area of the structure between building lines is a rectangle 66 feet wide and 398 feet long. On the basis of

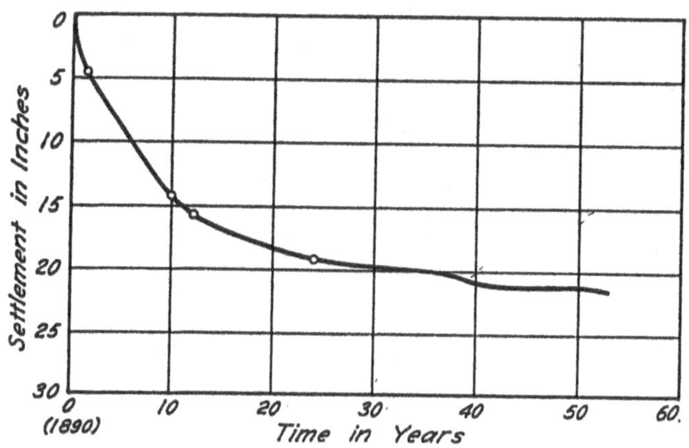

FIG. 23. TIME-SETTLEMENT CURVE FOR NORTH HALF OF MONADNOCK BLOCK

various surveys made from time to time, Mr. V. O. McClurg has reconstructed the settlement record shown in Fig. 23. The differential settlement of the structure in 1900 was of the order of 3 inches. Figure 24 shows a profile along the east side of the building, taken at that time. The profile extends through the south half of the building, built one year after the north half was constructed.

The Masonic Temple (Capitol Building), erected in 1892, was the first 21-story building in the world. It was rectangular in plan, 195 by 132 feet between building lines, and was designed for a soil pressure of 3200 pounds per square foot. In contrast to the heavy Monadnock Block, the Masonic Temple was of skeleton construction and exerted a load of only 1600 pounds per square foot over the gross area of the structure. Figure 25 shows time-settlement curves corresponding to two points in the building from the start of construction until 1913, twenty-two years afterward. The maximum settlement was ap-

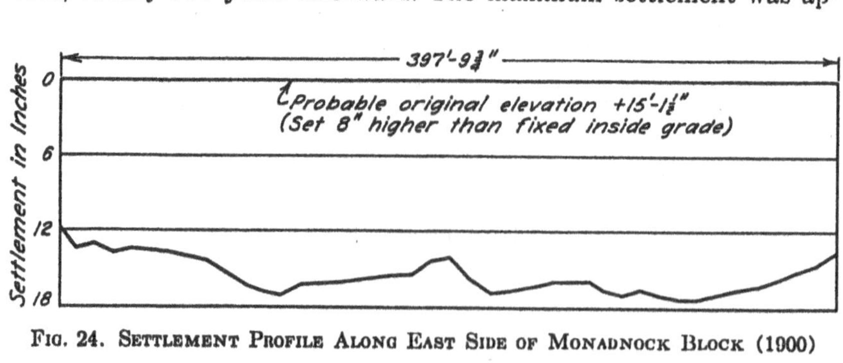

FIG. 24. SETTLEMENT PROFILE ALONG EAST SIDE OF MONADNOCK BLOCK (1900)

proximately one foot, and the greatest differential settlement was 7½ inches. Curves of equal settlement are shown in Fig. 26. They correspond to the year 1913. Settlement during the construction period was approximately 2½ inches.

The Great Northern Hotel, 16 stories, built of skeleton construction in 1892, settled between 9 and 10 inches during the first year after

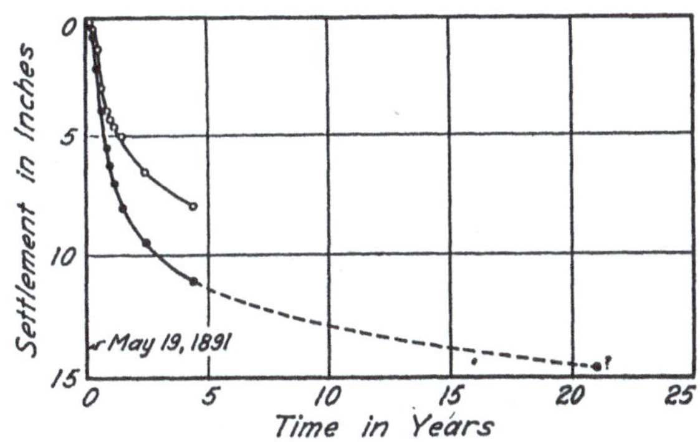

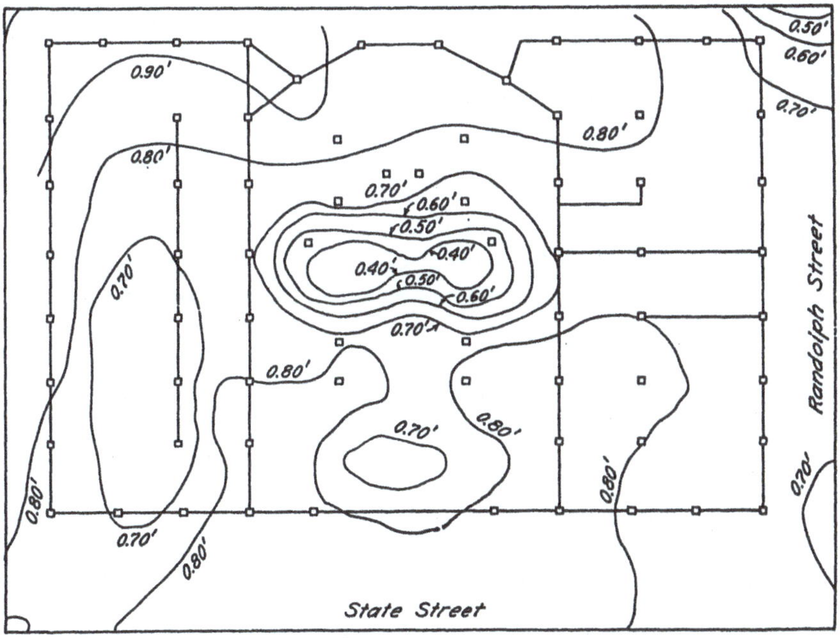

FIG. 25 (ABOVE). TIME-SETTLEMENT CURVES FOR MASONIC TEMPLE

FIG. 26. CURVES OF EQUAL SETTLEMENT FOR MASONIC TEMPLE (1913)

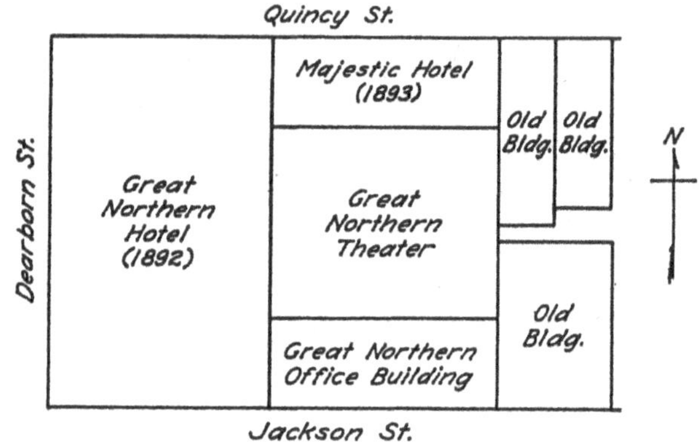

FIG. 27. LOCATION OF BUILDINGS, GREAT NORTHERN GROUP

construction. After this period three additions were constructed — the Great Northern Office Building, the Majestic Hotel, and the Great Northern Theater. The mutual relation of the four parts of the structure is shown in Fig. 27. One commentary on the new construction gives an interesting picture of the complacency with which Chicago engineers of this period regarded differential settlements:

"The new additional structures stand between the original hotel, which is of nearly equal weight . . ., and two other old style buildings which came to their permanent bearings many years since. Yet the new part fills the whole intermediate space, and it has been required to provide for the necessary settlement of the new building while regarding the safety and avoidance of disturbance of two classes of contiguous buildings. When it is understood that the estimated settlement of the new structure was nine inches, the difficulty in doing this can be readily appreciated. As a matter of fact, the Jackson street addition has settled nine inches and the Quincy street addition has settled six inches. But the former now has its full weight, while the latter has not yet attained its full weight, and the construction has been slower. The theater which stands between is therefore three inches lower on the south than on the north side, but as its whole construction is sufficiently elastic, nothing has to be feared while waiting for the north side to go down three inches more."[16]

During 1894, the Old Colony Building, a 17-story steel-frame structure, was completed. It was designed for a soil pressure of 3200 pounds per square foot. The average load over the gross area of the building was about 2500 pounds per square foot. The average settlement over its nine-month construction period was 4.2 inches, and the maximum value 5.1 inches. This settlement was considered quite normal.

Indeed, by 1895 most engineers and architects in Chicago felt themselves masters of their foundation problems. They accepted the idea that a certain amount of distortion of the building frames due to differential settlement was inevitable and, if anticipated, was not harmful. They did not consider it worth their while to investigate the possibility of less yielding foundations. As a matter of fact, the architect who contemplated a deep foundation for an office building would probably have failed to find any business, because Chicago business men were of a particularly utilitarian mind. Competition was keen among the architects to produce serviceable buildings at the lowest cost, and the expense of deep foundations seemed prohibitive and unnecessary.

VIII. Rise of the Long-Pile Foundation (1883-1903)

Earliest Uses of Piles

It is generally believed that pile foundations were first used in Chicago beneath the grain elevators along the waterfront. The earliest of these was built in 1848 on the northwest corner of Wells and S. Water Street (now W. Wacker Drive). In 1857, the first elevator constructed of 2-by-4 staves was erected, and from that date until the advent of reinforced concrete every elevator built in the city was of that type. All these structures were supported on long wood piles, loaded to about 20 tons per pile. The success of the foundations appears to have had little influence on the architects who constructed the tall buildings of a generation later, possibly because the construction of elevators was in the hands of a few specialists who did not engage in ordinary building construction. Furthermore, the apprenticeship of the skyscraper architects was usually in smaller buildings on spread footings, and as structures grew taller they were regarded merely as a development from the small structures.

The first use of piles beneath a conventional building was probably under the Sibley Warehouse, still in existence on the north bank of the Chicago River, on the east side of Clark St. Construction was started in 1883. This structure was of nine stories, each designed for a live load of 500 pounds per square foot. Only the river-front wall, 240 feet long, was supported on 30-ft. oak piles. The rest was carried on footings and masonry piers which, according to a contemporary account, "seem to nearly cover the whole area." The piles did not entirely prevent settlement or lateral movement of the river-front wall as the channel of the river was deepened from time to time, and the structure was tied together at an early date by iron tie-rods. In 1945, extensive alterations were made to the foundation along the river front. Systematic levels and alignment surveys have been made on the building for many years.

The first complete long-pile foundation under a building is that of the Grand Central Station, on the southwest corner of Harrison and Wells Streets, known formerly as the Wisconsin Central Station and as the Northern Pacific Station. It consists of a masonry tower, 27 feet square and 236 feet high, at one corner of the building, and adjacent portions only 8 and 5 stories in height. The piles are all about 50 feet long and loaded to 25 tons. This structure, built in 1890 by S. S. Beman, is still in use. A careful examination shows no signs of structural distress due to differential settlement, and the heavy tower and adjacent lighter portions are still perfectly bonded together.

The second long-pile foundation was that of the German (Garrick) Theater, built by Adler and Sullivan in 1892 at 64 W. Randolph St. The different parts of this structure range from

"a height of four stories for certain small portions of the structure to thirteen stories for the greater part of the same and from this to seventeen stories for the tower. . . . Borings made upon the ground show that the characteristic soft Chicago mud attains a depth here of from forty-two to forty-eight feet below the cellar floor, and that below this is found the well-known tunnel clay, so hard that it was impossible to bore through it with the ordinary boring tool used. Fifty-foot piles were driven through this soft mud until, as in the case of the Northern Pacific Station, their points strike the tunnel clay. Here they were merely given a sufficient number of blows to make sure that their points had penetrated the hard clay. . . . The heads of these piles are cut off at a uniform level of about three feet below . . . water line, and are then covered by a grillage of oak timbers running crosswise of each other and both layers drift-bolted into the heads of the piles. Upon this is formed a foundation of concrete and I-beams, of which the outer parts act as cantalevers [*sic*] for carrying the parts of the new enclosing walls, which are erected in immediate contact with the old party walls of the adjacent buildings."(17)

In itself, the foundation of the German Theater was an outstanding success. However, because of damage to adjacent structures due to the pile driving, the construction was followed by litigation.

In 1892 and 1893, three structures on long piles were completed. These were the Illinois Central Station, east of Michigan Avenue at Twelfth Street, Bradford L. Gilbert, architect; the Newberry Library, 60 W. Walton Street, Henry Ives Cobb, architect; and the Medinah Temple, northwest corner of Wells and Jackson, with Beers, Clay and Dutton as architects. Figure 28 shows a cross-section through one of the pile footings of the Illinois Central Station which is representative of the practice at this time.

In discussing this foundation, Mr. C. J. Mitchell, Assistant Civil Engineer for the Illinois Central Railroad, stated:

"Piles varied from 11 to 16 inches in diameter at the butt end, and were from 40 to 60 feet in length, averaging about 51 feet. . . . A cast iron cap was used in driving. . . . The drivers used, three in number, were of the usual pattern of drop hammer, the weights of which were

2800, 3200, and 3800 pounds respectively. . . . The fall of the hammer at the last blow was usually 35 to 40 feet, . . . [the] penetration under the last blow averaged about 3 inches, though a large number went less than 1½ inches. . . .

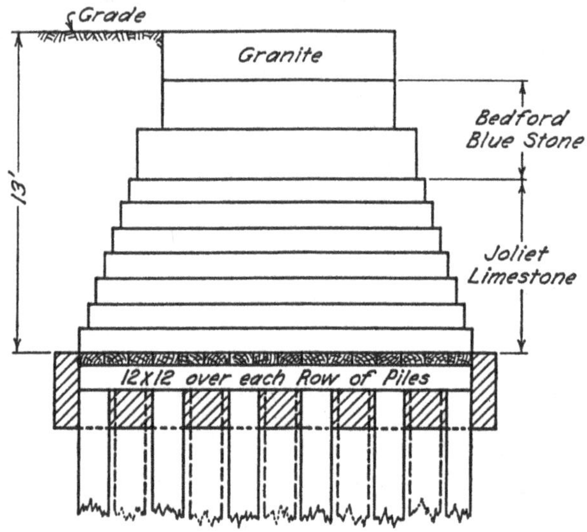

FIG. 28. PILE FOOTING OF ILLINOIS CENTRAL STATION

"After the piles had been driven, the tops were sawed off to a uniform height of 3 feet below datum in order that all timber should be below low water. . . . After the piles were cut off, the earth was excavated 18 inches below their tops and rich Portland cement concrete was tamped in even with the tops of the piles. Oak caps 12 x 12 inches were fastened on by a drift bolt in the center of each pile, and the space between the timbers was filled with concrete; on top of this came the masonry work."[(18)]

During pile driving, it was noted that piles previously driven were lifted as much as four inches. No mention is made of any redriving. Hence, the information given by Mr. Mitchell indicates that the piles were not driven to a particularly resistant layer and that the foundation did not attain present-day standards. Nevertheless, the structure, which consists of a 13-story tower, a 9-story office section, and a 3-story station structure, has not been damaged by settlement.

Pile Cut-off at Great Depth

In 1893, General Sooy Smith introduced a new conception in the design of the pile foundations for the Public Library on Michigan Avenue between Washington and Randolph Streets. For several years General Sooy Smith had expressed his fear that a deep system of sewers or the construction of a subway would lower the water table and expose the tops of wooden piles cut off only one or two feet below datum. In connection with the Auditorium, this concern was reflected by his provision for direct transmission of water from Lake Michigan to the gravel pockets around the timber grillages. In the Public Library, it led to the adoption of a pile cut-off twelve feet below datum. A cross-section through one of the bearing-wall foundations is shown in Fig. 29.

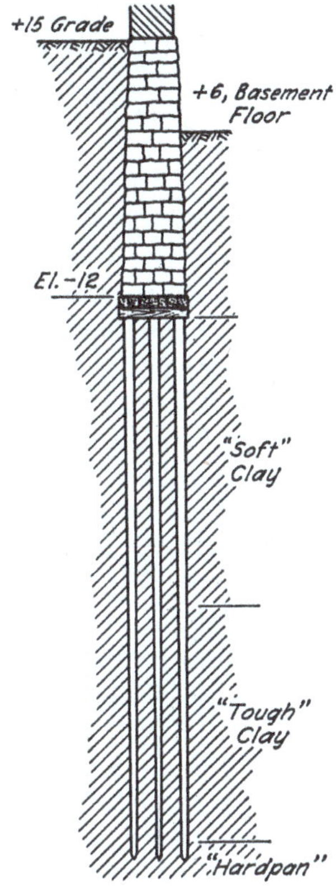

Fig. 29. Bearing-wall Foundation for Public Library

"The piles were driven [in the bottom of a sheeted trench] by a steam hammer . . . made by the Vulcan Iron Works; weight 4500 lbs.; fall, 42 ins., making 54 blows per minute. The last 20 feet were driven with a follower of oak. It was found that it required 48 to 64 blows to drive the last foot with the follower, and as the ratio of blows without follower to blows with follower is as one to two, it may be estimated that it would have required from 24 to 32 blows of the above hammer to drive the last foot directly, without follower. . . .

"The piles were driven 2½ feet between centers . . . deemed to be as close as they can be driven with ease. They were about 54 ft. long and were driven about 52½ ft. They had an average diameter of 13 in. . . . The heads of the piles, after being sawed off were 27 ft. below the street grade, and the tips were about 80 ft. below the same

[El. — 65]. They were driven about 27 ft. in soft, plastic clay, 23 ft. in tough, compact clay, and 2 ft. in hardpan."(19)

A load test was made on a group of four of the piles by Mr. N. E. Weydert, Superintendent of Buildings for the City. A platform seven feet square was constructed and placed on four piles. It was then loaded with pig iron to a height of 38 feet, subjecting each pile to a load of 50.7 tons. The load acted for eleven days, without producing measurable settlement. The design load was taken as 30 tons per pile.

Similar pile foundations were constructed by General Sooy Smith in 1894 for the Stock Exchange, still standing at the southwest corner of LaSalle and Washington Streets, and in 1897 for the Post Office, located between Clark, Dearborn and Adams Streets and Jackson Blvd. In the Stock Exchange, 30-ton piles were driven at three-foot centers and cut off at El. — 16′-10″; in the Post Office, piles were driven at 3½-foot centers and cut off at El. — 13′-8″. With a single-acting hammer weighing 4400 pounds, the average penetration per blow under the last 200 blows was less than one-half inch.

These foundations were practically unyielding and entirely satisfactory. However, Sooy Smith's fears of lowering the water table by drainage were based upon his failure to recognize the ability of the pervious sand and silt layer overlying the clay to furnish water more rapidly than it could be removed through the less pervious clay by any sewer or subway structure. Therefore, pile cut-off one or two feet below datum became the accepted practice, and no subsequent pile foundations were cut off at low level. In 1903 the LaSalle Street Station, located on the south side of VanBuren St. between LaSalle and Sherman Streets, was constructed, with E. C. and R. M. Shankland as engineers. It rests on 50-foot piles cut off just below datum. Hence, in this structure, the earlier practice became re-established; however, stone piers, such as were used in the Illinois Central Station, were superseded by grillages of I-beams and concrete.

A number of pile foundations have been constructed in the Loop, but as column loads grew heavier, piers to hard clay or bedrock became more popular. Piles are still the predominant foundation type for buildings north of the Chicago River, where a number of 15- to 25-story apartment houses have been built. In order to avoid cut-off at datum, concrete piles have been used extensively in this area.

IX. The "Chicago Caisson" to Hardpan

Limitations of Pile Foundations

In the congested Loop district, the pile foundation was not without its disadvantages. The vibrations due to the pile driving were injurious to adjacent structures, particularly if these were of old construction and founded on the silt layer, because under vibration the silt tended to compact. Furthermore, even newer structures were damaged by the heave caused by displacement of the soil by the piles.

Foundations for Stock Exchange

The litigation associated with the construction of the German Theater has already been mentioned. In 1894 the owners of the new Herald Building, adjacent to the side of the Stock Exchange, obtained an injunction preventing the driving of piles for the foundations of the wall of the Stock Exchange nearest their structure. To overcome the difficulty, General Sooy Smith introduced a line of cylindrical piers to replace the piles. These piers were the first to be constructed in shafts excavated by the so-called "Chicago caisson method." The term "caisson" was not used by Sooy Smith but was incorrectly introduced at a later date by construction men. Nevertheless, the term has persisted, and became accepted in Chicago to describe foundation piers of this type.

The Stock Exchange caissons were the outgrowth of several years' thought and experience on the part of Sooy Smith. The forerunner of the caissons was a group of wells sunk beneath the stage of the Auditorium in 1887 to house hydraulic cylinders for raising and lowering part of the stage. The procedure is best described in the General's own words in 1892.

"I sank these wells after the adjacent walls of the building were built. Some of them went to a depth of 24 feet below the footings of the foundations and were only four and one-half feet from them. The weights resting on the soil amounted to . . . 4320 pounds per square foot. . . .

"Extraordinary precautions were taken to prevent the movement of the earth around the walls, as such movement might have caused serious settlements of the walls. Strong steel rings were used, and tongued and grooved sheathing was driven outside of these rings, and kept well in advance of the excavation as this proceeded. We got the wells down without the slipping of a spoonful of earth outside of them, and this without the difficulty anticipated."[20]

In 1893, Sooy Smith prepared plans and estimates for underpinning the Board of Trade Building by an extension of this method, but the scheme was rejected by the owners in favor of dismantling part of the tower. Hence, the Stock Exchange represented the first full-scale trial of the method. According to a contemporary account:

"The clay at this point is as soft as has been encountered in any part of the city. The conditions were such that pile foundations seemed especially desirable, yet it was quite out of the question to drive piles so close to the Herald Building as would have been necessary in supporting the building wholly upon piles. The foundation plan finally adopted was that shown in [Fig. 30], all footings resting on pile

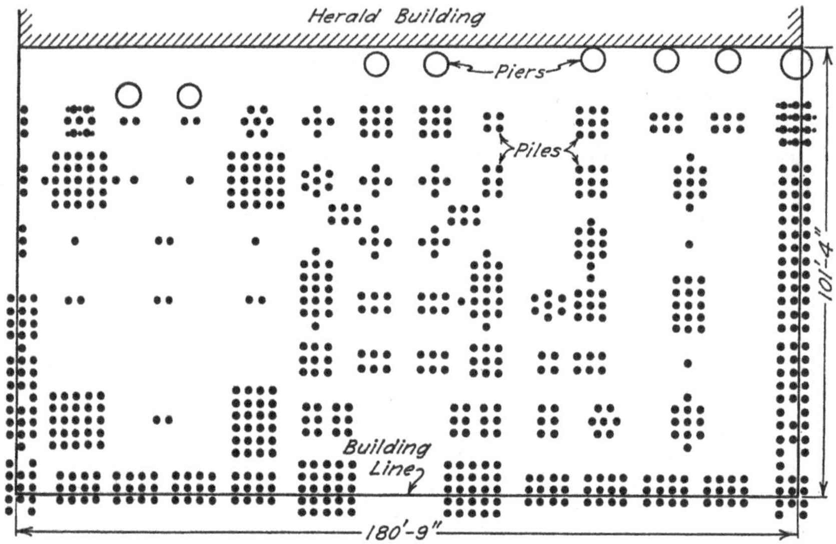

FIG. 30. FOUNDATION PLAN OF STOCK EXCHANGE

foundations except those near the Herald Building. These rest on foundations made by excavating wells down to the hard clay and filling the same with concrete. In making the excavations, the sides of the well were supported by oak sheathing held in place by steel rings, the sheathing being put in place as fast as the excavation progressed. These concrete foundations vary in diameter from 5 ft., to 6 ft. 4 in., increased at the base to 6 ft. 6 in. and 8 ft."(21)

Later Applications

The next use of the caissons was for underpinning the south wall of the Old Colony Building in 1896, three years after its erection. The footings for this wall had proven inadequate and excessive settlement occurred. The success of the operation, which was supervised by Sooy Smith, established the fact that the well method of construction was admirably suited for underpinning operations.

The first use of caissons under an entire new building probably occurred in 1899 at the Methodist Book Concern, now known as the Stop and Shop Warehouse at 12-14 W. Washington St. Within the next few years, many buildings were constructed on Chicago caisson foundations. Some 250 separate piers were constructed for the Cook County Hospital on Harrison St. between Wood and Lincoln Streets in 1912, and for the first time load tests were made to establish the bearing capacity of the hardpan at a building site. On the basis of the tests, the design load of 13,300 pounds per square foot at the bottom of the caissons was permitted.

X. The Controversy over Shallow and Deep Foundations

Before 1890, serious consideration had never been given to the use of deep foundations in Chicago. By 1892, the settlement of six prominent buildings on shallow foundations reached the values given in Table 3.

Of these buildings, only the Auditorium had been designed by Chicago engineers and architects. Dankmar Adler and his consultant on foundations, Sooy Smith, remained in Chicago, where they could not but observe the settlement of their structure. Further, Sooy Smith was at the same time observing the settlement of the Board of Trade in order to recommend remedial measures. Hence it is not surprising that these two men should have been the first to conclude that spread foundations above the soft clay did not represent sound engineering and should be abandoned. In June, 1891, Mr. Adler wrote:

"Neither the timber and concrete foundations of the Board of Trade and the Pullman buildings, nor the rail and concrete foundations of the Rialto, Phoenix and Rookery buildings, nor the timber, concrete, rail and I-beam foundations of the Auditorium, nor the concrete and I-beam foundations of the Tacoma and later buildings are the last word. . . . We should take into consideration the brilliant success of pile foundations as used under most trying conditions as supports for our grain elevators and as the substructure of the Northern Pacific Railway station."(22)

In October, 1892, General Sooy Smith, addressing the engineering students of the University of Illinois, said:

"It is the common practice of Chicago architects to load the soil at the rate of 3000 pounds per square foot. . . . Experience with heavy buildings shows that in the course of months and years they . . .

Table 3
Approximate Settlement of Buildings in 1892

Building	Age	Maximum Settlement	Differential Settlement
Post Office*	13 yrs.	14"	. . .
Polk Street Station	9 yrs.	8" (?)	. . .
Board of Trade	8 yrs.	14½"	7¼"
City Hall*	7 yrs.	. . .	14"
County Building†	7 yrs.	. . .	18"
Auditorium	5 yrs.	17⅜"	17"

* Mat.
† Short Piles.

continue to settle when the soil is not loaded to exceed the limit above stated. Experience also shows that while the initial settlements under a given load may be uniform throughout the area covered by a building, the progressive ones may eventually so differ as to cause serious cracks and demoralization in the structure. . . . [We] must, if practicable, secure an unyielding foundation. . . . Wherever it is practicable, the foundations of heavy buildings should rest upon the rock or upon the hard-pan immediately overlying it. . . . [Piles] can be driven . . . down to the rock, or into the hard-pan immediately overlying it. . . . And in cases in which the rock is within easy reach by wells, say, at a depth not exceeding 60 feet, they may be sunk to the rock and a pillar or column of rubble masonry or concrete may be carried up to the level of the basement or sub-basement floor.

"Careful estimates show that these foundations will cost less than those which are now generally employed; namely, platforms of steel and concrete, resting on compressible soil."(23)

Excerpts from Sooy Smith's speech were printed in Chicago papers; the reaction of the designers of office buildings was swift and unfavorable. Mr. Jenney, designer of the Home Insurance Building, responded:

"As to the deep piling . . . recommended by the General, . . . it is unnecessary in my estimation, and would be only a waste of money. The great high buildings put up on foundations supported by piers at datum are proving highly satisfactory. For instance, the Home Insurance Building has settled so uniformly that the greatest variation throughout is only three-fourths of an inch. . . . The present method of putting up buildings, if carefully and scientifically followed out, leaves nothing to be desired. Gen. Sooy Smith's ideas are above scientific criticism, but I am afraid it would be hard to persuade capitalists to pay the extra money to put them in practice. Nothing but an Eiffel tower would demand them."(24)

Mr. D. H. Perkins, manager for D. H. Burnham, whose works included the Rialto, Phoenix, Rookery and other buildings, remarked,

"I do not like to criticise so great an authority as Gen. Sooy Smith at a moment's notice. . . . As to the piling principle we do not use it and all our buildings stand on separate column footings which rest on the blue clay. We have not found that piles are necessary and our buildings have not suffered for lack of them."(25)

Evidently the architects and engineers involved in the controversy felt that there was no middle ground. Adler and Sooy Smith stood almost alone in 1892. Adler's belief in the lesson to be learned from the Northern Pacific Station was evidently not shared by Mr. Corydon T. Purdy, who remarked four months after Adler's article:

"If we could drive piles to bear upon a real hard pan or solid rock we could put a greater load per square foot over the area covered, than we can with the spread foundation. . . . Some claim that this can be done, but its practicability has yet to be demonstrated.

"We believe that the secret of equal settlements is very largely in the proper distribution of . . . loads. . . . I have yet to learn of a case where reasonable care was exercised to secure the proper proportions that settled unevenly enough to reveal the fact in the construction."(26)

The statement that the practicability of pile foundations had not been demonstrated was challenged by Mr. H. B. Herr, who drove the piles for the Northern Pacific Station. He rejoined:

"We have pounded down a good many piles for foundation work and other purposes. . . . I am well satisfied that if you drive a pile until it reaches the hard clay anywhere in this city, that is the best you can do for foundation support, and I believe it can be done anywhere in the city."(27)

Before 1900, Adler and Smith had demonstrated the practicability of both pile and pier foundations. Indeed, after 1892 there is no record that either man was ever again associated with a spread foundation for a building of any importance in Chicago. Mr. Purdy's confidence in the ability to proportion footings for equal settlement must have been shaken by the necessity for having Sooy Smith underpin the south wall of his Old Colony Building only three years after construction. In fact, by 1900 there was a slow trend toward the deeper foundations. By that time, the settlement even of some office buildings had become unpleasantly great. Table 4 lists the settlement of a few of the structures.

Hence, evidence in favor of deep foundations began to accumulate, and one by one the architectural firms began to adopt them. Adler and Sullivan built their first deep foundation in 1891. The firms of D. H. Burnham (successor to Burnham and Root) and Holabird and Roche

TABLE 4
APPROXIMATE SETTLEMENT OF BUILDINGS IN 1900

Building	Age	Maximum Settlement	Differential Settlement
Board of Trade	16	18″	9″
Auditorium	13	23½″	23″
Monadnock	9	14″	...
Masonic Temple	8	8″	...
Great Northern Hotel	6	10″	...
Old Colony*	6	5″	...

* South wall underpinned.

began their transition in 1896 and 1897, respectively. Mr. Jenney, the strongest supporter and probably the most successful user of the spread footing, did not alter his methods until 1904.

After 1904, contemporary accounts of Chicago foundation practice almost disappeared from engineering literature. Indeed, in 1904, an event occurred which made any future argument strictly academic. At that time the Chicago Tunnel Company began the construction of small freight tunnels along the principal streets of the Loop. The digging of the tunnels, about nine feet high and six feet wide and located fifty feet below the ground surface, caused the settlement of many structures and initiated a widespread need for underpinning. From that date, the possibility of settlement due to adjacent excavation could not be ignored, and very few new buildings were set on spread footings.

The transition from spread to deep foundations is illustrated by Fig. 31, which shows the height of the buildings erected during any one year and the number of buildings on spread footings, piles, or caissons. Only buildings of six stories or more are considered. The slow transition between 1890 and 1900 is significant, for during this period the average building height did not increase. The rapid change to almost no spread footings after 1905 is very likely the result of the combined influence of settlements due to excavation and of increased height of buildings.

Thus, the controversy which was so active in the decade from 1890 to 1900 resolved itself. The superior merits of deep foundations and the disadvantages of spread footings gradually forced themselves upon the attention of Chicago foundation engineers and architects, and nullified all arguments to the contrary.

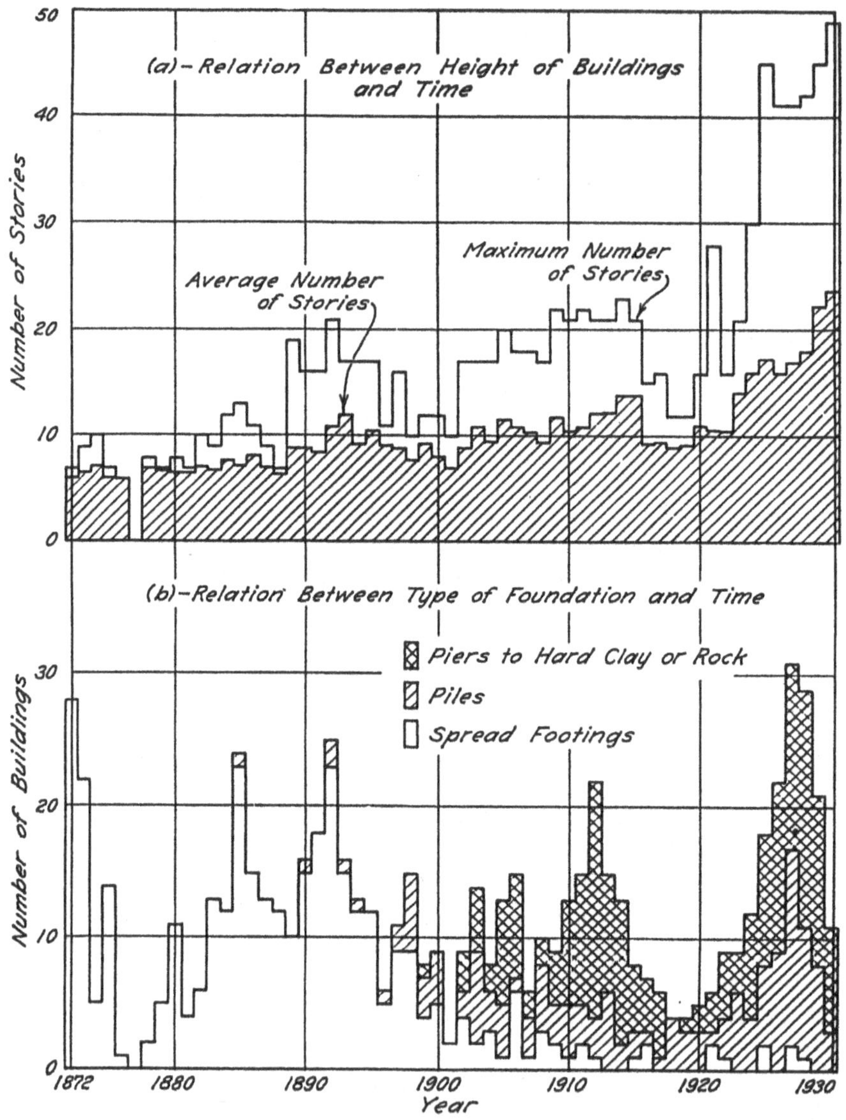

Fig. 31. Transition from Shallow to Deep Foundations. (a) Relation Between Height of Buildings and Time; (b) Relation Between Type of Foundation and Time

APPENDIX A

Selected References

The Inland Architect and News Record, published in Chicago from 1883 to 1908, contains many articles and news items concerning foundations for Chicago buildings. Among the more extensive are the following:

Recollections of Van Osdel, July 1883, p. 78; Aug. 1883, p. 89.

W. L. B. Jenney, *The Construction of a Heavy, Fireproof Building on a Compressible Soil*, Dec. 1885, p. 100. (Concerning the Home Insurance Building)

Dankmar Adler, *Foundations of the Auditorium Building, Chicago*, March 1888, p. 31.

Dankmar Adler, *Tall Buildings*, June 1891, p. 58.

W. L. B. Jenney, *The Chicago Construction, or Tall Buildings on a Compressible Soil*, Nov. 1891, p. 41.

Tall Buildings in Chicago, Railroad Gazette, Oct. 30, 1891.

Foundations of the Chicago Stock Exchange Building, Railroad Gazette, April 27, 1894, p. 302.

The Development of Shallow and Deep Foundations for Chicago Buildings, Engineering News, Dec. 22, 1904, p. 560.

Caisson Foundation Work in Chicago, R. C. Smith, Purdue Engineering Review, May 1922, p. 20.

The Building Problem in Chicago from an Engineering Standpoint, Wm. Sooy Smith, Illinois Technograph, 1891-1892, pp. 9-21. Summarized in Engineering News, Oct. 13, 1892.

The Foundations of the New Post Office and Government Building at Chicago, Engineering News, Jan. 27, 1898, p. 66.

Steel Skeleton Construction in Chicago, E. C. Shankland, Proceedings of the Institute of Civil Engineers (London), Vol. 128, Part II, 1897, p. 1.

Industrial Chicago, Goodspeed Publishing Co., Chicago, 1891, Vol. 1, p. 473.

APPENDIX B

Notes

1. *Inland Architect,* July 1883, p. 78.
2. *Ibid.*
3. E. C. Shankland, "Foundations," *A Half Century of Chicago Building,* Chicago, 1910.
4. *Op. cit.,* p. 7.
5. *Op. cit.,* p. 16.
6. *Op. cit.,* p. 14.
7. *Inland Architect,* Dec. 1885, p. 100.
8. *Ibid.*
9. *John Wellborn Root,* p. 114.
10. *Op. cit.,* p. 29.
11. *Journ. Assn. Eng. Soc.,* 1891, p. 569.
12. *Ibid.*
13. *Inland Architect,* Dec. 1898.
14. *Inland Architect,* Feb. 1894, p. 92.
15. *Inland Architect,* Feb. 1889, p. 7.
16. *Inland Architect,* Sept. 1896, special supplement.
17. For an earlier account, see *Am. Architect and Building News,* Oct. 24, 1891, p. 55.
18. *Technograph,* No. 7, 1892-93, p. 10.
19. *Eng. News,* July 6, 1893, p. 3.
20. *Technograph,* Oct. 1892, p. 11.
21. *Railroad Gazette,* April 27, 1894, p. 303.
22. *Inland Architect,* June 1891, p. 58.
23. *Technograph,* Oct. 1892, pp. 11, 14.
24. *Ibid.,* p. 20.
25. *Ibid.,* p. 21.
26. *Journ. Assn. Eng. Soc.,* Nov. 4, 1891, p. 571.
27. *Ibid.*

7
The Gothic cathedral: design and meaning

Otto G. von Simson

Within the last century the history of architecture has sought to explain the origin and meaning of the Gothic cathedral by singling out, one after the other, three of its main aspects: function (Gothic solution of statical problems), design (Gothic form as the expression of certain esthetic principles), and significance (Gothic form as the symbolic expression of certain ideas). The interpretations based upon each of these three aspects stand in a curious relationship to one another. Developed polemically and antithetically, each has sought to interpret Gothic architecture in terms of one of the three aspects, denying or belittling the significance of the two others. In point of fact, each of the three approaches has greatly advanced our knowledge; it seems to us today as if they not only complement, but actually point toward one another, encircling, as it were, a truth that lies in their midst.

Thus it is precisely to the history and analysis of Gothic form that we owe the knowledge that this form cannot be entirely understood in terms of stylistic development. Gothic architecture emerges suddenly and almost simultaneously with the great expressions of Romanesque art, not its heir or "logical sequel," but its rival and antithesis. The first Gothic art, moreover, is geographically so closely identified with one territory and its historical destinies that the late Henri Foçillon suggested, paradoxically but wisely, that Gothic be defined as the Romanesque of the Ile de France. The new style, finally, seems to have been the creation of a small group of men whose ideas we know and who were mutual friends; we are prompted to ask whether Gothic art must not be understood as an expression of these ideas.

In other words, the stylistic analysis of Gothic architecture has suggested its interpretation as the expression of certain ideas. And this second approach, the symbolic one, though long regarded with some suspicion, has recently attracted a good deal of attention. We have been reminded that the Christian sanctuary is, liturgically and mystically, an image of the heavenly Jerusalem, the eschatological vision described by the Book of Revelation. The medieval dedication rite establishes this relationship in explicit terms and the twelfth and thirteenth centuries appear singularly preoccupied with this symbolic significance of sacred architecture. May not this significance have actually determined the design of the cathedral? Indeed, a brilliant and provocative attempt has been made recently to interpret all the essential aspects of Gothic design as representations of the celestial city.[1] Such an attempt, to be sure, encounters serious difficulties.[2] But it is heuristically valuable in that it compels us to define the exact relation between the stylistic structure and the significance of the Gothic cathedral, throwing an entirely new light upon its esthetic as well as structural aspects.

The most striking feature of the new style is a new relationship between function and form, structure and ornament. In Romanesque (and Byzantine) architecture structure is a technical means to an artistic end; it remains concealed behind painted or stucco ornaments. Indeed, the entire edifice is often but an invisible scaffold for the display of great murals and mosaics. These compositions, especially the figure of Christ in Majesty surrounded by his heavenly court that usually adorns the Romanesque apse, evoke the symbolic significance of the sanctuary as an image of the celestial city. That the structure of the building is concealed by these images reveals the spiritual source of Romanesque "anti-functionalism": the celestial vision depicted is to make us forget that we stand in a building of stone and mortar. In the Gothic cathedral the relation between structure on the one hand and ornament and its symbolic function on the other is quite different. Here the design is entirely determined by the pattern of the structural members, vault ribs and shafts. It has been remarked that the flowering of the Romanesque mural was in large part due to the technical imperfections of the buildings it adorned, that wall painting declined in the measure in which these imperfections were overcome. In Gothic architecture, the wonderful precision with which every single block was shaped in the vault (leaving no ragged joints that it was necessary to conceal) suggests a new esthetic appreciation of the dignity of structural perfection.[3] This tectonic system is never concealed but rather underscored by Gothic wall painting. Even the stained glass windows submit, in composition and design, increasingly to the pattern of the stone and metal armature in which they are imbedded. The esthetic function of these windows is not only the creation of a new luminosity; the light they admit dramatically underscores the web of tracery, ribs, and shafts.

This new esthetic dignity ascribed to structure cannot be understood in terms of modern functionalism. Architectural form reveals function if it actually shows the physical interaction of weight and support as it does in the Greek temple. What concerned the Gothic builder was not such naked expression of statical function but rather the translation of function into an essentially graphic system. By concealing volume or "dissolving" it into a bundle of frail shafts, he obtained the visual effect of a geometrical grid on a two-dimensional surface.[4] With this qualification, however, Gothic architecture is indeed functionalist. And its emphatic recognition of structure is all the more remarkable if we recall the symbolic significance of the sanctuary. As we shall see, the "functionalist" aspect of Gothic art will help us understand the precise nature of its symbolic aspect, just as the latter is indispensable for a correct definition of Gothic functionalism.

It is hardly necessary today to stress the overwhelming importance of geometry in Gothic design. The reliance on geometrical formulae, apparent in every Gothic ground plan and elevation, is amply attested by medieval documents. So much research has been done in recent years on this question that I can limit myself to a brief summary. With but a single basic dimension given, the Gothic architect developed all other magnitudes of his ground plan and elevation by strictly geometrical means, using as "modules" certain regular polygons, above all the square. The knowledge of this way of determining proportions was considered so essential that it was kept a professional secret. Only toward the end of the fifteenth century—and of the cathedral age—was it made public by Matthew Roriczer, the builder of Regensburg cathedral. He teaches "how to take the elevation from the ground plan" by means of a single square. From this figure Roriczer derives all proportions of his edifice inasmuch as its dimensions are related to one another as are the sides of a sequence of squares the areas of which diminish (or increase) in geometrical progression. The proportions thus obtained the master considered to be "according to true measure."[5]

It was not only this late Gothic architect or the German lodges that made such modular use of the square. Perhaps the most important single piece of evidence regarding the principles of Gothic design is the famous model book by the Picard architect Villard de Honnecourt, who was active in the second quarter of the thirteenth century. He, too, demonstrates how to double (or halve) a square for the purpose of determining the proportions of a building, in this case the ground plan of a cloister. That this is no mere theory is shown by Villard's plan of one of the towers of Laon Cathedral, considered by him the most beautiful in the world. This plan indicates, as Ueberwasser has shown, that all horizontal subdivisions of the tower are recessed "according to true measure."

The square—along with the other polygons, such as the famous $\pi/4$ triangle, which the medieval architect derived from the square—and the proportion "according to true measure" have determined Gothic design to a remarkable extent. The façade of Notre Dame of Paris is composed of a sequence of four squares developed according to true measure. Of course geometrical formulae had been used by pre-Gothic architects too. Here, however, they were practical rather than artistic devices of which the observer usually remains unconscious. Nowhere do they determine the esthetic impression as they do in the Gothic system. One might almost say that the development of the style, from its origin to the classical maturity reached in the mid-thirteenth century, is marked by the gradual triumph of geometrical proportion. If we compare the façade of Notre Dame with the similar and earlier one of Noyon, we realize the increasing clarity with which the geometrical principle is realized in the Paris façade.

The same principle rules over all parts of the Gothic cathedral. F. Bond was struck by the tenacity with which the square is retained in the Gothic ground plan even after the advantages of oblong bays from the viewpoint of vaulting had been discovered. We shall soon return to this phenomenon. The square is equally apparent in the Gothic elevation. That of Noyon is a good starting point. The nave was built during the last third of the twelfth century. Thanks to the fine analysis of Charles Seymour, Jr., we know how the work progressed from East to West and how this progress involved stylistic changes that mark this transition from Romanesque to Gothic. In the proportions of Noyon the geometrical element is, as Seymour observes, not yet prominent. We

Noyon, elevation. (After Charles Seymour, Jr.)

Chartres, elevation and cross-section. (After Dehio-V. Bezold.)

encounter "shifts of design and alterations of proportions" in almost every bay. Even so, it is significant that only in the three Western bays, which were completed last, is the relation of the width of an aisle to that of the main vessel "brought to a ratio of nearly exactly one to two." In the elevation of the nave, moreover, the stringcourse under the galleries marks off a height equal to the distance between the main piers; and the square thus described occurs a second time since the distance from the stringcourse to the windowsills of the clerestory is again the same.

But the Gothic trend toward "geometrical functionalism" appears elsewhere at Noyon. He who designed this elevation perceived the relation of weight and support not as the interaction of physical bodies, but as an abstraction, as an interplay of lines. The columns under the nave arcades are the only architectural members designed and articulated according to their statical function; they alone unequivocally suggest volume. But they represent an old-fashioned element, doomed gradually to disappear from Gothic architecture. The main piers remain partly embedded within the wall which conceals their bulk. The intermediary shafts, slender like bamboo shoots, could not even maintain themselves were not they, in turn, braced by the walls between them. As the work progressed, the architects suppressed everything that might divert attention from the interplay of lines. Thus ornaments and even corbel rings disappear from the later sections of the nave, capitals become smaller and simpler until only the austere web of horizontals and verticals remains.

If, on the other hand, we compare Noyon with the first—and mother—of the classical cathedrals, that of Chartres, geometrical proportion suddenly seems to have come into its own. The ground plan of Noyon is Romanesque in that it suggests an additive process of composition: one could easily add or eliminate one or more of the rectangular units of which it is composed. The ground plan of Chartres presents the compact unity and cohesion of an organism. This unity is due to proportion. Dehio was the first to notice that this plan is based on the Golden Section, since the center of the crossing divides the entire edifice in the ratio 5:8. The main proportions of the ground plan, moreover, determine those of the elevation. The church is as high as is the distance from the center of the crossing to the end of the choir (excluding the apse) and of the transepts. And the great square described by the crossing and by each double bay of the nave also reappears in the elevation: the height to the first stringcourse being equal to the side of this square, the architect has created a series of "spatial cubes" that is quite noticeable to the observer. And this square reappears once again in the elevation, since its side is equal to the length of the slender shafts that support the vault ribs. If we increase this square "according to true measure" we obtain the height of the entire nave to the windowsills. Finally, the Golden Section appears in the elevation as it does in the ground plan, determining the proportion between the piers (to the arcade imposts) and the shafts above them.[8]

What matters is not only the existence of these measurements, but the way in which the architect has brought them to our attention. He has sprung his fourpartite vaults over transverse oblong bays, but has not entirely sacrificed the square: Chartres cathedral presents the alternating system of supports, in some respects an obsolete feature that occurred in Sens and Noyon but had already been dropped in Notre Dame of Paris. It is interesting to see how the master of Chartres employed alternation without sacrificing the homogeneity of identical supports. His piers consist, alternatively, of a cylindrical core surrounded by octagonal colonnettes and of an octagonal core surrounded by cylindrical colonnettes. The variation is just sufficient to induce the eye to see not one but two bays as one unit and thus to notice the square in the ground plan and the "cube" in the elevation. The main proportions of the elevation are made evident by the simple grid of intersecting verticals and horizontals. And our eye is similarly directed to notice the proportion (of the Golden Section) obtaining between the main parts of the supports; only one of the shafts in each compound lacks a capital under the nave arcade; it is the one facing the nave and just beneath the respond that rises to the springing of the transverse rib. We thus see the entire vertical member as one, harmonically subdivided, unit.

It is worthwhile to compare the elevations of Noyon and Chartres. The younger cathedral is nearly fourteen meters higher, yet it conveys the impression of far greater horizontal unification. The abandonment of the four-partite elevation in favor of the three-partite one is an important means to this end. The harmony of the Golden Section welds both the vertical and horizontal system of Chartres into an indissoluble unity. These proportions strike one as necessary and definitive, whereas the master of Noyon appears to be still groping for the right ratios. Equally increased is the tendency toward the two-dimensional in Chartres. With the elimination of the galleries, the side aisles have become narrower and lighter, a luminous membrane rather than a shadowy depth enveloping the nave. The *piliers cantonnés*—which the master of Chartres, as Panofsky observes,[7] employs for the first time—further accentuate the graphic tendency, seemingly reducing the solid core of the supports and dissolving their volume into the vertical rhythm of lines.

It ought not to be assumed that the line drawings we have been studying convey an inaccurate notion. In every other architectural style they would indeed give but a faint semblance of what the architect actually intended to build. Not so in Gothic. One has to look at the architectural drawings of the age—such as Villard de Honnecourt's or those of the contemporary Reims palimpsest, or the magnificent later collections from the cathedral lodges of Prague and Vienna—to realize that for their authors these drawings were not abstractions but the ideal which the completed edifice must seek to approximate. These wonderful systems of lines suggest neither space nor volume. Not until the end of the fourteenth century is there any indication of perspective.[8] It is as if these masters, as unconcerned with physical laws as with appearance, had been preoccupied solely with the reality of geometrical proportions.

Why this submission to geometry? One reason often given is a practical one: with measuring units varying from place to place, yardsticks were unknown or unusable; hence the use of proportions, in architectural drawings or models, that could be translated into large dimensions by geometrical means only. This explanation is but partially adequate, however. Villard de Honnecourt supplies numerical indications of size in a technical drawing, but always relies on geometry in his architectural designs. The proportion "according to true measure," whatever the facility of its practical execution, occurs, as Ueberwasser has shown, in Gothic paintings and engravings where the problem of translating one dimension into another did not enter. The Gothic artist would have overthrown the rule of geometry had he experienced it as a fetter. Yet he did not use geometrical formulae for purely esthetic reasons either; they occur in places where they are invisible to the observer. In short, the alternative "practical or esthetic" does not make sense in medieval terms. Happily, at least one literary document survives that explains the use of geometry in Gothic architecture: the minutes of the architectural conferences held during 1391 and the following years at Milan.

The cathedral of Milan was begun in 1386. After a few years difficulties developed and foreign advisors were called in from France and Germany. The minutes of the discussions between them and their Italian colleagues have survived. Two aspects of the deliberations are of importance in our present context: first, the reliance on geometric figures, attested by the German architect, Roriczer, of the fifteenth century, and the French architect, Villard de Honnecourt, of the thirteenth, is emphatically confirmed by the Italian document of the intervening century. The question debated at Milan is not whether or not the cathedral is to be built according to a geometrical formula, but merely whether the modular figure to be used is the square (which had already determined the ground plan) or the equilateral triangle.[9] The second and even more significant aspect of the Milan documents is that they suggest the reason for this use of geometrical formulae. The minutes of one particularly stormy session recall an angry dispute between the French expert, Jean Mignot, and the Italians. Overruled by them on a technical issue, Mignot remarks bitterly that his opponents have set aside the rules of geometry as if science were one thing and art another. Art, however, he concludes, is nothing without science, *ars sine scientia nihil est*. The terms art and science do not mean what they mean today. Art for Mignot is the practical know-how gained from experience; science the ability to account for the reasons that determine sound architectural procedure by rational and more precisely by geometrical means. In other words: architecture is scientific inasmuch as it is based on geometry, and unless he obeys the laws of geometry the architect must fail. This argument was considered unassailable even by Mignot's opponents. They hasten to affirm that they are in complete agreement and have nothing but contempt for an architect who presumes to ignore the dictates of geometry. It is taken for granted by both sides that the stability and the beauty of the edifice are not distinct, do not obey different laws, but are both comprehended in the perfection of geometrical forms.

Thus, the Milan document answers our question regarding the function of geometry in Gothic architecture. I think it also provides the clue to the reasons underlying what seems to us an almost superstitious belief in mathematics. Jean Mignot's juxtaposition of *ars* and *scientia* recalls, like a faint echo, the distinction that occurs almost a millenium before in the most influential esthetic treatise of the Middle Ages.

In the first book of his treatise, *De Musica*, St. Augustine defines music as the "science of good modulation." Before telling us what good modulation is, he explains why music, properly understood, is a science. He does not deny that music can be produced by instinct or practical skill, just as music can be appreciated by one who just "knows what he likes." Such understanding of music, however, creative or receptive, is but of a low order, according to Augustine. Vulgar performers and vulgar audiences have such an understanding; even a singing bird has. In fact, there is little difference between man and beast in regard to this kind of musical knowledge which Augustine calls contemptuously *art*. The true understanding of music, on the other hand, that knows the laws which are of its very essence, applies them in musical creation and discovers them in music, is what Augustine calls the *science* of music, and he goes on to explain the nature of this science as mathematical. The science of good modulation is concerned with the relating of several musical units according to a module, a measure, in such a way that the relation can be expressed in simple arithmetical ratios. The most admirable ratio, according to Augustine, is that of equality or symmetry, the ratio 1:1, since here the union or consonance between the two parts is most intimate. Next in rank are the ratios 1:2, 2:3, and 3:4, the intervals of the perfect consonances octave, fifth, and fourth. It is to be noticed that the pre-eminence of these intervals, for Augustine, is not derived from their esthetic or acoustic qualities. These are, rather, audible echoes of the metaphysical perfection which Pythagorean mysticism ascribes to number, especially to the four numbers of the *tetractys*. Without the principle of number, as Augustine calls it, the cosmos would return to chaos. Taking up the Biblical passage *Omnia in mensura et numero et pondere disposuisti*, Augustine applied Pythagorean and neo-Platonic number mysticism to the interpretation of the Christian universe, its creation, and its order. He shares with Plato both a distrust of the world of images and the belief in the absolute validity of mathematical truths. Platonic metaphysics is also the basis for Augustine's philosophy of art. The views he formulated, not only as regards the function of the arts in the Christian commonwealth, but also, one may say, as regards its style, have left their imprint on Christian art during a thousand years. This influence has three aspects.

1. The principles of good musical modulation and its appreciation which Augustine established in *De Musica* are mathematical principles and therefore apply, in his opinion at least, to the visual arts as they do to music. On the monochord, the musical intervals are marked off by the divisions on a string; the arithmetical ratios of the perfect consonances thus appear as the proportions between different parts of a line. And since Augustine adduces the musical value of the perfect consonances from the metaphysical dignity of the ratios on which they are based, it was natural for him to conclude that the beauty of certain visual proportions derives from their being based on the simple ratios of the *tetractys*. The place Augustine assigns to geometry among the liberal arts, like the place he assigns to music, is due to its "anagogical" function, that is, its ability to lead the mind from the world of appearances to the contemplation of the divine order. In the second book of his treatise *On Order* Augustine describes how reason, in her quest for the blissful contemplation of things divine, turns to music and from music to what lies within the range of vision: beholding earth and heaven, she realizes that only beauty can ever satisfy her, in beauty figures, in figures proportion, and in proportion number. The esthetic implications are clear. Augustine was quite as sensitive to architecture as he was to music. They are the only arts he seems to have fully enjoyed, and he recognized them even after his conversion, since he experienced the same transcendental element in both. For him, music and architecture are sisters, since both are children of number; they have equal dignity, since architectural mirrors eternal harmony as music echoes it.

2. Augustine uses architecture as he does music to show that number, as apparent in the simpler geometrical proportions that are based on the "perfect" ratios, is the source of all esthetic perfection. And he uses the architect, as he does the musician, to prove that all artistic creation follows the dictate of number, even though the architect, if he is a mere practitioner rather than a scientist of his art, may be unable to account rationally

for his instinctive use of mathematical rules. Such views, of course, confine artistic design and composition within the rigid limits of metaphysical doctrine. Along with a real appreciation of the abstract mathematical beauty that may and perhaps always does underlie artistic composition, Augustinian esthetics harbors a profound distrust and contempt of the image, the semblance of living form that may obscure the anagogical function of the work of art. And even the proportions he admits are limited to the "perfect" ratios of Pythagorean mysticism.

3. That Augustinian thought has profoundly influenced Western art during the Middle Ages, both in its recurrent iconoclastic tendencies and in its mathematical character, is beyond question. We are apt to underrate the positive consequences of this influence. While stripping the arts, and above all architecture, of much of their life, it also assigned to them an extraordinary dignity and mission. Because true beauty is, according to Augustine, anchored in a metaphysical reality, the contemplation of visual and musical harmony will actually lead the soul to the experience of the ultimate harmony and unity that is God.

The Middle Ages never questioned Augustine's authority. The passage from the Book of Wisdom "Thou hast disposed everything according to measure and number and weight," and the interpretation he had given to it, became, as has rightly been observed, the key word to medieval thought and learning until the advent of Aristotle. E. R. Curtius has recently shown how this world view, through number composition, has affected both the content and the form of medieval poetry. It has left an even greater impact on medieval art.

Augustinian esthetics were never forgotten during the Middle Ages. In the twelfth century, however, they gain an unprecedented importance in the Ile de France, under the influence of two movements, the first intellectual and speculative, the second spiritual and ascetical. The first centers in the group of Platonists assembled at the school of Chartres, the second in the monastic reform emanating from Cîteaux and embodied by Bernard of Clairvaux. French civilization in the twelfth century is in an important sense the synthesis of these two trends which, though distinct, are yet closely interconnected by personal and intellectual ties. Their common bond is the legacy of St. Augustine, their lasting achievement the creation of Gothic art.

The Platonism of Chartres was in many respects a true Renaissance movement. The group of men who gathered there in the second quarter of the twelfth century were primarily interested in theological and cosmological questions, to be solved by means of a synthesis of Platonic and Christian ideas. These early scholastics approached their task in a spirit of tolerance and respect with regard to the thought of antiquity that often reminds one of the "universal theism" of the fifteenth century; yet, theirs was a strange Platonism indeed. It was almost entirely based on one single treatise, the *Timaeus*. Of this treatise but a fragment was available; of this fragment not the Greek original, but only a garbled translation along with two commentaries—by Chalcidius and Macrobius—that viewed Plato's cosmology through the lenses of an eclectic and confused neo-Platonic mysticism. The Platonic fragment (and the two mediocre commentaries) were approached by the theologians of Chartres with nearly the same awe and respect as was the Book of Genesis. Both works, it was believed, were in substantial agreement in what they revealed about the creation of the universe, indeed, about the Creator himself. If one considers that the theology and cosmology of Chartres resulted largely from the interpretation of two documents as different as Plato and the Bible, but approached with the notion that they must not contradict each other and that the interpreter must not contradict either, one can but marvel at the wonderful and daring speculative system that resulted.

The aspects of the theology and cosmology of Chartres that interest us most in our present context are, first, the emphasis on mathematics, particularly geometry, and, second, the esthetic consequences of this thought.[10] The masters of Chartres, like the Platonists and Pythagoreans of all ages, were obsessed with mathematics; it was considered the link between God and world. The most influential exponent of the system, Thierry of Chartres, hoped to find, with the help of geometry and arithmetics, the divine artist in his creation; he went further and sought to explain the mystery of the Trinity by geometrical demonstration. The equality of the Three Persons is represented, according to him, by the equilateral triangle; the square unfolds the ineffable relation between Father and Son. Thierry recalls that Plato "like his master Pythagoras" identified the metaphysical principles of monad and dyad with God and matter, respectively. God is thus supreme unity, and the Son is unity begotten by unity as the square results from the multiplication of a magnitude with itself. Rightly, Thierry concludes, is the Second Person of the Trinity therefore called the first square. It has been said, that, under Thierry's influence, the school of Chartres attempted to transform theology into geometry. The attempt, which appears so strange to us, conveys a glimpse of what geometry meant to the twelfth century.

More daring than this theology, more dubious from the standpoint of orthodoxy, and more significant for the art historian is the cosmology of Chartres and the philosophy of beauty which it engendered. In the *Timaeus* Plato describes the division of the world soul according to the ratios of the Pythagorean *tetractys*. The esthetic, especially musical, connotations of this idea, barely hinted at by Plato, are underscored by Chalcidius who points out that the division is effected according to the ratios of musical harmony. He, as well as Macrobius, insists that the Demiurge, by so dividing the world soul, establishes a cosmic order based on the harmony of musical consonance.

It was easy to fuse this notion with the Augustinian idea of a universe created "according to measure and number and weight." As a result the creation appeared as a symphonic composition. It is so described in the ninth century by John Scrotus Erigina, and the idea was seized upon by the school of Chartres. William of Conches, the teacher of John of Salisbury, and Abelard, who seems to have studied mathematics under Thierry and whose cosmology is that of the school of Chartres, both identify the Platonic world soul with the Holy Ghost in its creative and ordering effect upon matter; and they conceive this effect as musical consonance. The harmony it establishes throughout the cosmos is represented, however, not only as a musical composition but also as an artistic one, more specifically, as a work of architecture. The ease with which the transition from the musical to the architectural sphere is here effected must not surprise us in view of the sistership of the two in Platonic and Augustinian thought. But for the theologians of Chartres, the notion of the cosmos as a work of architecture and of God as its architect has a special significance, since they assume a twofold act of creation: the creation of chaotic matter, and the creation of cosmos out of chaos. The Greek word *cosmos* signifying ornament as well as order, it was plausible to view matter as the building material, the creation proper as the "adorning" of matter by the artful imposition of an architectural order. In the Platonic cosmology, moreover, the masters of Chartres could detect the design and method according to which the divine architect had built the universe, the cosmic temple as Macrobius calls it.

In the *Timaeus* the primary bodies of which the world is to be composed are conceived as building materials ready to be put together by the builder's hand. This composition is effected by means of fixing the quantities in the perfect geometrical proportions of squares and cubes (1:2:4:8 and 1:3:9:27) —the same proportions that also determine the composition of the world soul. According to this composition, the world's body, consisting of the four primary bodies, whose quantities are limited and linked in the most perfect proportions, is in unity and concord with itself and hence will not suffer dissolution from any internal disharmony of its parts; the bond is simply geometrical proportion.[11] In this view, the perfect proportions, the beauty of which we may admire in musical and in architectural compositions, also acquire an explicit technical or tectonic function: these proportions chain and knit together the different elements of which the cosmos is composed. William of Conches quite correctly interprets the Platonic passage in this sense. Here, then, perfect proportion is thought to account for both the beauty and the stability of the cosmic edifice.

The significance of these ideas for the history of architecture is very real. Some years ago N. Pevsner pointed out that the term architect is rarely used in the Middle Ages and, if it is, denotes either clerics interested or experienced in architecture, or masons.[12] Pevsner concluded that the professional architect, in the classical sense (which is also the modern one), hardly existed in the Middle Ages, and suggested that the revival of the term in the mid-thirteenth century coincides exactly with the change from the humble master mason to the architect of the thirteenth century, no longer considered as a craftsman but as the principal artist and a "theoreticus" or scientist. There may be a good deal of truth in all this, but Pevsner is surely wrong when he seeks to connect this sociological and philological development with the introduction of Aristotle's *Metaphysics* (where *architect* is defined in our sense) to Western thought after 1200. Quite apart from the writings of Vitruvius, known and studied since Carolingian times, it was Augustine who kept alive the classical definition of the architect. His distinction between the mere practitioner and the true architect who deliberately applies scientific principles occurs in at least three different treatises, all studied and admired throughout the Middle Ages. While this definition permitted the application of the term architect even to the mere craftsman, it left no doubt that only the "scientist" schooled in the liberal arts was truly entitled to it. And since knowledge of the *quadrivium* was generally the privilege of clerics, it is not surprising to find so many ecclesiastics among medieval builders and the term architect so often applied to them.

But it was the school of Chartres which dramatized the image of the architect in the classical sense (more than a century before Aristotle's *Metaphysics* could have done so) by depicting God as a master builder, a *theoreticus* creating without instrument or effort by means of an architectural science that is essentially mathematical. And not only Augustine but Boethius, the greatest mathematical authority of the Middle Ages, taught the school of Chartres how to visualize in geometrical terms the perfect consonances. He points out that the proportions of double, half, triple, and third—those, in other words, that marked the perfect ratios on the monochord—are as readily perceived visually as they are acoustically, for, he continues, echoing the *Timaeus*, "the ear is affected by sound in quite the same way as the eye is by sight." And Boethius confines this doctrine of synesthesia not only to the proportions of line or surface; he discovers "geometrical harmony" in the cube since the number of its surfaces, angles, and edges—6:8:12—contains again the ratios of the consonances.

Toward the end of the twelfth century—at a time when the first Gothic cathedrals were nearing completion—

Alanus ab Insulis described the creation of the world. To Alanus, the *doctor universalis*, the Platonism of Chartres owes probably its widest influence and diffusion. To him, God is the artful architect *(elegans architectus)* who builds the cosmos as his regal palace by composing and harmonizing the different genera of created things with the "subtle chains" of musical consonance.

The impact of these views on the architecture and architectural procedure of the twelfth and early thirteenth centuries cannot be overestimated. Since art is an image of nature, Professor de Bruyne asks, "must not the ideal church be constructed according to the law of the universe?" We suddenly understand why the high Middle Ages defined and practiced architecture as applied geometry; why the experts at Milan pay such astonishing tribute to this discipline. And we also understand why the great lay architects of the Gothic period have themselves invariably depicted, ruler and compass in hand, as geometricians. With these same attributes the creator himself was represented. Only by observing geometrical principles did architecture become a science in Augustine's sense; by submitting to its laws, the human architect imitated his divine master and in doing so his calling acquired metaphysical significance.

In order to understand this significance of Gothic architecture we must bear in mind that the musical harmony which the Christian Platonists beheld in the cosmos is primarily not a physical but a metaphysical principle. Medieval man looked upon the creation as the first of God's self-revelations, the Incarnation of the Word being the second. Between the two the twelfth century perceived innumerable mystical correspondences. The theological meaning of the universe had been obscured through Adam's fall on earth. It still appeared clearly in the perfection of God's celestial palace. Hence the tendency, familiar to every reader of Dante, to link the realm of the stars with the celestial mansions; hence also the seemingly dual symbolism of the cathedral which is at once an image of the cosmos and of the celestial city. If the Gothic architect designed his sanctuary according to the laws of harmonious proportion, he did not only imitate the perfection of the visible world but also created an image, in as much as that is possible to man, of an invisible one.

The symbolic concatenation is well explained in a passage of Abelard. After identifying the Platonic world soul with world harmony, he first interprets the ancient notion of a music of the spheres as referring to the "heavenly mansions" where angels and saints "in the ultimate sweetness of harmonical modulation" render eternal praise to God. Then, however, Abelard transposes the musical image into an architectural one. He relates the celestial Jerusalem to the terrestrial one, more specifically to the Temple built by Solomon as God's "regal palace" and at once a model for the Christian sanctuary and a mystical image of Heaven. This temple, Abelard remarks, was pervaded by the divine harmony as were the celestial spheres. The passage reflects the influence of Platonic cosmology upon Christian eschatology and symbolism in the twelfth century, the notion of an ineffable harmony gradually subdues the imagery by which the celestial city had formerly been depicted. This notion goes far to explain the transition from Romanesque to Gothic, the striking rule of harmonious proportion in the new style that emerges around 1140.

It must be emphasized at this point that the musical mysticism of the Platonic tradition was by no means the exclusive property of the School of Chartres. Embodied in Augustinian thought, it influenced the spiritual formation of the monastic movement that centered in Cîteaux and Clairvaux and is personified by St. Bernard. As I remarked earlier, this movement contributed as much to the civilization of the twelfth century as did the Platonism of Chartres, and both trends are intimately related. It is time to consider Bernardine thought in its impact upon Gothic art.

Bernard's artistic views are usually described as those of a Puritan. They are in point of fact Augustinian. No other author has had greater influence upon Bernard's theological formation than Augustine. He considered the Bishop of Hippo the greatest theological authority after the Apostles; with Augustine, Bernard writes, he wants to err, as well as to know. And Augustine musical mysticism could claim as its greatest spokesman. The following passage gives an idea of the place and function of music in Augustine's theological experience. In his treatise *De Trinitate* he meditates on the mystery of Redemption by which the death of Christ atoned for man's twofold death of body and, through sin, of soul. As the Bishop of Hippo ponders this "congruence," this "correspondence," this "consonance" of one and two, musical experience gradually takes hold of his imagination, and suddenly it dawns upon him that *harmony* is the proper term for Christ's work of reconciliation. This is not the place, Augustine exclaims, to demonstrate the value of the octave which seems so deeply implanted in our nature—by whom if not by Him who created us?—that even the musically and mathematically untrained immediately respond to it. Augustine feels that the mystery of Redemption is conveyed to human ears by the consonance of the octave, the musical expression of the ratio 1:2. The remarkable passage conveys an esthetic experience radically different from our own. It was not the primary enjoyment of musical consonances that led Augustine to interpret these as symbols of metaphysical or theological truth. On the contrary, the consonances were for him echoes of such truths and the enjoyment which the senses derive from musical harmony (and its equivalent, geometrical proportion) is our intuitive response to an ultimate reality that may defy reason but to which our entire nature is wonderfully attuned.

This experience determines the medieval attitude towards music. It accounts for the emphasis on musical studies even, and especially, in the monasteries of strict ascetical observation. To take a typical example, Othlon of St. Emmeram (1032-70), in embracing the most austere monastic ideal, renounced all his former humanistic interests. But arithmetic and music retain their mystical function for him; he uses them in his writings to convey divine secrets to his fellow monks, to prepare them for the life in a world to come. Even the order prevailing among the heavenly hosts, he writes, corresponds to the intervals of the perfect consonances.

Bernard's attitude towards music was quite similar. He was profoundly musical, and, as Father Luddy observes, an Augustinian even in musical matters. Something of a composer himself, he was once invited by the abbot of another monastery to compose an office for the feast of St. Victor. Bernard's reply is noteworthy. What he demands of ecclesiastical music is, above all, that it "radiate" truth, "sounding" the great Christian virtues and kindling the light of truth. Music, Bernard thinks, should please the ear in order to move the heart; it should, by striking a golden mean between the frivolous and the harsh, wholesomely affect man's entire nature.

These are not the views of a Puritan. Bernard must have responded to musical experience with unusual sensitivity. In demanding that music be attuned to the great metaphysical and ethical experiences of Christian life, he confronted music, not with an attitude restricting its creative scope, but with a challenge. The importance of Bernard's views on music for our present inquiry lies in the fact that they also provide an indispensable clue to his convictions regarding the visual arts. That the laws of music, generally understood, "embrace everything," that they extend to all the arts is a view frequently expressed during the Middle Ages.[13] And to a man steeped, as Bernard was, in the Augustinian tradition, not only the metaphysical dignity of the perfect consonances but their presence in the sister art of music, architecture, must have been self-evident.

The appraisal of Bernard's views regarding religious architecture must rely not only on the opinions he expressed in literary form but on the testimony of Cistercian architecture, the design of which was certainly determined by his views. Bernard's demand that the "monstrous" imagery of Romanesque art—and in fact all images besides the crucifixus—be banished from the Cistercian cloister and church; his attack upon the "immense" height, the "immoderate" length, the "supervacuous" width of the Cluniac churches are but a negative statement of his views. To understand its meaning we must bear in mind that the sumptuousness of Cluniac sanctuaries was considered incompatible with monastic humility. More important than this ethical consideration, however, was a spiritual one. To Bernard the life of the Cistercian cloister—the *paradisus claustralis*—was an image and foretaste of Paradise. He sought to prepare his monks, even while in this life, for the fruition of an ultimate truth which the relatively crude imagery of Romanesque art could never convey.

Its elimination from the Cistercian monastery was therefore inevitable. Significantly, however, music maintained its place. And the disappearance of figurative sculpture and painting from Cistercian churches cleared the way for a purity of proportions that is all the more remarkable in view of the role which sacred architecture played, in the mystical contemplation of the Order, as an image of Heaven. For Bernard, as for Augustine, the perfect consonances, visible or audible, were not delusions of the senses, but echoes of transcendental reality.

We do not yet know all the geometrical modules used by the Cistercian builders. Yet the use of such modules is strikingly obvious in their churches. Augustine's "perfect" ratio 1:2 generally determines the elevation. In the abbey of Fontenay (1130-47), the best surviving example of early Cistercian architecture, this octave ratio determines the proportions of the ground plan as well. Moreover, the bays of the side aisles being of equal length, width, and, up to the stringcourse, height, we obtain in each of these cells a spatial cube—the "geometrical harmony" of Boethius which, later, we also encounter in Chartres. The austere façade, if we include the buttresses and the upper stringcourse, describes a square. The distance between the upper and lower stringcourse is determined "according to true measure." Medieval preference for this proportion appears in a new light if we recall Augustinian preference for the octave and the preoccupation with the octave and with the square in the thought of Augustine on the one hand and of the school of Chartres on the other: the proportion "according to true measure" may be defined as the geometrical expression of the ratio of the octave based on the square as module. It can hardly be a coincidence. Nowhere in Western architecture are the Augustinian consonances as present as they are in the simple and solemn proportions of this Cistercian church. Its design, in any event, is singularly attuned to the musical mysticism of the twelfth century.

Thus we have seen that the use of geometrical formulae in sacred architecture—so emphatically evident and yet so surprising to the modern observer—is rooted in the very world view of the twelfth century. Such interaction between the spheres of thought and of art is, of course, most evident in the case of Cistercian architecture.

But what renders this evidence important in our present context is the close relationship between Cistercian and early Gothic architecture. This relationship has often been described. It would be incorrect to define Gothic as the daughter of Cistercian architecture. In criticizing the art of the Cluniac order, St. Bernard himself had made it very clear that his views as regards religious architecture applied to monastic buildings but not to the secular cathedrals. He readily conceded that the latter, "since they cannot excite the devotion of the carnal populace with spiritual ornaments, must employ material ones," in other words, that cathedral art had to make concessions to sensuous experience which the mystic no longer required. On the other hand, St. Bernard's insistence that all religious art and music must be attuned to spiritual experience, that they are justified only inasmuch as they lead man to ultimate truth, confronted cathedral builders with the same basic challenge which monastic architecture had answered with the creation of the Cistercian style. Cistercian and Gothic may be described as two branches of religious architecture—the one monastic, the other episcopal—growing out of the same basic experience.

There is nothing surprising about this close relationship. St. Bernard was the most forceful exponent of a vision of art, the basic tenets of which had their champions in cathedral schools like that of Chartres. It would indeed be a serious mistake to consider the two intellectual movements we have studied as entirely distinct. The Platonists of Chartres and the Augustinians of Cîteaux and Clairvaux were united by innumerable personal ties as well as by a common heritage. Of particular interest in this connection are the relations between St. Bernard and the three men who commissioned the three monuments with which Gothic art begins: Suger, Abbot of St. Denis, and the Bishops Henry of Sens and Geoffrey of Chartres.

Suger's place in the history of art is mainly due to the importance of the west façade and choir of St. Denis which, for the development of architecture as well as sculpture, mark an epoch. But his fame also rests on the fact that, unlike any other medieval builder, he has, in two treatises, rendered account of the significance of his art and indeed of the spiritual motivations that prompted him to adopt a style the dramatic novelty of which he himself seems to have sensed.

The interpretation of Suger's writings—and of his art as well—has generally suffered from the assumption that "an irreconcilable contrast" existed between his artistic views and those of the Abbot of Clairvaux. In point of fact, Bernard's views were from a twelfth-century viewpoint far less extreme than is usually believed and, as I hope to have shown, offer many points of contact with the esthetic convictions of his contemporaries. Suger in particular, steeped in the neo-Platonism of Pseudo-Dionysius, must have been singularly attracted by Bernard's insistence that religious art is admissible only inasmuch as it guides contemplation toward the transcendental source of all beauty. This thought is the continued theme of Suger's treatises in which he interprets his artistic creations as he believes they ought to be understood. We have no reason to doubt his sincerity in this regard. In fact, there is good reason to think that the art of St. Denis may actually reflect Bernardine influence. The same influence, as I can but mention in passing, may have inspired the building activities of Henry of Sens and Geoffrey of Chartres.

St. Denis, Sens and Chartres West mark a beginning. But the "classical" Gothic cathedrals that follow their lead embody the same metaphysics of beauty. During the second half of the twelfth century the Platonism of Chartres and the spirituality of St. Bernard lose nothing of their influence. On the contrary, this influence is consolidated through a union of the two movements that is often as intimate as is, during the same period, the fusion of Cistercian and Gothic architecture.

This fusion, and its origin in the metaphysics of musical harmony, is most clearly embodied in the work of a professional architect. I have already mentioned the model book of Villard de Honnecourt. One of his designs represents the ground plan of a Cistercian church drawn *ad quadratum*, i.e., the square bay of the side aisles is used as a module, as a measuring unit from which all proportions of the ground plan have been derived. And these proportions, as E. de Bruyne observes, correspond in each case to the ratios of the musical consonances. Thus the length of the church *in opere* is related to the transept in the ratio of the fifth (2:3); the ratio 1:2 determines the relation between side aisle and nave, length and width of the transept, and, we may assume on the basis of Cistercian practice, that of the interior elevation as well. The 3:4 ratio of the choir evokes the musical fourth; the 4:5 ratio of nave and side aisles taken as a unit corresponds to the third, while the crossing, liturgically and esthetically the center of the church, is based on the 1:1 ratio of unison, most perfect of consonances.

Villard's testimony is of great significance. He seems to have received his architectural training at the Cistercian monastery of Vaucelles and certainly was employed as an architect by the Order. His design embodies the esthetic principles of the Cistercian tradition. While it does not seem to represent any church actually built, it shows sufficient resemblances to Pontigny and other churches of the Order to warrant Hahnloser's suggestion that Villard's ground plan was designed as the ideal type of a Cistercian church.

Villard's authority and interest was not confined to

the sphere of monastic architecture. Trained during the first third of the thirteenth century and active during the second, he saw, completed or in the process of building, nearly every one of the classical cathedrals of the Ile de France. He was a distinguished Gothic architect in his own right, and it is all the more significant that, not only in his Cistercian ground plan, but on nearly every page of his model book he stresses the importance of geometry for the realization of harmonical proportion.

But it is the Gothic cathedral itself that bespeaks this artistic conviction. The proportions of the great sanctuaries of the thirteenth century are based on the simple ratios of the musical consonances that also determine Gothic music. This affinity between architecture and music was an esthetic truism to an age which perceived in the "science" of music the all-embracing principle of the universe. Abbot Suger, in the opening passages of his treatise on the consecration of the church, describes the universe as a symphonic composition in terms strikingly similar to those used by Alanus ab Insulis in the passage on the work of the divine architect. Suger invokes this vision as the sublime prototype of the sanctuary he is going to erect. But if his musical phraseology suggests that he considered his architecture a "frozen music," musical writers of the High Middle Ages, conversely, compare musical composition to architecture.

To sum up: the metaphysics of "measure and number" explains the design as well as the symbolic significance of the cathedral and influenced architectural procedure itself. The observation of geometrical canons was imposed by an endeavor that had less to do with artistic invention than with science (as these terms are understood today). Designed in an attempt to reproduce the structure of the universe—not unlike the great scientific experiments of our time in this respect—the cathedral is perhaps best understood as a "model" of the medieval universe. It is the theological transparency of this universe that transformed the model into a symbol.

1. H. Sedlmayr, *Die Entstehung der Kathedrale*, Zuerich, 1950. See my review in *Kunstchronik*, IV, 1951, pp. 78 ff.

2. The Book of Revelation is neither the only source on which the Middle Ages based their eschatological vision, nor is Gothic architecture a better monumental "illustration" of that source than other styles. The temple of Solomon, especially as described by Ezekiel, probably influenced ecclesiastical symbolism as much as did St. John's description of the Heavenly-City. And if a contemporary writer compares the Romanesque *Trinité* of Fécamp to the celestial Jerusalem, the sanctuary evokes for him the splendor of the biblical vision by means very different from those of the Gothic design and yet quite as effective. See V. Mortet, *Recueil de Textes*, I (Paris, 1911), p. 345.

3. See the remarks of F. Bond, *Introd. to Engl. Church Arch.*, I (London, 1913), p. 321 on French and English methods of web filling.

4. See J. Bony, *French Cathedrals* (Boston, 1951), pp. 9 ff., who defines Gothic "functionalism" much as I do.

5. Cf. esp. W. Ueberwasser, "Nach rechtem Maasz," *Jahrb. d. preuss. Kunstslgn.*, 56, 1935, and P. Frankl, "The Secret of The Medieval Masons," *Art Bulletin*, 27, 1945.

6. The most exact measurements of Chartres Cathedral are still those given by Lassus, *Monographie de la Cath. de Chartres* (Paris, 1867). Measurements vary from bay to bay and the proportions given above are but approximations. Variations seem to be due partly to changes of plan and considerations imposed by the earlier building, and partly to the rather crude methods of setting out—by means of pegs and strings—employed by the medieval builder. See now J. Harvey, *The Gothic World* (London, 1950), pp. 16 ff. and L. F. Salzman, *Building in England* (Oxford, 1952), p. 17.

7. *Gothic Architecture and Scholasticism*, Latrobe, 1951, pp. 74 ff.; Panofsky's analysis of the elevation of Chartres is illuminating even if one does not accept the author's interpretation.

8. See O. Kletzl, *Planfragmente aus d.d. Dombauhuette von Prag* (Stuttgart, 1939), pp. 19 f.

9. *Annali della Fabbrica del Duomo di Milano* (Milan, 1877-85), esp. I, pp. 68 ff. and 209 f. The best discussion of the subject is that of J. S. Ackerman, "Ars sine scientia nihil est," *Art Bull.* 31, 1949.

10. The first comprehensive interpretation of the esthetic system inherent in the thought of the school of Chartres we owe to E. de Bruyne, *Études d'Esthétique médiévale* (Bruges, 1946), II, pp. 255 ff.

11. See F. M. Cornford, *Plato's Cosmology* (London, 1948), pp. 59 ff.

12. "The term 'architect' in the Middle Ages," *Speculum*, 17, 1942. On the entire question see now the very sensible remarks of Salzman, *l. c.* pp. 1 ff.

13. See M. F. Bukofzer, "Speculative thinking in medieval music," *Speculum*, 17, 1942, and L. Spitzer, "Classical and Christian ideas of world harmony," *Traditio*, II, 1944, and III, 1945.

8
Stability concepts from the Renaissance to today

Rowland J. Mainstone

Today, structural stability is usually understood as the capacity of a structure to return to a previous state of equilibrium after imposed small displacements. In framed buildings, the displacements likely to be critical are usually the result of elastic flexural deformations of relatively slender beams and columns and elastic shearing deformations of associated walls, brought about by applied loads such as the wind. In shells they are more likely to be displacements normal to the surface and liable to cause this to buckle. The term may also, by extension, be used to denote the ability of the structure to remain standing after local permanent damage. But it seems preferable to use a different term — such as structural robustness — for this.

Stability proper is therefore closely linked now with elastic stiffness in flexure and shear, both of which are comparatively recent concepts. It will also be influenced, and may be jeopardized, by inelastic deformations. But, in a good design, these will have a significant adverse influence only when the imposed loading exceeds the maximum that is to be expected.

The structures of the ancient buildings we are considering on this occasion were of a different kind. They were mostly built of masonry, with arched spanning elements when the spans exceeded the capacity of single lintel, and vaulted roofs or ceilings. The proportions of their columns, arches, vaults, and walls were less slender, in relation to heights ans spans, than the proportions typical today. Where there was no seismic risk, self-weight greatly exceeded any intermittent load. Elastic deformations in flexure and shear under loads such as the wind were thus too small to pose, in themselves, a threat of collapse. Where such a threat did exist, it arose either from a barely stable initial state of equilibrium or from gross long-term inelastic deformations due largely to continued yielding and cracking under self weight and repeated temperature change. Seismic loading could pose a further threat to some buildings. But this, like self-weight loading but unlike many of the critical imposed loads on modern buildings, was also directly related to the structural geometry.

The concepts of stability which guided the design of these buildings were also different. Essentially they were concepts linked to the assurance of comfortable equilibrium under self-weight loading, and it is such concepts on which I intend to concentrate, though I shall, as I was asked, attempt to carry the story up to the present day. Even without attempting more, it will be difficult to do justice to them in the limited time at my disposal, and necessary to be highly selective in describing them.

Stability concepts in the Renaissance

Some insight into the ways in which stability was envisaged well into the Renaissance, and in which designers normally sought to achieve it, is given by the records of discussions at Milan in 1400 and by a number of short manuals written subsequently to illustrate contemporary design procedures.[1]

By 1400, construction of the new (essentially Gothic) cathedral in Milan had advanced to the point where a decision on the heights of the cross section and the adequacy of the piers and buttresses to carry the main arches and vaults was becoming imperative. The sometimes-heated debates between the Milanese masters and others from across the Alps (whose assistance they had sought) was conducted almost wholly in terms of different controlling geometric figures and different proportionate relationships governing the cross sections of the supports. Some recognition of the tendency of arches to thrust outwards on their supports was implicit in the discussions of the required provisions for buttressing the thrusts, even in the wilfully perverse assertion of the Milanese at one juncture that pointed arches do not thrust in this way – "non dant impulsam contrafortibus". But nowhere was there any attempt to consider quantitatively the balance of force and counter-force.

On a theoretic level, stability was thought to be ensured by adherence throughout the design to a geometric discipline whose ultimate authority derived from the cosmic ideas of Plato (figure 1). On a more practical level, as seen more fully in some of the later design manuals, it was ensured by following certain geometric or numerical rules for determining the necessary cross sections.

All these disciplines and rules were simply ways of codifying or systematising experience gained previously from building other similar structures. To that extent, stability was simply an attribute of those structures that has succesfully stood the test of time, and it was seen to reside primarily in an appropriate choice and use of materials and in the way in which the principal elements were proportioned. Seeing it in this way was justifiable to the extent that the structural organism was virtually identical with the visible geometric form – as indeed it largely was when the entire construction was of a particular type of stone masonry and when this masonry carried nearly all the loads by direct internal compressions.

The chief shortcomings of the codifications were that the more generalised geometric disciplines contained, within themselves, no clues to the ways in which they should be interpreted and used, while the more specific rules might no longer be valid if there was a significant departure from precedent in the overall design. It was these shortcomings that were really at the root of the disagreements and arguments in Milan. They were best expressed at the time when — after the principal northern master had justified his advocacy of one geometric discipline by asserting that "ars sine scientia nihil est" – the Milanese countered by rightly pointing out that it was equally true that "scientia sine arte nihil est". This indispensible "ars" was, in fact, a body of unformulated practical knowledge and operational rules, all learnt through the direct experience of building in a particular way.

Underlying the more innovative design through which this experience was gained, there must, however, have been some wider tacit or intuitive understandings of the conditions of static equilibrium, which might be regarded as other concepts of stability.

Study of the ways in which new forms were developed, coupled with a certain amount of imaginative introspection involving putting oneself in the place of their designers, strongly suggests that this understanding had two components. The best names I can think of for them are those I first gave them – spatial and muscular.[2]

The spatial component was that which recognised that all things have a tendency to fall unless prevented by the presence of some obstacle. This obstacle could be directly beneath if there was no possibility of rotation about it. But, if rotation would otherwise be possible, there would also have to be at least one other obstacle that would, by itself, rotate in the opposite sense – the two rotations then being incompatible so that neither was possible. More generally, it recognised that any movement that would threaten stability could, in principle, be prevented by interposing some suitable obstacle or obstacles.

The muscular component sprang from an ability to feel, by analogy with our direct awareness of pushes and pulls exerted on and by our own muscles, the similar pushes and pulls exerted on such obstacles to threatened movements, and the compressions and tensions that must, therefore, exist within them. By extension, it would also have embraced some intuitive apprehension of the balance of forces that guaranteed stability once a potentially stable spatial configuration had been achieved. Thus, where the feet of an arch were prevented from spreading by something linking them directly together, this link would be seen as being subjected to a pull at each end, and hence to be in tension. And where this spreading was prevented by an obstacle on each side, these obstacles would be seen as being subjected to outward thrusts, which would put them in compression.

The best instance of structurally innovative design in the early Renaissance, and of the role of these more intuitive understandings, was the construction of the dome of Florence Cathedral.

As in Milan, construction had begun much earlier. The supporting piers had already been completed before any detailed consideration was given to the way in which the dome was to be built. They seem to have been proportioned, in relation to the spans of the dome, solely on the basis of earlier comparable, though smaller, structures. If anything, they were cautiously made substantially more massive, and initially they were tied together by iron rods at the springing levels of the arches. They were amply strong and rigid and gave no subsequent trouble. Above them, an octagonal drum had also been built.

The octagonal dome of the adjacent baptistery furnished a possible prototype for the dome itself, and the initial plan seems to have envisaged it simply as a scaled-up replica of this earlier dome. A major problem nevertheless confronted those responsible for the completion of church a decade and a half after the discussions in Milan – that of actually building such a huge dome at an unprecedented height above the ground.

From the present point of view, this was chiefly a problem of achieving and maintaining stability during construction, while the constituent radial arches of the full dome were, as yet, incomplete. It was compounded by the fact that the plan was octagonal and not circular. In constructing the much smaller part-octagonal domes over the three eastern arms of the church, stability had been assured by erecting timber centering to give temporary support. Figure 2a shows, in principle, where the need for support arises. But, on the scale of the main dome, the construction of this centering, construction of the dome upon it with operations such as the lifting of materials impeded by its presence, and the necessary subsequent decentering, would have presented virtually insuperable problems at the time.

STABILITY CONCEPTS 5

Brunelleschi's genius lay first in recognising that a circular dome can be built with little or no temporary support if it is constructed in horizontal rings (because each ring constitutes a horizontal arch capable of keying together all the incomplete radial arches), and that, even though the cathedral called for an octagonal dome, this also could (on account of the thickness of the drum) be constructed as if it were a circular one (figure 2b). After this it lay in devising the detailed procedures that made construction in this manner possible.

There is no space here to describe these procedures in detail.[3] What is chiefly relevant to the present theme is that Brunelleschi seems both to have recognised the possibility and elaborated the detailed procedures wholly in terms of the spatial and muscular intuitions just referred to, in all probablity seeing them latent in certain earlier Roman forms and types of construction. His concepts were largely spatial or geometric, but with the further component that he must have seen that, irrespective of what happened as construction advanced further, each successive ring would be subject initially to compression only. There is no hint that he concerned himself with stress levels or, in any quantitative way, with the internal forces, and there would have been no need for this. Indeed it would not have been possible at the time in the absence of any suitable theoretic basis.

It had long been possible to measure vertically-acting weights by means of a balance arm, and a quantitative appreciation of the relevant conditions of balance had been available at least since the time of Archimedes. But it was more difficult to quantify and measure forces acting in other directions, and the ability to do this was won only over a long period from late mediaeval times onwards. Tentative attempts to do so by reference to the action of weights descending obliquely down imaginary frictionless inclined planes are found in works attributed to Jordanus written in the 13th century. But only at the end of the 15th century do these attempts first seem to have led to a more developed concept of stability in terms of an overall equilibrium of forces.

Several sketches by Leonardo at this time suggest that he was then seeking a fuller understanding by considering the forces in weighted cords passed over frictionless pulleys (figure 4a). Other sketches show that he was exploring such practical consequences of the conditions of equilibrium as the horizontal thrusts exerted by an arch (figure 4b) and the directions of the forces acting through an arch rib or vault (figure 5a). Others again show that he was carefully noting or visualising likely modes of instability or collapse that were to be guarded against (figure 5b).

Yet, even then, his approach was far from typical. The more typical Renaissance concept remained a purely intuitive and predominantly spatial or geometric one. Most design seems to have relied fairly directly on past experience for an assurance of stability, either through the "know-how" of the builders, through direct reference to earlier designs as exemplars, or through standard geometric or numerical procedures for the detailed sizing of particular elements such as the piers supporting an arch.

Even in the 17th century, reliance on such procedures was still the normal practice. It is still rules of this kind that are given in texts such as Derand's "L'architecture des Voûtes", 1643; Garcia's "Compendio de arquitectura", c. 1680 (derived from Rodrigo Gil, c. 1550); and the 1698 edition of Blondel's "Cours d'architecture" (figures 6a and 6b).

Developments of these concepts in the 17th and 18th centuries

Such purely empirical rules were, however, criticised by Wren in his unfinished "Tract on architecture", written at about the same time.[4] And, though Wren advocated in their place an erroneous application of Achimedean statics (figure 6c), he was attempting to envisage the conditions of stability more as Leonardo had done. He seems even to have gone somewhat further in one of his sketches for the dome of St Paul's in London. For, superimposed on possible cross sections, are lines that look very much as if they were meant to be catenary curves (figure 7).

Nevertheless, the full change to the use of a precise quantitative concept based on the equilibrium of internal and external forces can be documented only in the middle of the 18th century. The occasion was renewed concern for the stability of the dome of St Peter's, whose dome, and the drum on which it stood, were badly cracked. Le Seur, Jacquier, and Boschovich, three mathematicians consulted by Pope Benedict XIV, took, as the starting point of an analysis of the overall stability, the visible mode of potential collapse – in which the sections of the drum between the windows would rotate about their feet, allowing the orange-slice sections into which the dome above had cracked to open out also by pivotting about their bases on the drum and about their common crown region (figures 6d and 8). A calculation, equating the virtual-work done in a further small displacement by the relevant weights and by the tension in a circumferential tie at the dome springing level, was then made to show what tie force would be required to prevent such further displacement. Their conclusions were published in 1743.[5]

In parallel with their investigation, another mathematician, Giovanni

Poleni, made a further analysis.[6] On the basis of the analogy between an arch and a hanging chain with a similar distribution of weight along its lenght that had already been noted in Wren's time, he derived experimentally a possible line of thrust acting through a typical slice of the dome (figure 9) – though, surprisingly in the view of the fact that the obvious weakness lay in the drum, he terminated his analysis at the dome springing level instead of carrying the chain and the thrust line down to the base of the drum to explore the overall stability.

Shortly afterwards, in 1771, the French engineer Gauthey first analysed the static equilibrium of a structure still to be built – the new church of St Geneviève in Paris.[7] Soufflot had designed the church without, apparently, any such quantitative consideration of the overall stability. Yet, largely ignoring the obvious precedents of earlier and otherwise similar domed churches and relying more on his knowledge of Gothic architecture, he had made the supports immediately beneath the dome unusually slender (figure 10), leaving it to the outer walls to resist the thrusts. Gauthey's analysis, undertaken after the design had been strongly criticised, demonstrated that the proposed structure was indeed safe, by considering the equilibrium of the relevant weights and internal forces for an assumed mode of action of the dome thrusts (figure 11).

All these analyses drew on the considerable development in structural mechanics that had taken place in the late 17th and early 18th centuries.[8] The idea that the hanging chain could be considered as a analogue for an arch had been formulated by Hooke abd further developed by Gregory. A completely general definition of force had been provided by Newton. And numerous analyses had been made of the stability of the arch and of its requirements for stable support. Initially (in 1695) it was assumed that each voussoir was held in equilibrium by thrusts acting at right angles to the faces abutted by adjacent voussoirs, and then (in 1712) by considering the equilibrium of larger rigid blocks consisting, respectively, of the upper part of the arch on each side of the crown and the lower part plus its support. La Hire, who was responsible for the first analyses of both kinds, based the latter one on the observation that when the supports of an arch moved apart, separations usually occurred only in the crown and haunch regions.[9] Conformatory tests on model voussoir arches (figure 12) were reported by Danisy in 1732, and first published by Frezier.[10]

A further new feature of the analyses of the dome of St Peter's, and of some additional analyses of the piers of St. Geneviève made when these had, after all, begun to show signs of distress, was an explicit consideration of strengths. This was necessary for St Peter's in order to determine the sizes of tie necessary to develop the calculated tension. For St Genevieve, it was done in order to seek the reason for observed cracking and spalling of the facings. In both cases it was made possible by newly developed testing techniques. Previously all sizing of structural elements seems to have been based solely on experience of what had previously proved adequate in similar situations.

Changes in concept

Looking back, we can now see the changes that had taken place since the beginning of the Renaissance. They are well illustrated by comparing, on the one hand, Cesariano's Renaissance depiction of one of the proposals for the cross section of Milan Cathedral (figure 1) and typical Renaissance rules for determining necessary depths of pier to support an arch (figures 6a and 6b) with, on the other hand, the explicit considerations of the necessary equilibrium of the forces brought into play by the three mathematicians, Poleni, and Gauthey. The former were purely geometric, though in slightly different senses; whereas it would be truer to describe the latter as mechanistic.

What has been said about Brunelleschi's scheme for constructing the dome of Florence Cathedral, about some of Leonardo's explorations and projects, and about one of Wren's drawings for the dome of St Paul's, has shown, however, that the change was a gradual one, and less a complete supplanting of one concept by another than a sharpening of the intuitive mechanistic concepts — both spatial and muscular — that lay behind the more innovative earlier design. Putting this in other words, it might be said that the typical earlier concept of a stable structure was essentially one of a particular disposition of elements built in a particular manner to provide all necessary mutual support without excessive deformation. While largely spatial or geometric, it was closely tied, in practice, to particular manners of construction learnt from experience. The underlying muscular intuition gave some feeling for the internal forces called into play and for their balance, but not one that could be dissociated from the particular physical embodiment of the structures. The later concept was a far more abstract one. It was focussed primarily on the balance of these forces and freed from particular physical embodiments by a much more powerful abstract concept of a force acting in any direction.

19th century developments

The 19th century saw chiefly a further development of this later more abstract concept. Though the concept remained essentially unchanged, consideration of the conditions of equilibrium was greatly facilitated by new graphical techniques of analysis and the matching of strengths to calculated forces was similarly facilitated by a vast amount of testing of materials and structural elements.

Allied with the testing of strengths — particularly testing of the strengths of new iron and steel columns and beams — was an increased emphasis on internal stress distributions to facilitate sizing members by calculation.

In the analysis of the overall stability of arched masonry structures, there had hitherto been some uncertainty about where the forces necessary for equilibrium should be considered to act within the relevant cross sections. La Hire implicitly assumed that the arch thrust would act at the intrados at the sections where the haunches split, and the analysis of the impending collapse mode at St Peter's by the three mathematicians also implictly assumed that the forces would finally act at the extremities of the sections)figure 8). Poleni (closely echoing a similar earlier statement by Gregory) stated that it was a sufficient guarantee of stability for the thrust line to lie wholly within the section – implying again that it could safely move over to the intrados or extrados.

There was, however, some recognition that, if the thrust acted right at the edge of a section, local crushing of the masonry, and possibly slip, would result. Possibly for this reason, Belidor, who first publicised La Hire's later theory and showed its practical application to design, usually placed the thrust at the centre of the section in his examples[11], and Gauthey also conservatively assumed that it would act here in his analysis of the stability of St Genevieve (figure 11).

Simple bending theory now suggested that, to be sure of safety, the resultant force should act within the middle third of the section, to ensure that the compression was distributed over the whole depth so that there could be no opening of the joints. Guadet, in a suggested redesign of the external flying buttresses of the church of St Ouen in Rouen, even took the calculated thrust line as the centreline of the new flyer (figure 13), and this seems also to have been Gaudi's intention in his designs for the chapel of the Colonia Guell and the Church of the Sagrada Familia in Barcelona on the basis of three-dimensional modelling of the structures as suspended-cord analogues loaded with suitable weights.

More recent developments in relation to the assessment of the margins of stability of ancient masonry arched and vaulted structures

For structural design purposes, what Guadet and Gaudi did was, of course, entirely reasonable, even if questionable architecturally or on grounds of construction economy. But if the masonry is sound there is certainly no need, when assessing the stability of an existing structure, to expect the thrust lines to do more than stay within the middle third of the cross section. Even this may be unnecessarily cautious. Can we now do better?

The opposite extreme of the 18th century approach of La Hire, the three mathematicians and Poleni, has recently been advocated afresh as a preferable one.[13] The simplifying assumptions that they made have been subsumed under more generalised assumptions about possible failure states, and the whole stability problem has thereby been brought within the framework of "plastic" limit-state theory – a theory which ignores internal stresses and strains and assumes that instability can result only from the formation of a kinematic mechanism, neither elastic nor plastic deformations between the "hinges" of this mechanism having any adverse effect.

The resultant "plastic" theory goes even further than Poleni in asserting that, if we can find any equilibrium state in which all loads are transmitted to the ground wholly within the masonry, the structure is safe, regardless of how the loads may actually be transmitted – obviating, thereby, the problems that arise from multiple possible load paths in most structures actually encountered.

Unfortunately it tells us nothing we do not already know about a structure that is already standing and that is unlikely to be subjected in the foreseeable future to loads significantly greater than those it has been carrying in the immediate past. From the mere fact that the structure is still standing, we already know that at least one such state must exist, irrespective of whether we are able to find it. We learn nothing about the adequacy of the margin of safety beyond what we might be able to judge for ourselves, independently of the theory, from the positions of the estimated lines of thrust.

Local stresses are indeed important in making such a judgement, as may be seen from actual failure sequences (figure 14a).[14] If the thrust comes too close to the edge of the section, local splitting or spalling does occur and does precipitate slips. These reduce the effective depths and make the situation worse until collapse occurs. Deformations under stress also affect the actual load paths.

These shortcomings of analyses based solely on statics have suggested the use of modern techniques of stress analysis, using either photo-elastic models or numerical finite-element models.[15] The difficulties of adequate modelling and then of making realistic and useful analyses are, however, many. If, for instance, the actual structure is represented in a photoelastic model as an isotropic elastic continuum equally strong in tension and compression and is implicitly assumed to carry no load until complete, not one of the assumptions will match the true situation, and the divergences are of such importance that the predicted stresses will bear little relation to those that will exist at any particular time in practice. Finite-element modelling is more versatile, but still calls for simplifications that may render the analyses invalid or misleading.

In assessing the usefulness of any concept, criterion, model or analytical technique for the purpose of estimating the margins of stability of existing ancient structures, we should not, of course, demand answers to the questions we should like to ask that are more precise than those we should expect if we were designing a new structure. Simplifying assumptions are always essential, and the real test of their admissibility is whether an adequate basis is provided for a final judgement, or at least a better one than would otherwise be available. But simplifications do seem out of place where, simply by inspecting the structure itself, we can deduce important aspects of its strengths and weaknesses and modes of action that, for a new structure, we could only estimate.

In this situation I strongly feel that, of the basic approaches that have been considered so far, the most helpful one is essentially that adopted by the three mathematicians in their indentification, in the first place, of a possible collapse mode. In this way, account is taken from the outset of what the structure has itself disclosed about its capacities and weaknesses instead of ignoring this. Analysis of safety margins will thus be more realistic. Any exploration of possible equilibrium systems of internal thrusts should, I feel, similarly take fully into account the actual deformed geometry and any actual restrictions on possible load paths due to such things as significant discontinuities at cracks.

This is not to dismiss the usefulness of all modern theoretic developments and insights. We have seen that developments in theory in the 17th and 18th centuries transformed earlier primarily geometric concepts into more abstract and powerful mechanistic ones. Subsequent developments have not brought about a further change of this magnitude from the present point of view. But they have further clarified the spatial and muscular intuitions through which we can understand and interpret what the structures are able to disclose to us. We can now see as much and more in far less obvious indications that were present at St Peter's. And, when we have read the evidence, we have been provided with a much wider range of tools for futher analysis if this is called for. The plot of increase in a critical deformation reproduced in figure 14b is one instance of what may now be done to provide a further basis for judgement.[16]

I link this emphasis on first studying as it is with a plea for respecting the original organism where possible. Any approach that starts only from the geometric form is more likely to result in a desire to modify this organism to make it conform with the analysis.

Setting aside the special case of structures liable to seismic loading,[17] margins of safety will ultimately be measured by tolerable further increases in deformation and by the nearness with which the resultant loads approach the boundaries of critical cross sections. The relevant concepts of stability are therefore, in the first instance, a reasonable ability to accept further deformation and, in the second, a "comfortable" state of equilibrium. "Comfortable" is not, of course, a precise term. The need for judgement is inescapable. But the better we understand the existing structural organism, the better fitted we shall be to make this judgement.

[1] Mainstone, R J, "Structural theory and design before 1742", *Architectural Review*, vol. CXLIII, 1966, 303-10, discusses this period more fully with fuller references to the documentary sources. See now, in addition, Boucher, F, "Design in Gothic architecture, a preliminary assessment", *Journal of the Society of Architectural Historians*, vol. XXVII, 1968, 49-71; Mainstone, R J, *Developments in structural form*, London and Cambridge, Mass, 1975; Shelby, L R, *Gothic design techniques*, Carbondale, 1977; and Boucher, F, *Architector*, vol. 1, New York, 1979, each with further references. Ackerman, J S, "Gothic theory of architecture at the Cathedral of Milan", *Art Bulletin*, vol. XXXI, 1949, 84-111, still gives the best account of the discussions there in 1400.

[2] Mainstone, R J, "The springs of structural invention", *Journal of the Royal Institute of British Architects*, vol. LXX, 1963, 57-71; and idem, "Intuition and the springs of structural invention", *VIA* (Graduate School of Fine Arts, University of Pennsylvania), vol. 2, 1973, 42-63 and 192-5.

[3] Mainstone, R J, "Brunelleschi's Dome", *Architectural Review*, vol. CLXII, 1977, 156-66, gives a fuller account. Idem, "Le origine della concezione strutturale della cupola di Santa Maria del Fiore", *Filippo Brunelleschi: la sua opera e il suo tempo*, Florence, 1980, 883-92, concentrates on the development of Brunelleschi's ideas; and a preliminary paper, idem, "Brunelleschi's dome of Santa Maria del Fiore and some related structures", *Transactions of the Newcomen Society*, vol. XLII, 1969-70, 107-26, includes a statical analysis of the development of stress in the circular dome contained within the thickness.

[4] Published in Wren, C, *Parentalia of the family of the Wrens*, 1750, London, 356-8.

[5] Le Seur, T, Jacquier, F, and Boscovich, R G, *Parere di tre mattematici sopra i danni, che si sono trovati nella cupola di S. Pietro*, Rome, 1742.

[6] Poleni, G, *Memorie istoriche della gran cupola del Tempio Vaticano*, Padua, 1748.

[7] Gauthey, E M, *Memoire sur l'application des principes de la mechanique a la construction des voutes et des domes*, Dijon, 1771.

[8] The best summary of this development (with full references) is to be found in Heyman, J, *Coulomb's memoire on statics*, Cambridge, 1972, 82-8 and 168-83. Only a few of the more important publications are cited below.

[9] La Hire, P de, "Sur la construction des voutes dans les edifices", *Memoires de l'Academie Royale des Sciences, 1712*, vol. 69, 1731.

[10] Frezier, A F, *La theorie et la pratique de la coupe des pierres ... ou traite de stereotomie*, vol. 3, Paris, 1739. (Figure 12 is reproduced from pl. 111 of the 2nd edn., 1769).

[11] Belidor, B F, *La science des ingenieurs*, Paris, 1729, bk. 2.

[12] Guadet, J, *Elements et theorie de l'architecture*, Paris, nd., (Figure 13 is reproduced from the account in the 4th edn., c. 1909, vol. 3, 342-8).

[13] Heyman, J, "The stone skeleton", *International Journal of Solids and structures*, vol. 2, 1966, 249-79.

[14] A corresponding photographic sequence is reproduced in Mainstone, R J, (for working Group F of UNIDO/UNDP Project RER/79/015), *Repair and strengthening of historical monuments and buildings in urban nuclei: Building construction under seismic conditions in the Balkan region*, vol. 6, Vienna, 1984, fig. 3.37.

[15] See especially Mark, R, *Experiments in gothic structure*, Cambridge, Mass, 1982.

[16] Mainstone, R J, *Hagia Sophia*, London, 1987, gives the derivation.

[17] Mainstone, R J, op. cit., note 14, deals most comprehensively with this aspect in relation to ancient buildings.

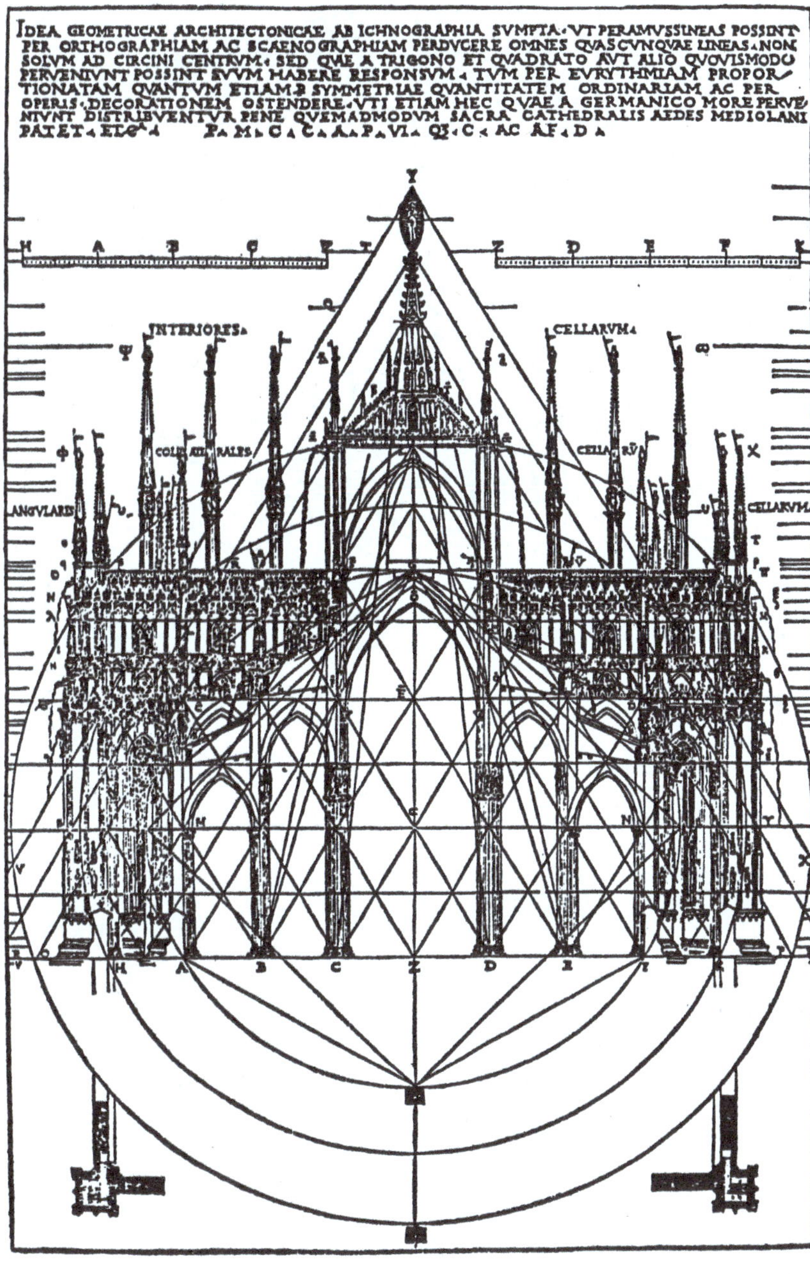

Illustration I. Cross section of Milan Cathedral showing a controlling geometric grid "ad triangulum" (from Cesariano's edition of Vitruvius, 1521).

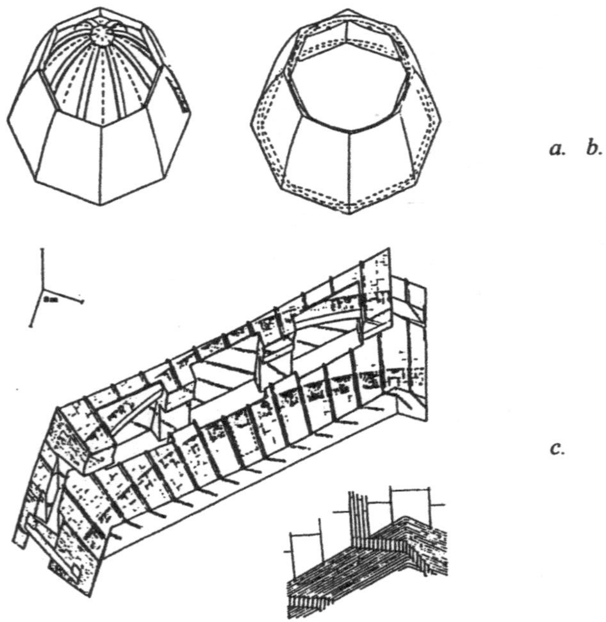

Illustration 2. At a) normal centering requirements for a dome of octagonal plan; b) a thicker octagonal dome containing within its thickness a circular one which allows construction without centering if provision is made to prevent those sections inside the circle (lettered a, a, a, a) from falling away (Author). At c) details of the construction devised for this purpose by Brunelleschi for Florence Cathedral, the effective circular dome being shaded (Author).

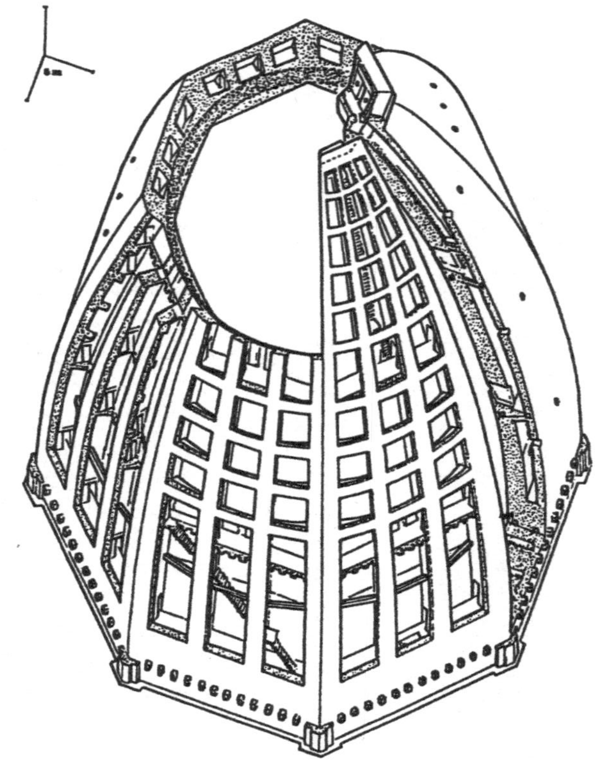

Illustration 3. Florence Cathedral dome: partly cut-away drawing of the whole construction (Author).

a.

b.

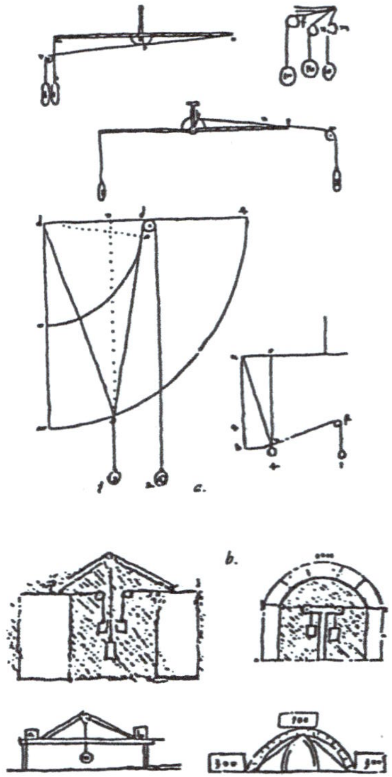

Illustration 4. Sketches of Leonardo illustrating: at a) possible ways of envisaging the action of non-vertical forces and the associated conditions of balance and at b) the outward thrusts exerted by coupled rafters and arches, and ways of resisting these thrusts.

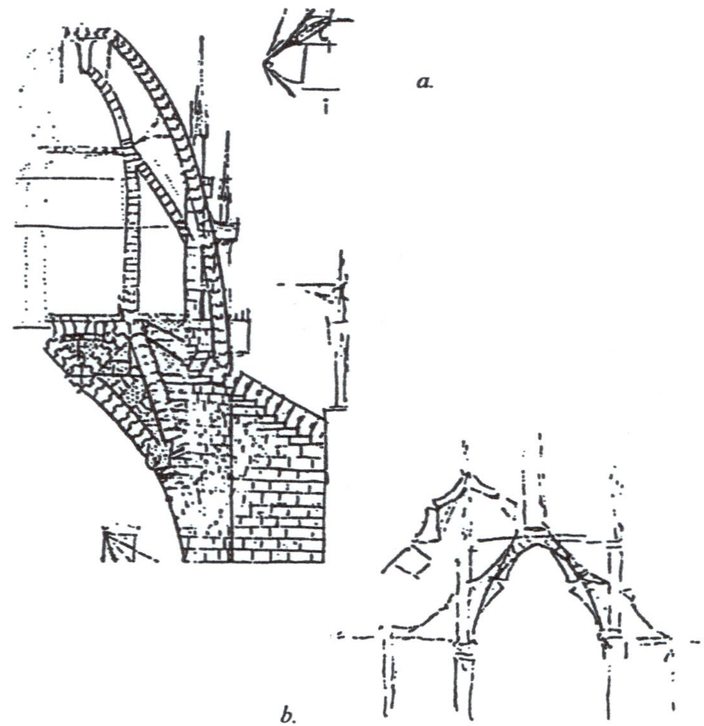

a.

b.

Illustration 5. Sketches by Leonardo illustrating: at a) (especially at the upper right), the action of forces in the arched ribs of a proposed vault and at b) two modes of collapse of a similar vault.

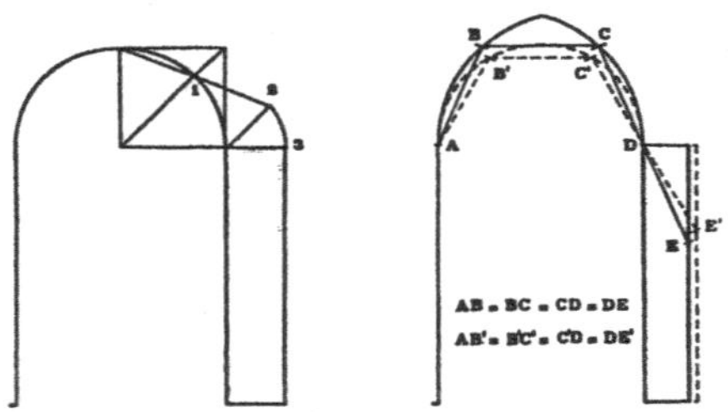

a. b.

Illustration 6. Empirical geometric rules for determining the sizes of piers needed to support an arch: a) from Derand and b) from Garcia (after Rodrigo Gill).

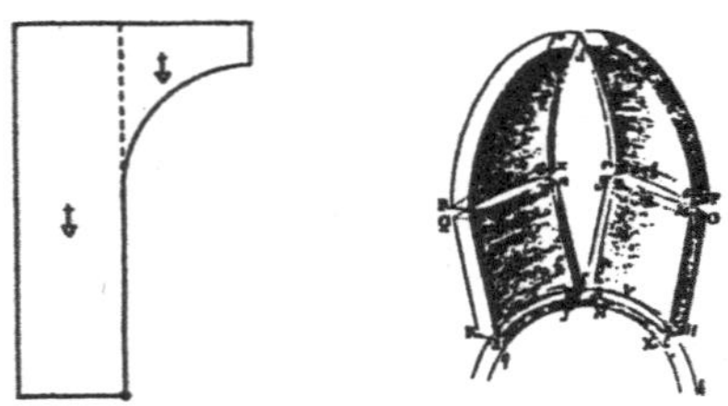

c. d.

At c) Wren's suggestea Archimedean statical determination, assuming that collapse could occur only through each pier rotating about the inner edges of its base. At d) the collapse mode of the dome of St Peter's deduced from the observed pattern of cracking by Le Seur, Jacquier and Boscovich (a, b, c, Author; d from Poleni).

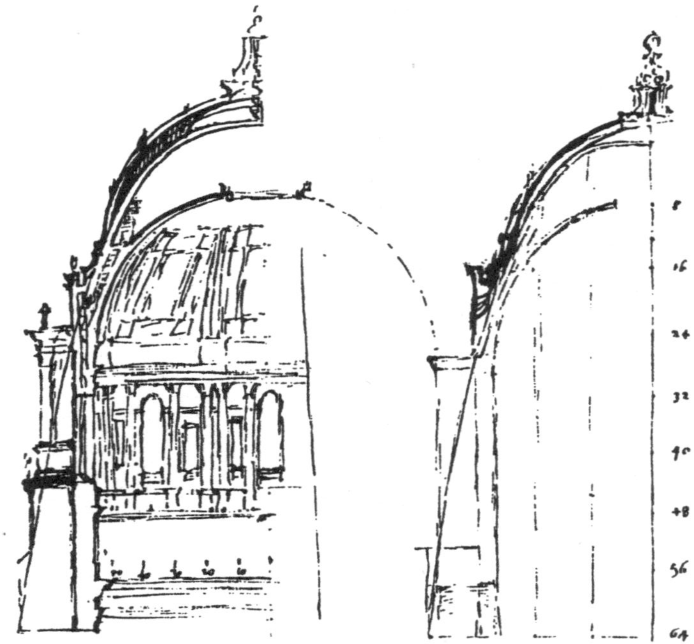

Illustration 7. Sketches by Wren of possible cross sections for the dome of St Paul's.

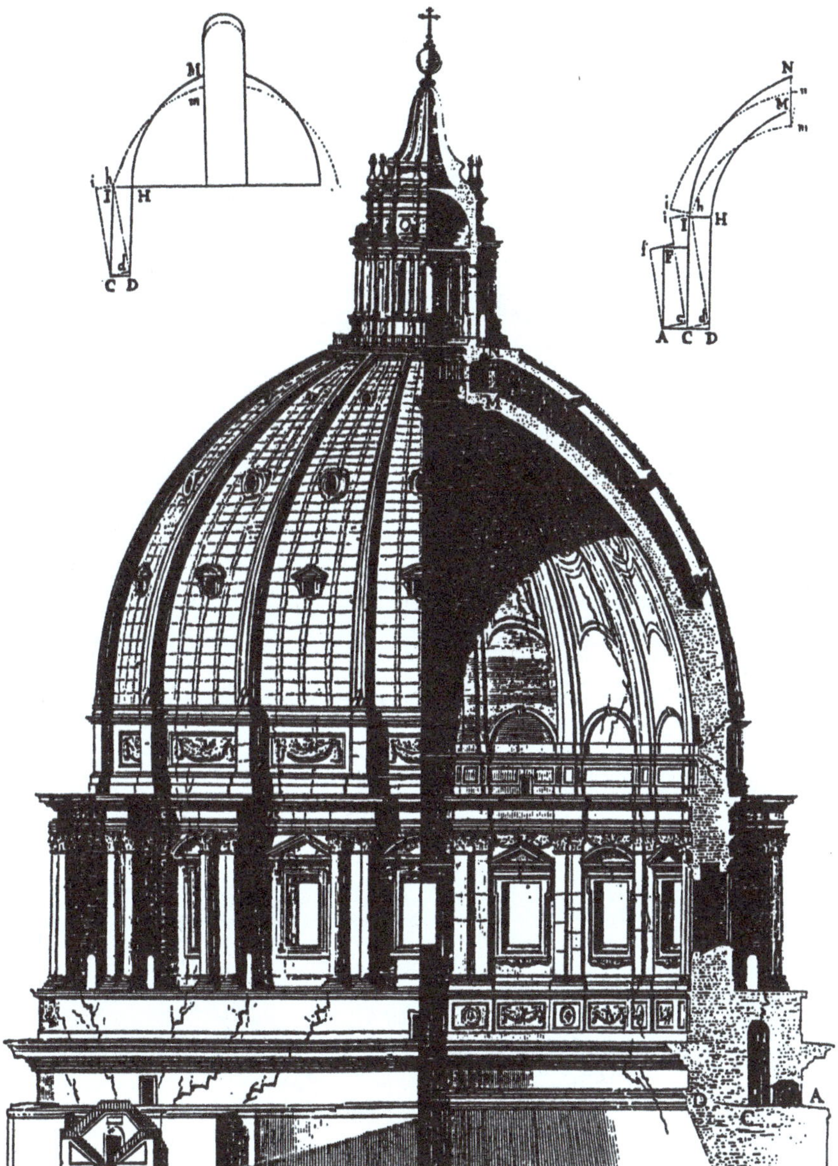

Illustration 8. Part sectional elevation of the dome of St Peter's showing the observed extensive cracking and (at the top) sketches of the deduced mode of collapse (from Le Seur, Jacquier and Boscovich).

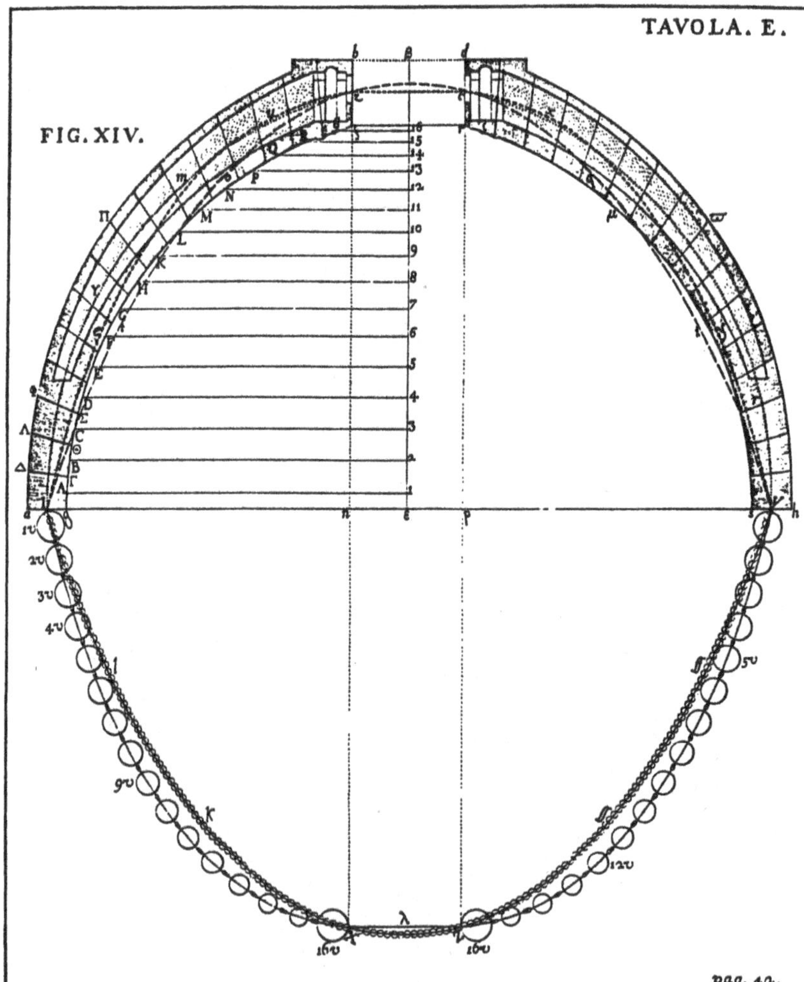

Ilustration 9. Experimental analysis of the dome of St Peter's on the basis of analogy of a similarly weighted hanging chain (from Poleni).

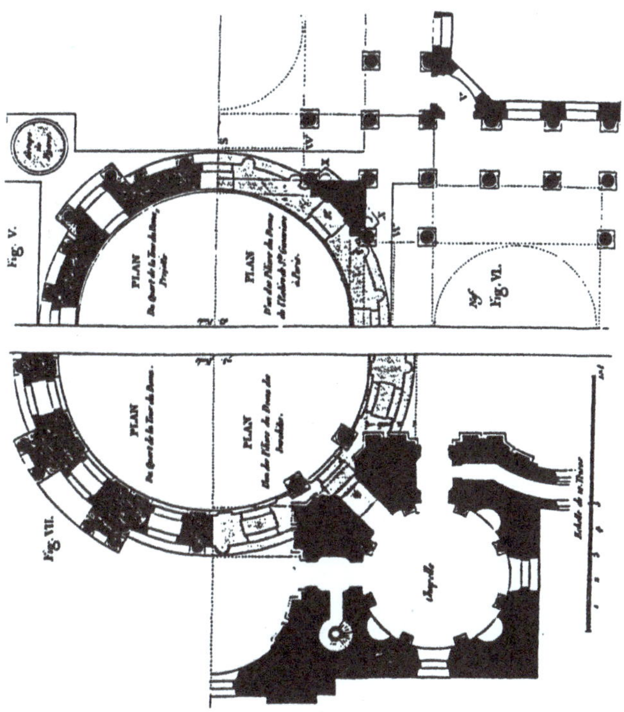

Illustration 10. Comparative part-plans of the Invalides, Paris, and the projected church of St Geneviève, and their supports.

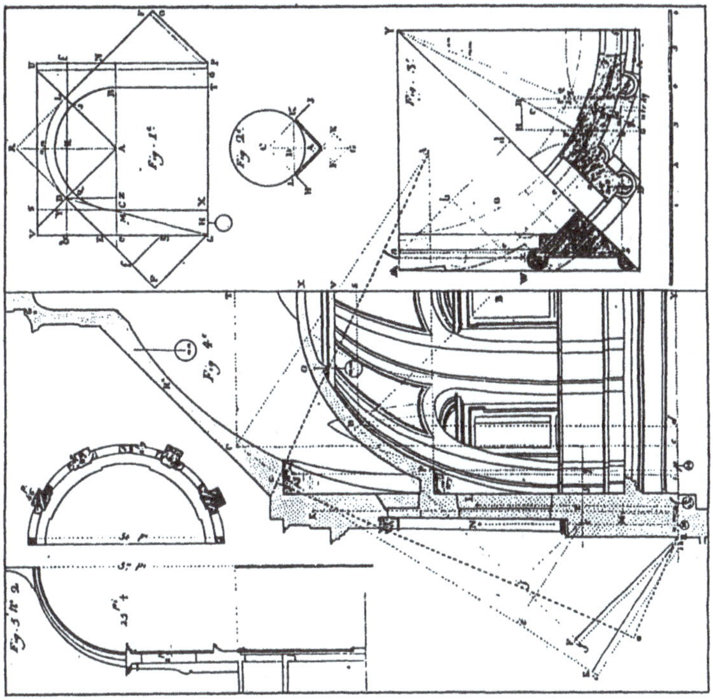

Illustration 11. Analysis by Gauthey of the stability of the dome of the projected church of St Geneviève (from Gauthey).

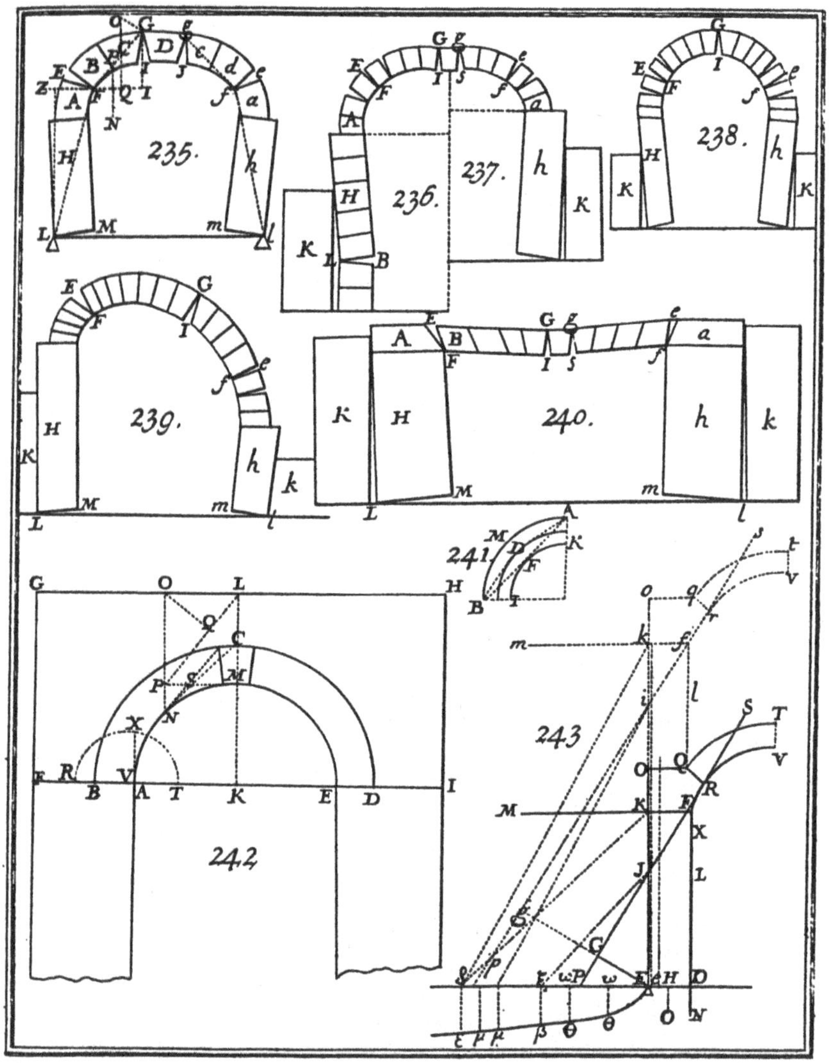

Illustration 12. Tests on model voussoir arches (figures 235-240) and ways of ensuring their stability and that of supporting piers (from Frezier and Danisy).

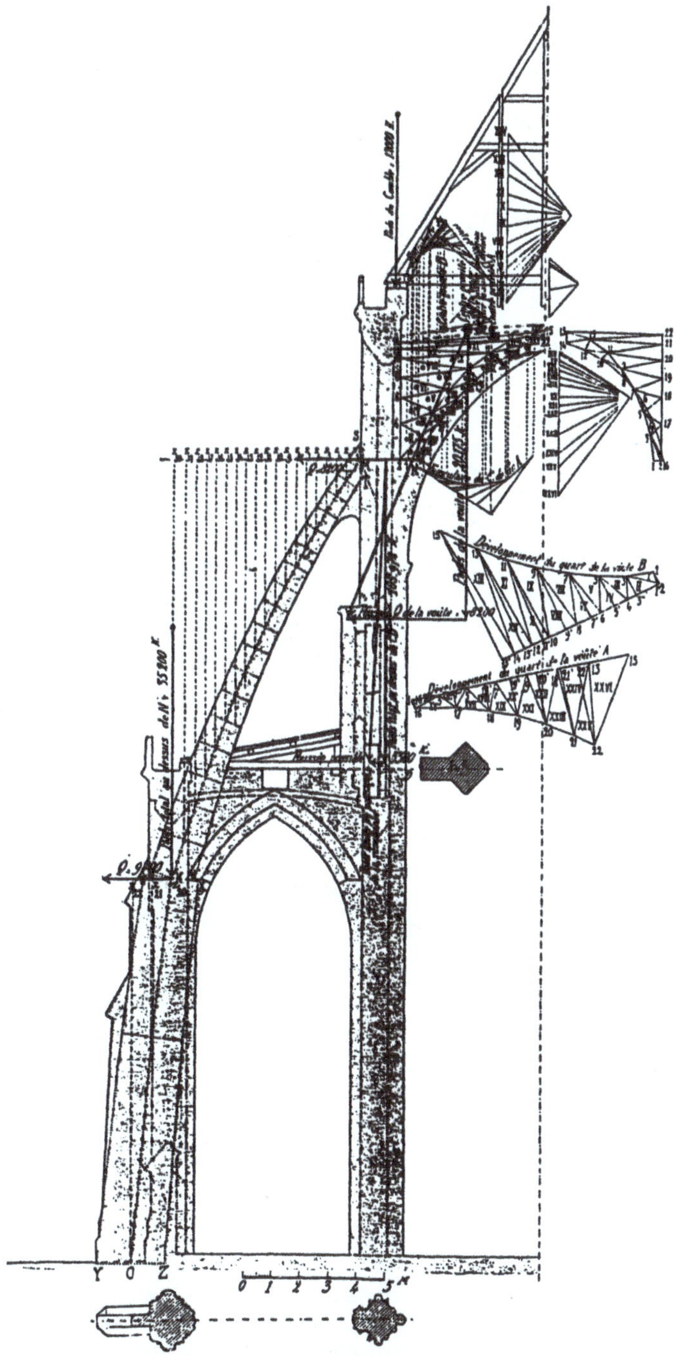

Illustration 13. Analysis by Guadet of the lines of action of the thrusts generated by the existing roof and vaults of the Church of St Ouen, Rouen, together with his suggested re-design of the outer buttress flyers (from Guadet).

a.

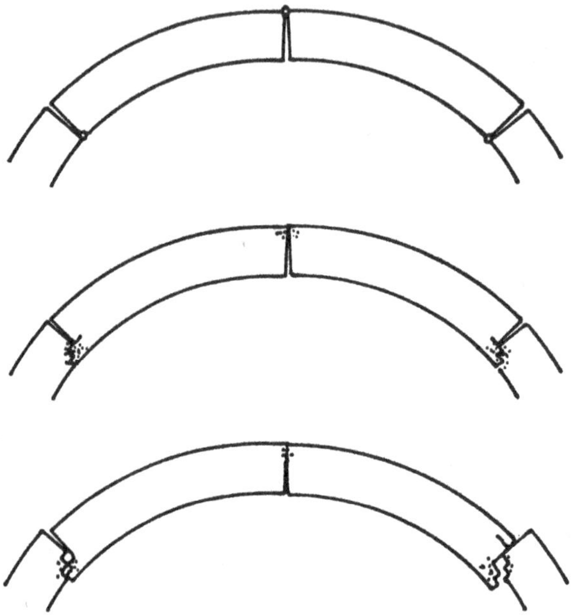

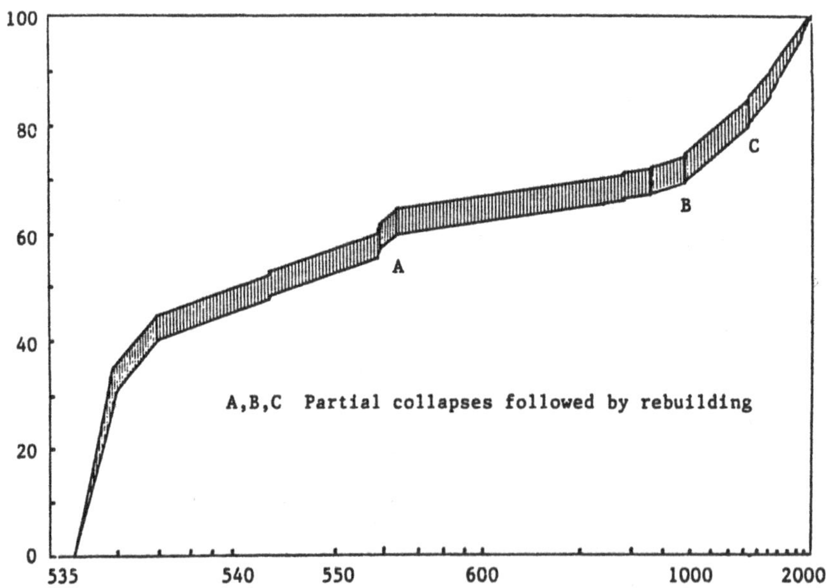

Illustration 14. At a) typical failure sequence for a masonry arch carrying a normally distributed top load (for unsymmetrical loading the sequence will be similar but with different hinge positions; where the top loading is lighter and the supports move apart collapse by simple slips of the voussoirs is more usual). At b) increase in inclination of the principal piers of the Church of Hagia Sophia, Istanbul, since construction, as a percentage of the present inclination. Partial rebuilding of the arches and vaults ocurrs as noted (both Author).

9

The place of Sir Christopher Wren in the history of structural engineering

Stanley B. Hamilton

Much has been written during the last few years on the work of Sir Christopher Wren, as artist, man of science, and public servant. He is entitled, however, in addition, to a place of honour in the history of structural engineering, and the purpose of this paper is to exhibit the evidence which should enable us to assess his merit in that respect.

Prior to Wren's time the only scientific work on the strength of materials of which any record exists was contained in the unpublished notes left by Leonardo da Vinci (1452-1519) upon his experiments on the strength of beams and struts; and in the first two " Dialogues " in Galileo's *Two New Sciences*, published in 1638, when Wren was six years old. Galileo's work was discussed by mathematicians and physicists on the Continent, and introduced to the notice of the Royal Society in 1662 by correspondence received from Marin Mersenne, who had translated Galileo's work into French. Experiments were devised by certain members of that body to test Galileo's theories, but no results of value were achieved at the time; Wren, although an active founder member of the Royal Society (1662), does not appear to have interested himself in these experiments, nor, indeed, to have shown any marked interest in architecture, until the King, at John Evelyn's instigation, created for him the office of Assistant to the Surveyor General in 1662. Evelyn could hardly have foreseen that the young anatomist-astronomer on whose behalf he sought the Royal favour would prove, after one brief tour of study in Paris and a few years of subsequent practice, to possess the talents which, in due course, marked Wren as one of the greatest exponents of late Renaissance architecture. No one, indeed, could have foreseen that the disastrous fire which swept the City of London in 1666 would find in Wren the one man eminently capable of seizing so unique an opportunity of raising from the ashes the magnificent Cathedral and the many lesser gems of beauty which are still the city's pride.

It seems strange, and is unfortunate, that Wren wrote nothing on the subject of strength of materials, since he made a profound study of early buildings and exercised exceptional ingenuity in the combination of structural forms.

WREN AND THE HISTORY OF STRUCTURAL ENGINEERING

His opinions on many phases of structural design, however appear in the comments and criticisms contained in the reports he submitted in his official capacity on the condition of several famous public buildings.

OLD AND NEW ST. PAUL'S.

In describing the ruin of the roof of Old St. Paul's, for instance (the building subsequently destroyed by the Great Fire), Wren wrote:[1] "The work was both ill-design'd and ill-built from the Beginning; ill-design'd, because the Architect gave not Butment enough to counterpoise, and resist the Weight of the Roof from Spreading the Walls; for the eye alone will discover to any Man, that these Pillars as vast as they are, even eleven Foot diameter, are bent outwards at least six inches from their first position; which being done on both Sides, it necessarily follows that the whole Roof must first open in large and wide Cracks along by the Walls and Windows, and lastly drop down between the yielding pillars. This bending of the Pillars was facilitated by their ill Building; for they are only cased without, and that with small stones, not one greater than a Man's Burden; but within is nothing but a Core of small Rubbish-Stone, and much Mortar, which easily crushes and yields to the Weight."

It was one thing, however, to criticise adversely the methods adopted by early builders: quite another to devise a system of construction which should not, in the course of centuries develop similar defects. The new St. Paul's, like the old, under stress of false economy imposed by the owners, was carried on piers, ashlar-faced and rubble-hearted. In our day these have had to be grouted with Portland cement and secured by heavy steel reinforcement to remedy the defects which two and a half centuries have rendered only too apparent. A section of the main piers which carry the dome of St. Paul's Cathedral is shown in Figure 1. It will be noticed that the whole weight of the superstructure comes on the rubble, and not on the ashlar, at ground floor level. Many of the facing stones were also hollow bedded.

What calculations Wren made of loads and stresses will never be known, for none have been preserved, even in the case of St. Paul's. That they were the simplest and the fewest possible, made hastily on scraps of paper, is not improbable, for he had an enormous practice. Exact calculations, moreover, would have been useless to a designer who repeatedly changed his mind, even when construction had reached an advanced stage. No original drawing of St. Paul's, of the many still extant, not even the accepted design, shows the building as it was ultimately constructed. The only accurate

[1] *Parentalia of the Wren Family*, by Stephen Wren, 1750, p. 275.

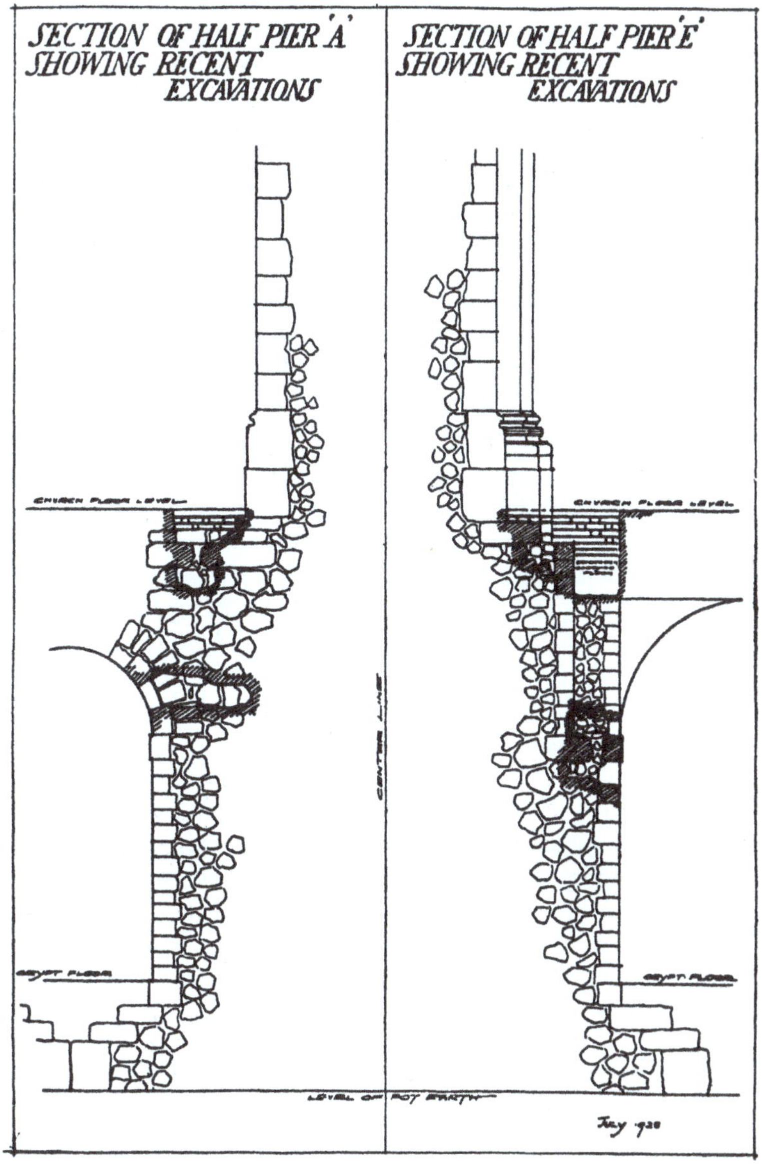

FIG. 1. ST. PAUL'S CATHEDRAL: BASE OF MAIN PIER IN SECTION.
From a drawing lent by Mr. W. Godfrey Allan.

drawings of the building as completed, and all known estimates of loads and stresses, are the outcome of comparatively recent measurements.

According to Gwilt[2] the stress in the main piers beneath the dome is 17.1 tons per square foot. This is, presumably, the average at the base. Local stresses at the pendentives must be enormously higher: hence the spalling of the masonry at these points. In St. Peter's at Rome the corresponding stress is given by the same authority as 15.0, and in the Invalides in Paris, 13.6 tons per sq. ft.; but in neither is the local concentration so acute. The Pantheon in Paris is worse, 26.9 tons per sq. ft., and it is not surprising that the condition of that building was the subject of grave anxiety to those responsible for its maintenance at a much earlier stage. A later estimate of the average load per area on the piers under the dome of St. Paul's given in the 1925 report prepared by the Commission of eminent engineers and architects appointed to investigate the structural condition of the dome is 15 tons per square foot.[3]

WREN'S CRITICISM OF SALISBURY CATHEDRAL AND WESTMINSTER ABBEY.

An illuminating report on Salisbury Cathedral was compiled by Wren in 1669.[4] After commending the original architect for the proportion, restraint and stateliness of his work which was described in some detail, Wren criticised the inadequate spread of foundation under the piers, the choice of a site too low for effective drainage, and the "Poise of the Building"; "generally," he wrote, "the Substructions are too slender for the Weights above." "The Pillars," he continued, "appear small enough, and yet they show much greater than they are; for the Shafts of Marble that encompass them, seem to fill out the Pillars to a Proportionable Bulk; but indeed, they bear little or no weight, and some of those that are Pressed, break and Split: if those Ornaments should be taken off, the Pillar would then appear too little for its Burthen; but this is no where so enormous as under the Steeple, which, being four Hundred Feet in Height is borne by four Pillars, not much larger than the Pillars of the Ailes. . . As this addition of a Spire was a second Thought, the Artist is more excusable for having omitted Buttresses to the tower; and his ingenuity commendable for supplying this defect by bracing the Walls together with many large bands of iron within and without, keyed together with much Industry and Exactness; and besides those that appear, I have reason to believe that there are Divers other Braces concealed within the thickness of the Walls; and these are so essential to the Standing of the Work that, if they were dissolved, the Spire would spread open the Walls of the Tower, nor could it stand one Minute."

[2] *Encyclopaedia of Architecture,* 1881, Edn., p. 406.

[3] See Joint Paper by Capt. C. Stanley Peach and Godfrey Allen, on the Preservation of St. Paul's, in the *Journal* of the R.I.B.A., for 9th August, 1930.

[4] Milman's *Wren,* pp. 332 et seq., from *Parentalia,* pp. 303/6.

WREN AND THE HISTORY OF STRUCTURAL ENGINEERING

" But this way of tying Walls together with Iron, instead of making them of that substance and Form, that they shall naturally poise themselves upon their Butment, is against the Rules of good Architecture; not only because it is corruptible by rust, but because it is fallacious, having unequal veins in the metal, some pieces in the same bar being Three Times stronger than the other, and yet all sound in Appearance."

In this report the first feature that strikes a modern structural engineer is Wren's keen eye for the anatomy of a building; the second, that a pillar would " appear too little for its Burthen " unsupported by any figures to show whether such was in fact the case: the third is the reference to the liberal use of iron in thirteenth century structures, not as a casual addition, but as an inherent element and a necessary corrective to the defect in that very balance for which, in popular esteem, the mediæval architect is famous. The unreliability of such iron due to its frequent welds is also indicated.

Wren's report on Westminster Abbey in 1713 displayed the same features.[5] He described the history of the building; the origin and development of various faults, and how they might be remedied. He referred at length to the durability or otherwise of the materials selected, how this had been affected by their mode of working and their position in the fabric. The usual fault of lack of balance was again in evidence in the four pillars which had been intended to carry a central tower and steeple, but which had been " bended inward considerably "[6] by the vaulting they carried.

" I conceive the Architect knew very well," he wrote, " that the four Pillars above the Intersection of the Cross nave would not prove a sufficient Butment to stand against the Pressure of so many Arches, unless they were very much bigger than the other Piers; but that could not be without cumbering up the principal part of the Church; but though these angular Pillars could not be made bigger, yet they could be made heavier to stand against the Pressure of the several Rows of Arches which might prove an equivalent."

He proceeded to show that, if a pier could not be made wider to withstand a horizontal thrust it must be more heavily loaded from above.

" This is the Reason," Wren continued, " why in all Gothick Fabricks of this Form, the Architects were wont to build Towers or Steeples in the Middle, not only for ornament, but to confirm the middle Pillars against the Thrust of the several Rows of Arches, which force against them every Way. The Architect understood this well enough, but, knowing that it might require Time to give such a Butment as the Tower to his Arches, which was to be last done; and lest there should be a Failing in the mean Time, he wisely considered

[5] Milman's *Wren,* p. 347 et seq., from *Parentalia,* pp. 295-303.
[6] loc. cit., p. 300.

WREN AND THE HISTORY OF STRUCTURAL ENGINEERING

that, if he tied these Arches every Way with Iron, which were next to the Middle of the Cross: this might serve the turn, till he build the Tower to make all secure, which is not done to this day. These Irons which were hooked on from Pillar to Pillar have been stolen away; and this is the Reason of the four Pillars being bent inward, and the Walls above cracked."

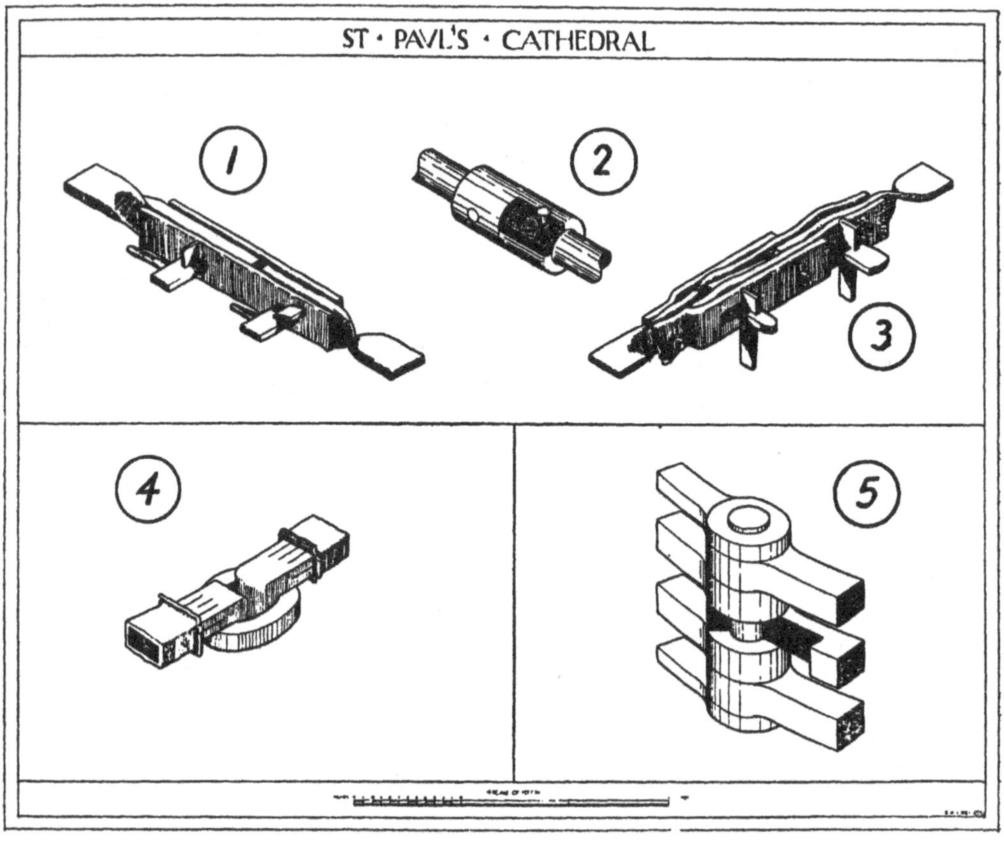

FIG. 2. NOS. 1, 3 AND 4: OLD IRON TIES. NOS. 2 AND 5: NEW NON-CORROSIVE STEEL TIES INSERTED BETWEEN 1925 AND 1930.
Measured and Drawn by R. A. Kirby.

Wren referred to a model, prepared to show how the necessary strengthening should be executed. He would then add a tower, and advises a Spire, which " will give a proper Grace to the whole Fabrick, and the West-end of the City, which seems to want it."

The West End of London still wants Wren's tower and spire to Westminster Abbey, which no modern architect would care to mount on the lofty slender columns, manifestly bent, even to casual observation, in spite of plentiful iron bars at the springing of the vaults, where Wren replaced the original ties, which had, he stated, been " stolen away."

WREN'S USE OF IRON IN ST. PAUL'S CATHEDRAL.

Wren, although he recognized the dangers of exposed and buried iron-work, made extensive use of both in his own St. Paul's. Many of the minor defects of that building, indeed, were due to its subsequent corrosion. Rustless alloy-steel embedded in Portland cement has now taken its place. In Fig. 2 are shown five isometric views of tie connections from St. Paul's Cathedral.

No. 1 shows one of the original wrought-iron ties which span between the eight main piers under the dome and the four bastions which share with the piers the horizontal thrust from the arches under the dome. (See Fig. 3.)

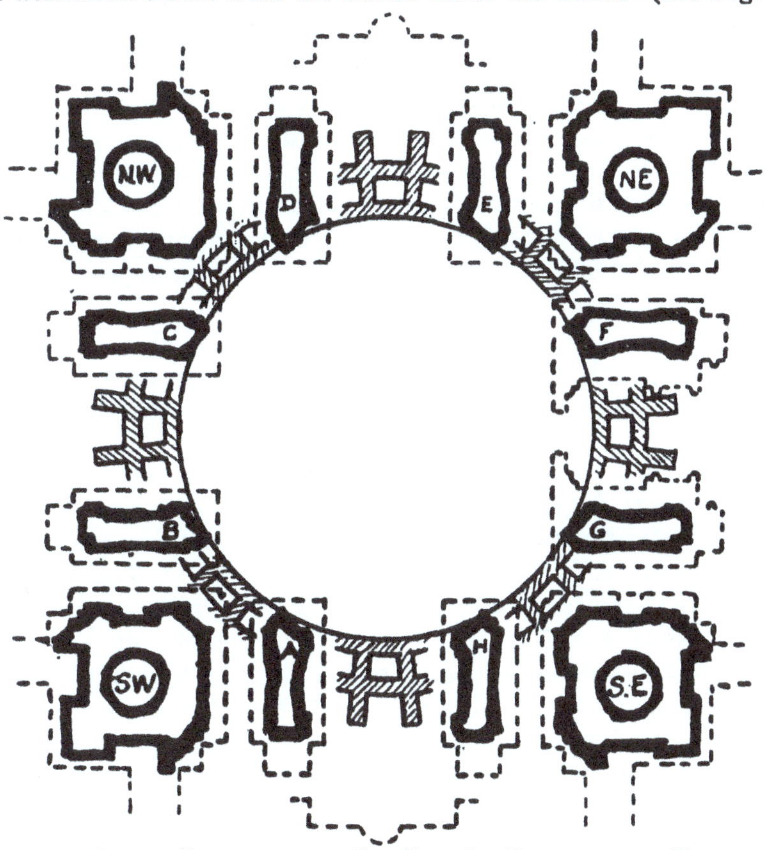

FIG. 3. THE DOME SUPPORTS OF ST. PAUL'S CATHEDRAL. PIERS IN CRYPT SHOWN BY DOTTED LINES. PIERS ABOVE GROUND SHOWN BY FULL LINES. DRUM WALLS AND BUTTRESSES PARTIALLY SHOWN BY FINE FULL LINES AND HATCHED. THE LOAD FROM THESE WALLS WAS CARRIED DOWN TO THE GREAT ARCHES BELOW BY A SERIES OF INVERTED ARCHES. (*See* Plate III).

WREN AND THE HISTORY OF STRUCTURAL ENGINEERING

No. 2 shows a typical joint in one of the non-corrosive alloy steel bars, which have replaced those shown in No. 1.

No. 3 shows one of the old tie joints repaired about 1873.

No. 4 shows a connecting ring in the " great chain " at the base of the brickwork cone, between the inner and outer domes. Each link was 10 ft. long, formed of five bars, 1 in. square in section, welded together at both ends, and bent to engage the ring. The lead sheath protecting the links is shown rolled back.

No. 5 shows the new chain.

The positions of the old and new chains are indicated on the Section, Plate I.

The new ties were inserted during the restoration work carried out between 1925 and 1930. The views of these ties were prepared by Mr. R. A. Kirby.

WREN ON THE EQUILIBRIUM OF THE ARCH.

From the foregoing reports it is evident that Wren was highly dissatisfied with the Gothic treatment of the vault and its supporting columns. Nor was he better pleased with the current graphical solution of the necessary size of abutment for end-piers or bridge spans, which he criticised adversely in the second of two unfinished tracts on architecture, which appear to have been the commencement of a treatise on that subject.[7]

" If the butments be more than enough," he wrote, " 'tis an idle expense of materials; if too little it will fail; and so for any vaulting; and yet no author hath given a true and universal rule for this, nor hath considered the various forms of arches."

Wren proceeded to criticise in detail a rule which may be described as follows (see Fig. 4): A Q B C is the soffit of an arch of any customary form—semicircular, elliptical or pointed. Trisect the curve of this intrados at Q and B. Join B C and produce to D such that CD = BC. Then the vertical through D gives the correct position of the back of the arch abutment.

Wren demands why GF should determine the abutment. " For add to the bottom as KL the arch must then certainly press more upon the higher part than the lower; or if some additional weight be added above the arch, that must still press more than before this was added. So this rule, (if it were built upon any sure geometrical theorem, as it is not) is neither true nor universal: and what is true will be shown to be only determinable by the doctrine of finding the centres of gravity in the parts of the proposed design. . . Fig. 5: Let a stone be cut in this form: FB is a parallelogram: CD a semicircle added,

[7] See *Parentalia.* Also Appendix to *Memoirs of the Life and Works of Sir Christopher Wren,* by James Elmes, London, 1823, p. 128.

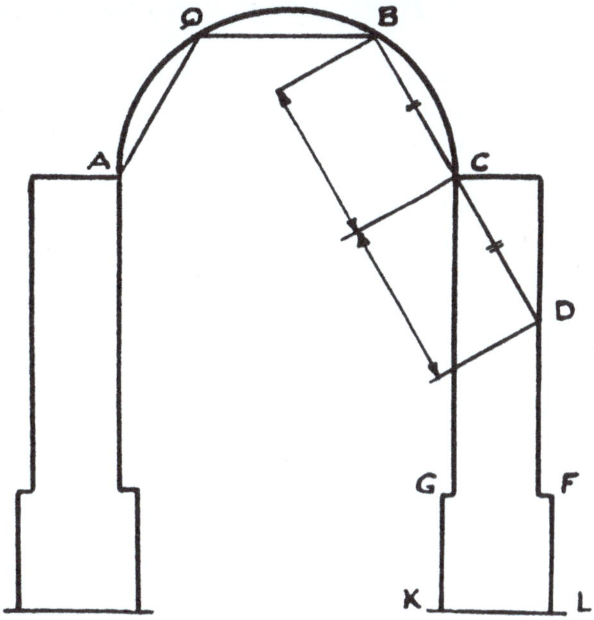

FIG. 4. CONSTRUCTION TO OBTAIN THE THICKNESS OF AN ARCH ABUTMENT.
From Father Derand's *L'Architecture des Voûtes*, 1643.

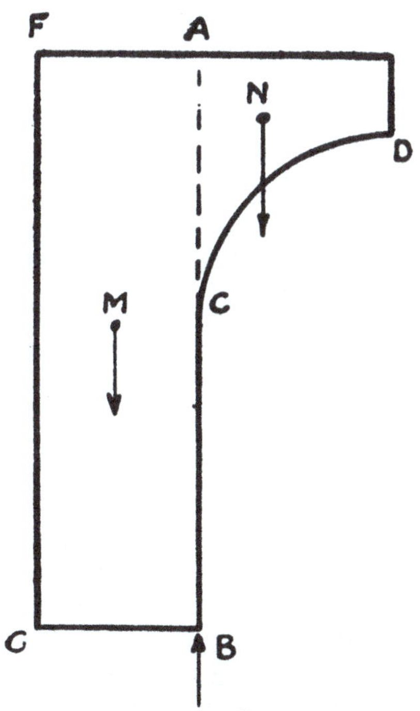

FIG. 5. WREN'S CRITERION FOR THE STABILITY OF AN ARCH ABUTMENT.
From *The Parentalia*, 1750.

WREN AND THE HISTORY OF STRUCTURAL ENGINEERING

AB a perpendicular, M the centre of gravity of FB, and N of ACD: now if N be equiponderant to M on each side of the perpendicular AB, it is certain the whole stone will stand immovable upon the base at B, although it be but half-an-arch: add the like stone on the opposite side, till the horns meet in an entire arch, so the whole will stand as well as the halves. If anything be added without M, that alters nothing, only 'tis an useless expence; but if any thing be added above N, that alters the centre of gravity, which, therefore, must be provided for by adding more weight to M; and the same may be shown in all kinds of vaulting."

The errors in this explanation would appear at once, if sections of the material used were proportioned to the stresses that such a system would apply. By ignoring the horizontal thrust at D it is made to appear (1) that the section at D might be vanishingly small; actually it must transmit both the whole horizontal thrust and any shearing force due to unsymmetrical loading of the two halves of the rib. The tensile stress at A calculated on the cantilever basis, would be excessive, and Wren gave no indication as to how it should be carried in masonry. (2) The counterpoised centilever would produce maximum pressure on the work below at B, whereas if BC were at all lofty the horizontal thrust would cause the centre of pressure to shift across towards G, if not beyond.

It need hardly be said that in actual design Wren's sound structural sense did not allow him to be seriously misled by his own inadequate theory, although his arch-work was in general inferior to the Gothic he criticised adversely.

A final reference to Wren and the arch must be made. When Robert Hooke announced in January, 1671, that he could demonstrate the correct form of an arch to carry any assigned load, Oldenburg, the secretary of the Royal Society, mentioned that Dr. Wren also had such a demonstration, and advised that both Hooke's and Wren's solutions should be delivered, and opened together by the President. There is no record as to whether this was done. Hooke's opinion that the correct form of an arch was that of a hanging chain inverted was only expressly stated in papers published after his death, although he openly announced to the Royal Society that the correct form for a cupola was a "cubico-paraboloidal conoid."

Gregory's Catenaria and Wren's Dome.

In spite, however, of the care with which Hooke appears to have attempted simultaneously to let everyone know that he held the secret of the balanced arch, and to preserve the actual secret inviolate, it became known to others. Whether it was discovered independently, leaked out, or was guessed from the hints he dropped, is not known; but in 1697 it was published without acknowledgment to Hooke in the *Philosophical Transactions* of the Royal Society, in a paper by David Gregory on "The Catenaria."

Gregory's statement that any arch must contain a catenary within its depth in order to be stable is not literally true; but, nevertheless, appears without modification in most essays on the subject until well into the nineteenth century.[8] The catenary is the ideal form only when the load to be carried is proportional to the length of the arch-rib itself. The catenary is merely a special case of the general solution by means of the link polygon.

Mr. W. Godfrey Allen, F.R.I.B.A., and the late Captain C. Stanley Peach, F.R.I.B.A., in presenting their paper to the Royal Institute of British Architects in June, 1930,[9] on "The Preservation of St. Paul's Cathedral," exhibited several interesting slides containing catenaries. One of these was a photograph of a drawing by Wren's own hand in which the proposed line of the Cathedral dome, not, however, the line ultimately adopted, formed part of a catenary which is shown continued through the supporting piers. Another slide showed a light chain suspended over an inverted drawing of the section through the existing structure. In order to keep the catenary within the lines of support it was here necessary to load the chain at the vertex, where the heavy concentrated load of the lantern is borne. A repetition of this procedure made by the present author on the lines of Captain Peach's view is shown in Plate I.

Had Wren in his mature years completed the fragments of his earlier projected treatise on architecture, his standing in relation to the catenary theory would doubtless have been clarified. When he retired from the Surveyorship, however, he was already in his eighties, and wrote no more.

"Si Monumentum Requiris, Circumspice."

Wren's palaces and private houses were elegant and dignified, but called for no extraordinary structural skill. His city churches are of varying merit, and must to a great extent have been designed by assistants, or "ghosts." Much of the supervision of his widely scattered works must also have been carried out by subordinates. Nor was his work free from trouble and failure. St. Paul's, standing as it does on a thin bed of clay over running sand, settled seriously. Before the dome could be built, and at several stages, splayed courses were laid to even up the beds for subsequent work. A large amount of ashlar spalled and had to be replaced during Wren's lifetime, leading to

[8] e.g. in Edward Cresy's *Encyclopaedia of Civil Engineering*, 1856 edn., vol. II, p. 1428.
[9] *Journal of the R.I.B.A.* vol. XXXVII (3rd Series) No. 18, pp. 664, 665.

accusations of delay and unsound work. Accidents occurred at Kensington Palace and Hampton Court; the latter was serious: a wall collapsed, several roof-trusses fell in; two men were killed, and a number more injured. The Controller of the Works blamed Wren's design, doubtless to protect himself against charges of incompetent supervision. A committee was appointed by the Treasury to report; but the technical record is not extant. The two men killed were buried in the garden; the injured were attended to in a hospital extemporised on the estate, and the Surveyor's loyal staff appear to have seen that the incriminating report in due course disappeared, since only the Treasury minute has survived.

Undeterred by such incidents, undismayed by constant over-load of work and worry, Wren continued to raise his great Cathedral, adapting the design as defects occurred, or his widening experience suggested improvements. He appears, according to the " Warrent Design," to have intended to construct a heavy stone dome. The dome, as it was built, is much lighter and more graceful; the brick cone was a stroke of genius; and it is on the superstructure of this superb building that his enduring fame as a structural engineer must rest. The weakness of his work lay in the omission adequately to provide for shear; the counterforts between the inner and outer drums which should have distributed the weight from above on to the backs of the main piers and arches ultimately cracked so badly as to be unable to do so. Failure on a large scale was probably within sight when the famous " dangerous structure notice " was issued, and the great engineering work undertaken which has saved the edifice.

The marvel, however, is not that defects occurred, but that such a stupendous mass of intricately balanced masonry was ever made to hold together. The way in which the weight was shared and distributed, especially if done with little or no calculation, showed a marvellous sense of balance and nice adjustment of load to support. This is suggested by the plans shown in Fig. 3, but is far more evident in the beautiful isometric views measured and drawn by Mr. W. Godfrey Allen, and by his generous permission reproduced as Plates II and III in the present paper.

Wren made full use of every method and material available in his day: he tried them all, and added innumerable improvements. Well versed in the history and tradition of the building arts, he applied both Gothic and Classic styles as seemed most suitable; but above all he applied them with the skill, knowledge, imagination and courage which marked him not only as the master builder of his own age; but as a shining example to artists in construction for all time.

PLATE I.

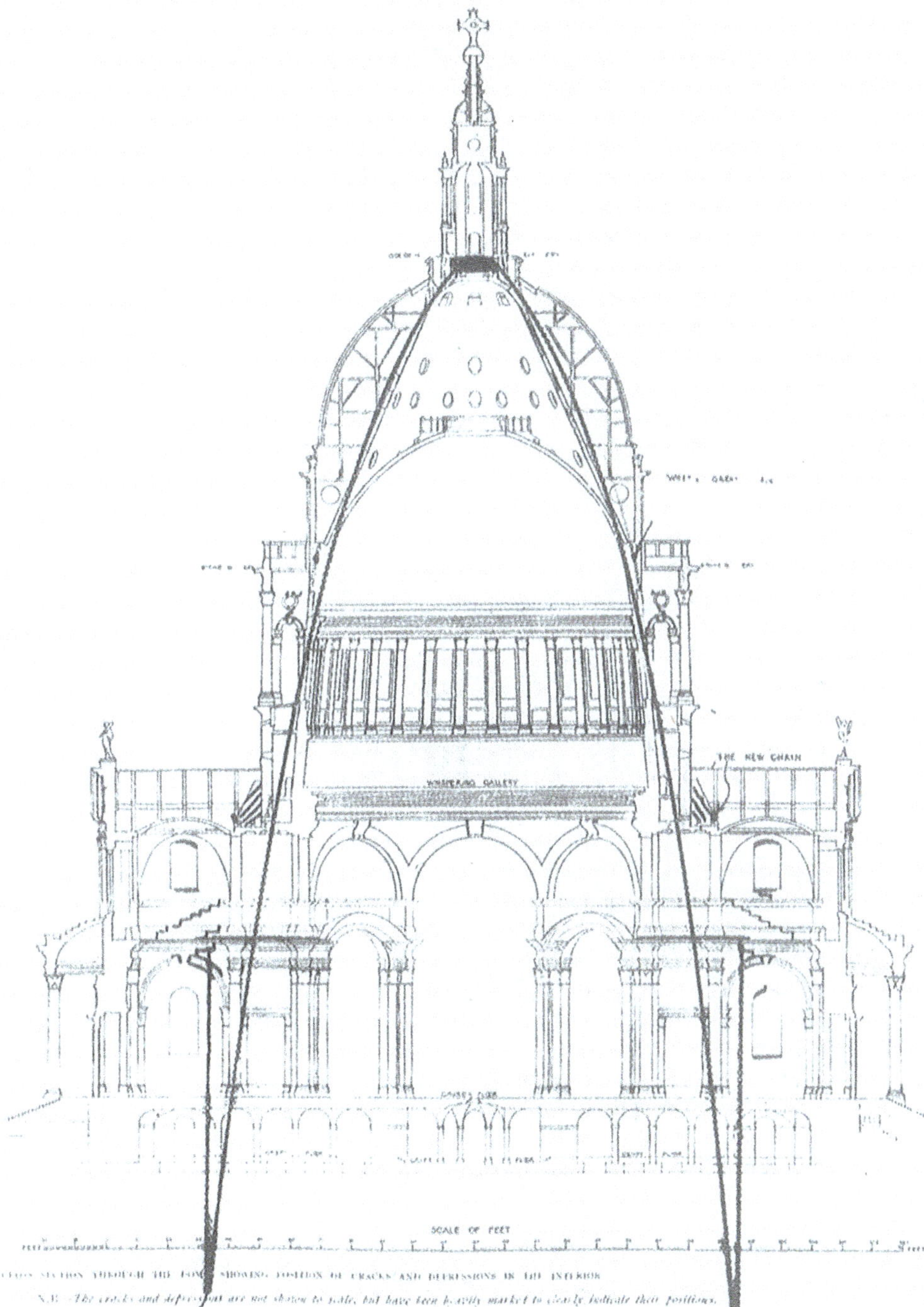

CROSS SECTION OF ST. PAUL'S CATHEDRAL,
showing the loaded catenary form of the Dome support after the manner devised by the late Capt. C. Stanley, Peach, F.R.I.B.A.

PLATE II.

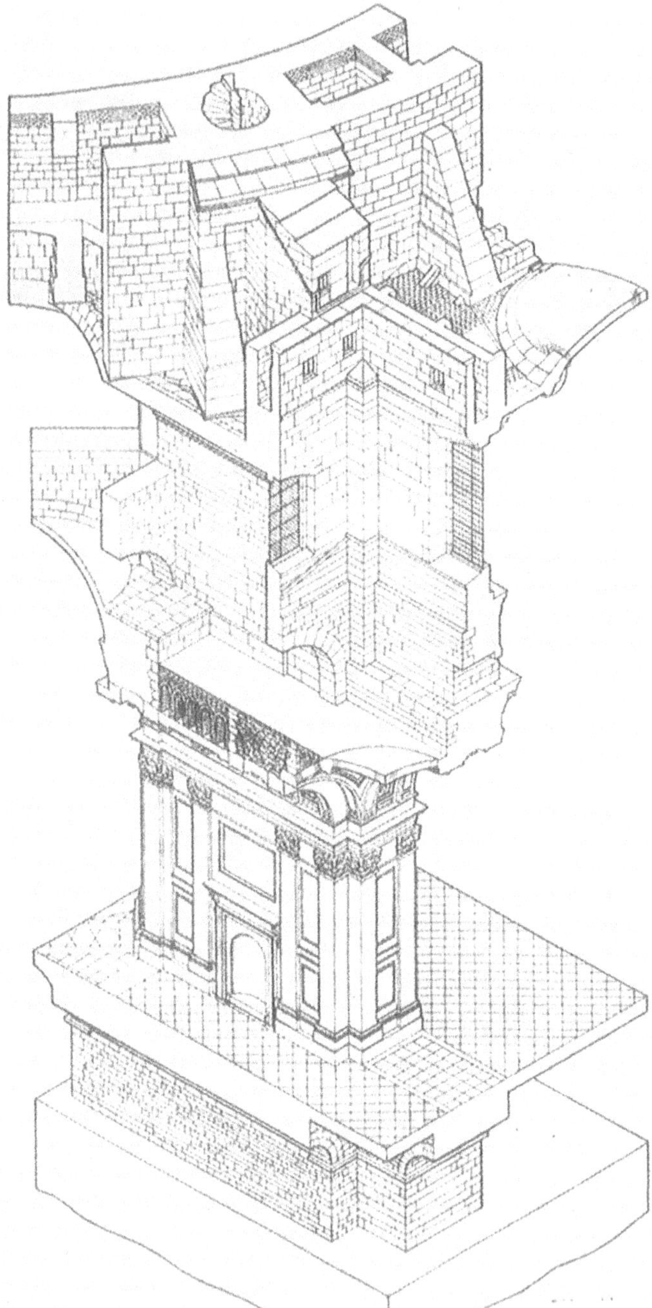

ISOMETRIC OUTSIDE VIEW OF A PIER
UNDER THE DOME OF ST. PAUL'S CATHEDRAL.
Measured and Drawn by W. Godfrey Allen, Esq., F.R.I.B.A.

PLATE III.

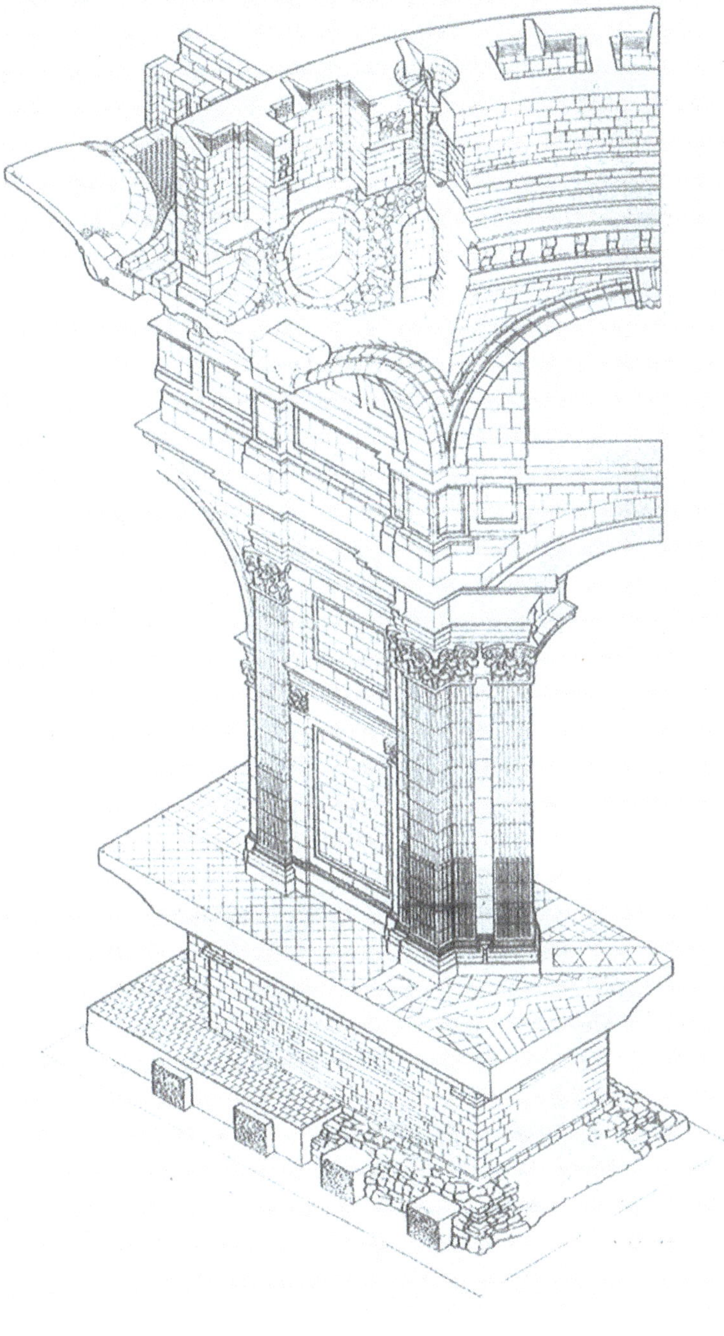

ISOMETRIC INSIDE VIEW OF A PIER
UNDER THE DOME OF ST. PAUL'S CATHEDRAL.
Measured and Drawn by W. Godfrey Allen, Esq., F.R.I.B.A.

10
The origin of iron beams

A.W. Skempton

Introduction.

The steel beam, introduced about 1885, is one of the most important components of modern architecture and structural engineering. It took the place of the wrought-iron beam which, in its turn, replaced the cast-iron beam around 1850. But the cast-iron beam was the first metallic beam to be used consistently as a major element in building construction. Previously timber had been employed for this purpose. Consequently in tracing the origin of iron beams we are concerned also with the origins of a basic change in construction practice.

From the historical and technical points of view the following questions require considerations:

I – When, and for what purpose, were iron beams first used?

II – How was their shape decided upon?

III – By what stages did they evolve towards a standard form?

IV – What tests and calculations (if any) were used in their design?

The first iron beams.

Wrought-iron beams were used experimentally in a few Parisian buildings in the decade preceding the Revolution (1), and cast-iron was employed for the lintel beams in several furnaces in England,

(1) Bannister (1950). This paper also contains the first systematic study of the early iron-framed mills in England. The conclusions, however, have had to be modified in the light of recent research.

an example dating back to 1658, at Coalbrookdale, Shropshire (2). But the first iron beams from which a continuous structural development can be traced are those used by Charles Bage (1752-1822) (3) in a 5-storey flax mill built a Shrewsbury in the years 1796-97.

The early textile mills adopted the traditional timber floor construction. This consisted typically of heavy timber beams supporting light wooden joists, on which were laid the wooden boards of the floor. Although economical, these floors could burn only too easily and, as mills became bigger, the machinery more complex and the number of work people greater, the risks both to life and capital became intolerable.

The first successful solution to this problem was achieved by William Strutt (1756-1830) in a 6 storey calico mill, built by him at Derby in the years 1792-93 (4). This mill no longer exists, but it was identical in structure with a 4 storey warehouse at Milford, near Derby, also completed in 1793, and his 6 storey water-powered cotton mill (known as the West Mill) at Belper in Derbyshire, erected between 1793 and 1795.

These buildings at Milford and Belper still stand essentially in their original form. The floors consist of brick arches spanning about 9 ft. and springing from 12" square timber beams which are supported, internally, by cast-iron columns and externally by the masonry walls of the building. The spandrel space above the arches is filled with sand, making a level surface on which brick tiles are laid. The top of the timber beam and both its sides are therefore completely protected against fire, and the underside of the beam is covered by about an inch and a half of hard plaster. The arch bears upon wooden skewbacks, which are themselves protected by sheet metal. The beams are continuous over the columns and, in fact, consist of one great piece of timber stretching from one wall to the other. A sketch of a beam in the Milford Warehouse is given in Fig. 1 (a). The width of this building is 27 ft. and two rows of columns reduce the unsupported span of the beams to 9 ft.

With this construction, William Strutt achieved the first multi-storey fire-proof framed buildings. The presence of timber, however, even though well protected, was not altogether satisfactory.

In February 1796 a large flax mill in Leeds, completed only 5 months before, was completely destroyed by fire. This mill belonged to Marshall and Benyons, a firm who were rapidly expanding their business and had already decided to establish a branch at Shrewsbury (5). The Leeds mill was rebuilt with all possible speed, still

(2) Raistrick (1953), pp. 103-104. The section of these beams is not known.
(3) A brief account of Bage's life is given by Gameson (1954).
(4) Strutt's mills at Derby, Milford and Belper are fully described by Johnson & Skempton (1956) with full references to sources.
(5) Letters from Marshall to Boulton & Watt, and Marshall MSS. I am debted to Mr. W. G. Rimmer of Leeds University for his great help in stu' the Marshall MSS.

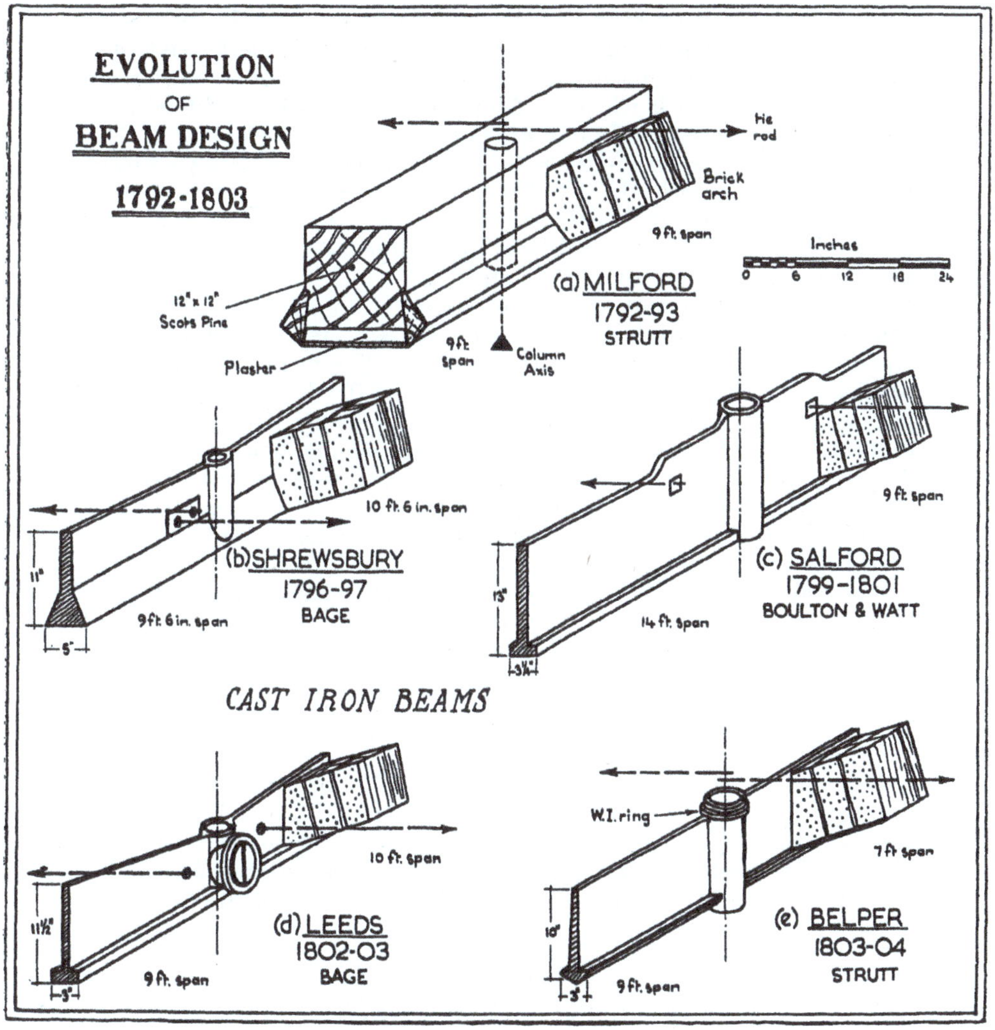
EVOLUTION
OF
BEAM DESIGN
1792-1803
tie rod
Brick arch
9 ft. span
Inches
0 6 12 18 24
12" x 12" Scots Pine
Plaster
9 ft. span
Column Axis
(a) MILFORD
1792-93
STRUTT
10 ft. 6 in. span
(b) SHREWSBURY
1796-97
BAGE
11"
9 ft. 6 in. span
5"
9 ft. span
(c) SALFORD
1799-1801
BOULTON & WATT
13"
14 ft. span
3¼"
CAST IRON BEAMS
10 ft. span
11½"
(d) LEEDS
1802-03
BAGE
9 ft. span
3"
W.I. ring
7 ft. span
10"
(e) BELPER
1803-04
STRUTT
9 ft. span
3"

with timber floors [6], but no risks were taken with the new mill at Shrewsbury.

Construction of this building began in May 1796 and it was in operation by September 1797. Charles Bage was responsible for its design. From a letter written by him to Strutt at this time, we know that he received much help from the designer of the Derby mill, and most of the features of Strutt's fire-proof mills were incorporated at Shrewsbury, including the brick arch floors. But Bage substituted iron for timber in the beams. The Shrewsbury beam is shown in Fig. 1(b) and we see at once how closely this was modelled on Strutt's timber beam with its skewbacks [7]. These first iron beams are extemely primitive, but this is hardly surprising as they had no precedent in building construction. William Hazledine cast the columns and beams at his foundry in Shrewsbury [8].

The Shrewsbury mill has an internal width of 36 ft. and three rows of columns. Each beam is cast in two equal lengths, bolted together on the centre line of the building. The beams are therefore continuous over the columns. The profile of the beam between each pair of supports is turtle-backed; that is to say its depth is greatest at mid-span. This shows an appreciation of the bending moment diagram, but it shows also that Bage had not realised the existence of negative moments over the supports. Indeed, with its comparatively low tensile strength, cast iron is unsuited as a material for continuous beams.

Nevertheless, in spite of these technical drawbacks, the beams at Shrewsbury have continued to serve their function for 160 years, and the building is in a good state of repair.

Improvements in beam design.

The next structure with iron beams was a 7 storey cotton mill built for Philips and Lee at Salford in Manchester [9]. The foundations were laid late in 1799 and the building was completed towards the end of 1801. The iron framing was designed by Boulton and Watt, who installed the steam engine for the mill, as they had also done at

---

([6]) The valuation of the mill after rebuilding was almost identical with that of the original structure, namely 2/- per ft. super. Fire-proof mills with iron-framing cost nearly double this figure. Marshall MSS.

([7]) PEVSNER (1949) p. 68 and BANNISTER (1950) give short descriptions of the Shrewsbury mill, but the shape of the beams remained unknown before the investigations of Mr. Johnson and the author in April 1956. Plans of the mill are in the Boulton & Watt Collection, where letters from Marshall, Benyons and Bage are also to be found which give the dates of construction etc. I am grateful to Mr. H. L. Hobbs and Mr. R. E. James for help at Shrewsbury, and to Miss W. R. N. Payne at Birmingham.

([8]) MINSHULL (1803), p. 74.

([9]) BRITTON & BRAYLEY (1802), p. 364, list the first four fire-proof mills as being those at Derby and Belper, Shrewsbury and Manchester.

Shrewsbury. Originally the span was to be 12 ft. 10 ins. and the beams were similar to those at Shrewsbury. It was decided, however, to increase the span to 14 ft. and the beams were made correspondingly larger [10]. At the same time the section was altered to the more economical flanged (or inverted *T*) section [11] shown in Fig. 1(c). This was an important advance, but the beams were still continuous over the supports whereas, logically, a cast-iron beam should be simply supported in order to eliminate the negative bending moments.

This feature was incorporated by Charles Bage in Benyon's flax mill built at Leeds completed in July 1803 [12]. Only the foundations of this building now exist, but in August 1804 Simon Goodrich visited the mill and made a sketch of the beams from a descripition given to him by the foreman of Mathew Murray, who had installed the engine and almost certainly cast the iron columns and beams at his neighbouring foundry in Leeds [13]. From Goodrich's sketch the drawing in Fig. 1(d) has been prepared. The beams are flanged and strongly turtle-backed, which is now a very appropriate form corresponding closely to the bending moments for simply supported beams. The flange accounts for 40% of the total cross section, a marked improvement on the Salford beam where only 20% of the metal is in the flange. The method of connecting the beams together was somewhat clumsy, although a quite practical solution.

In this respect an improvement was made by William Strutt in his 5 storey North Mill at Belper, on which work commenced in 1803 and which was completed late in 1804 [4]. Goodrich made careful sketches of the beams, which he saw when the building was under construction, and his details have been confirmed by recent investigations made on the actual structure. Referring to Fig. 1(e) it will be seen that the connection is much neater than at Leeds, and in shaping the flange in an arris continuing the line of the arch soffit, Strutt has produced an elegant design.

The typical iron beam of the succeeding 25 years combined the rectangular flange and Strutt's type of connection; and attained spans of 17 ft. in the South Mill at Belper in 1811 [4] and 21 ft. in Wood's mill at Bradford in 1825 [14].

---

(10) Letters from George Lee to James Watt Jnr. and many drawings of the mill in the Boulton & Watt Collection. Two of these drawings are reproduced by GIEDION (1941), Figs. 64, 65.

(11) FAIRBAIRN (1854) p. 3. The first cast-iron flanged engine beam was also used at the Salford mill, DICKINSON & JENKINS (1927) p. 202. But according to BANKS (1803) p. 97, flanges were considered only as stiffeners to the beam.

(12) Evidence given before the *Factories Inquiry Commission* of 1834 (C. 1, p. 37).

(13) HAMILTON (1941), describes a later addition to this mill and, in discussion, Mr. A. E. Forward drew attention to the account given by Goodrich of his visit to Leeds.

(14) FAIRBAIRN (1854), pp. 6-8.

Bage's theory for the strength of flanged iron beams.

It may well be questioned whether these early beams were calculated or whether they were designed by eye to what seemed to be reasonable proportions .It is one of the most interesting features of the whole investigation that we know that the beams were calculated, and proved by full scale tests.

When Strutt was beginning to build the North Mill at Belper, Bage wrote to him from Shrewsbury, in a letter dated May 19th 1803:

« If on a balance of advantage and disadvantage you should at length prefer Iron beams, I should be glad to submit my reasoning on the strength and shape of them to your examination ».

And in a letter dated August 29th 1803 he says:

« My conscience has long upbraided me for having delayed to send you some theorems for ascertaining the strength of cast iron beams, but with the care of building two Mills [15]... I really find but little time ».

These « theorems » are enclosed in the letter and may be summarised as follows.

Starting with a rectangular beam, Bage assumes that at the instant of failure the « cohesion of the particles » is fully mobilised over the whole section. If *d* is the depth of the beam and *b* its breadth, then the « strength », or moment of resistance, will be proportional to:

(1)*a*. $$\frac{1}{2} b\, d^2$$

or:

(1)*b*. $$M = k \cdot \frac{1}{2} b\, d^2$$

where *k* is a constant equal to the tensile strenght of the marial. This is the classic result first obtained by Galileo in 1638, modified by Mariotte in 1686, but accepted throughout the 18th century as a practical rule for brittle materials, even by such able theoreticians as Coulomb (1773) and Girard (1798) [16]. It must be emphasised that we are here dealing with the failure condition. Coulomb himself was the first to treat correctly the beam in the elastic state, the moment of resistance then being $\sigma \cdot \frac{1}{6} bd^2$ where σ is the fibre stress. But before the work of Thomas Young in 1807 it was difficult to check

[15] These were at Leeds and a second mill at Shrewsbury, both for Benyons.
[16] For the work of Galileo, Mariotte, Coulomb and Girard, see TIMOSHENKO (1953), pp. 12, 21, 42, 49.

experimentally any results pertaining to the elastic state, and engineers therefore concerned themselves with the condition of failure. And, in fact, tests by Coulomb and others had shown that the Galilean theory was not unacceptable for masonry beams.

Bage, however, used this theory only as a basis for comparing the strength of an actual beam with that of a small test beam. He was faced with the problem of the inverted *T* beam, a form never before considered. His solution was simple yet, as we shall see, amply justified in practice.

The tensile force was assumed to be mobilised only in the flange, but the « cohesion of the particles » in the flange was taken as developing a moment about the top of the web. Thus, if the dimensions of the beam are as shown in Fig. 2, the moment of resistance is proportional to:

$$\text{(2)}a. \qquad \frac{c+b}{2}(d-g)g+\frac{2c+b}{6}\cdot g^2$$

or, if A is the area of the flange and y is the distance of its centre of gravity from the top of the beam, we may write:

$$\text{(2)}b. \qquad M = k\,.\,A\,.\,\bar{y}$$

The expressions (1) a and (2) a are as given by Bage, and he goes on to say:

« From these comparative strengths, and from the knowledge that a Cast Iron Bar 1 inch square and 3 feet long will sustain 8 cwt. laid on the middle, the weight to break a beam of any other size or length may be calculated. N. B. Though I have broken a bar of the above dimensions which sustained above 8 cwt. I always calculate as if it sustained only 6 cwt. ».

He was writing to Strutt, an able engineer to whom the practical conclusions of this theory would be obvious.

For, if we test a simply-supported rectangular beam of span l_1, and dimensions $b_1 \times d_1$ and if the central breaking load in this test is W_1, then the failure load in any other simply supported rectangular beam of span l and dimensions $b \times d$ will be:

$$\text{(3)} \qquad W = W_1\cdot\frac{l_1}{l}\cdot\frac{b\,d^2}{b_1\,d_1^2}$$

and similarly the breaking load of a simply-supported flanged beam will be:

$$\text{(4)} \qquad W = W_1\cdot\frac{l_1}{l}\cdot 2\cdot\frac{A\,.\,\bar{y}}{b_1\,d_1^2}$$

But for l_1 = 36ins. $b_1 = d_1$ = 1in. Bage states that $W_1 = 0.4$ tons (8cwt.). Hence:

$$\text{(5)} \qquad W = 14.5\,.\,b\,d^2/l \qquad \text{(rectangular beam)}$$

$$\text{(6)} \qquad W = 29\,.\,A\,\bar{y}/l \qquad \text{(flanged beam)}$$

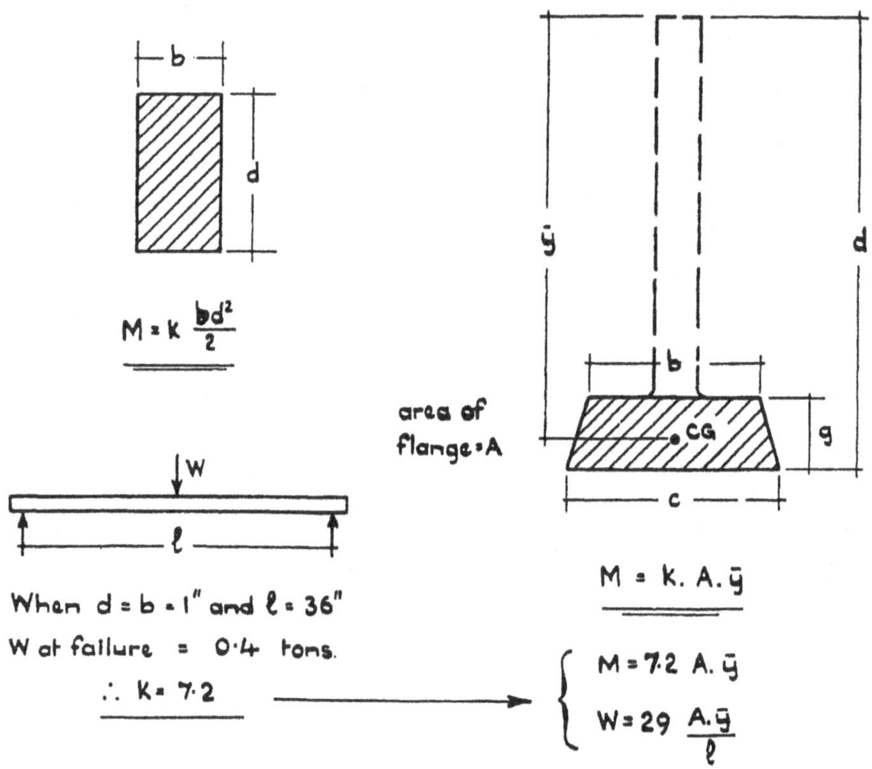

BAGE'S THEORY FOR THE STRENGTH OF CAST-IRON FLANGED BEAMS (1803)

Further, it is clear that the uniformly distributed load causing failure would be double, and that in a cantilever the failing load would be one quarter, the values given by the above equations; and that, in equns. (1) *b* and (2) *b*, $k = 7.2$ ton/sq.in. (with which the tensile strength of 8.5 ton/sq.in. obtained by Rennie (1818) may be compared).

There is nothing of any great originality in equn. (5); it merely compares the strength of two similar beams. Rules given by Banks in 1803 for the design of cast-iron beams for steam engines can be reduced precisely to this form, the numerical coefficient lying between 12 and 16 depending on the quality of the iron ([17]). Bage's real contribution lies in equn. (6) which is the first attempt to express the strength of a flanged beam. We cannot consider here the developments in theory which occurred after 1803, but we may note that equn. (6) is almost identical with the well-known formula $W = 26. Ad/l$ derived by Hodgkinson in 1830 from his tests on cast-iron beams.

As would be expected, Bage did not trust to theory alone. From Telford ([18]) we know that Bage tested a beam with a span of 9 ft., an overall depth of $11^1/_2$ ins., and a flange 3 ins. wide and $1^1/_2$ ins. thick. Since these dimensions coincide with those of the beams in the mill at Leeds, built in 1802-03 ,we can conclude that he made the test for this mill. The beam broke under a central load of 14 tons, and Bage's formula gives 13.0 tons. Bage then tested an identical beam with a uniformly distributed load and found that failure occurred at 26 tons, which is in exact agreement with his theory.

To be on the safe side, however, Bage says that he takes W_1 as 6 cwt., instead of 8 cwt. as in his test. This allows for the lowest result obtained by Reynolds, at Ketley, in five tests on inch bars in 1795 ([19]). Thus Bage's calculated failing load of the Leeds beams, for design purposes, would be 20 tons.

Now these beams had a span of 9 ft. and were placed 10 ft. apart. The weight of brick arch floor supported by each beam would be about 4 tons, and a live load of 100 lb./sq. ft. brings the total distributed load to 8 tons ([20]). It therefore appears that Bage was designing with a factor of safety of $2^1/_2$.

([17]) Banks (1803), pp. 94-103. Banks deals only with engine beams and ignores the contribution of the flanges to bending resistance.

([18]) Telford's Pocket Memorandum Book, published in the *Life of Thomas Telford*, p. 682, contains, in an undated entry details of the tests by « Mr. Bage of Shrewsbury ». The flange width is given as 3 ft.; an obvious printer's error for 3 ins.

([19]) Banks (1803), pp. 89-90. Reynold's average result of 6 1/2 cwt. is decidely low. Banks himself reports values of W_1 from 7 3/4 to 8 3/4 cwt. and Rennie (1818) obtained 8 cwt.

([20]) This figure is confirmed by data relating to the Shrewsbury mill, given in a letter of 1796 written by Bage to Strutt. The dimensions at Shrewsbury and Leeds were very similar.

It is highly probable that Bage also tested the 1796 beams, before daring to use them in such a large building as the Shrewsbury mill. But we know of no tests other than for the Leeds beams until those of Fairbairn [14] and Hodgkinson. Their results on inverted *T* section beams are tabulated below and show good agreement with Bage's formula.

TEST ON CAST-IRON FLANGED BEAMS

	Span	Flange	Overall Depth	Central Breaking Load	
				Calculated	Test
Bage, 1802	9'	3" × 1½"	11½"	13.0 tons	14. tons
Fairbairn, 1824 (I)	14'	5" × 1"	15"	12 5	12.5
Fairbairn, 1824 (II)	16'	6" × 1⅜"	15"	17.8	23
Fairbairn, 1825	20' 9"	6" 1½"	18"	18.1	19
Hodgkinson, 1830	4' 6"	2¼" ½"	5"	3.2	3.9

Thus we see that by 1803 cast-iron beams had reached a satisfactory form, and could be designed with sufficient accuracy. Their use then became widespread.

Acknowledgments.

The field investigations were carried out jointly with Mr. H. R. Johnson of Sheffield University. I owe much to him, to Dr. S. B. Hamilton and Professor Turpin Bannister.

REFERENCES

MANUSCRIPT SOURCES

Letters from Charles Bage to William Strutt. *Shrewsbury Public Library* MSS. 2657.

Letters from Marshall, Benyons and Bage to Boulton & Watt and Drawings. *Birmingham Public Library*, Boulton & Watt Collection.

Account Books and other MSS relating to John Marshall. *Leeds University Library*.

Journals and Drawings of Simon Goodrich. *Science Museum*, London.

PRINTED SOURCES

J. Banks, *On the Power of Machines*. Kendal 1803.

T. Bannister, « The first iron-framed buildings » *Architectural Rev.* vol. 107, pp. 231-46, 1950.

J. Britton & E. W. Brayley, *The Beatuies of England and Wales*. vol. 3, London 1802.

H. W. Dickinson & Rhys Jenkins, *James Watt and the Steam Engine*. Oxford 1927

Factories inquiry Commission, Supplementary Report, Part 2, vol. 20, 1834.

W. Fairbairn, *On the Application of Cast and Wrought Iron to Building Purposes*. London 1854.

F. R. Gameson, « The story of Charles Bage », *Shropshire Mag*. January 1954.

S. Giedion, *Space, Time and Architecture*. Harvard 1941.

S. B. Hamilton, « The use of cast iron in buildings », *Trans. Newcomen Soc.* vol. 21, pp. 139-55, 1941.

E. Hodgkinson, « Theoretical and experimental researches to ascertain the strength and best form of iron beams », *Mem. Manchester Lit. & Phil. Soc.* Series 2, vol. 5, pp. 407-544, 1831.

H. R. Johnson & A. W. Skempton, « William Strutt's fire-proof and iron-framed buildings 1793-1812 ». *Trans. Newcomen Soc.* vol. 30, 1956.

T. Minshull, *The Shrewsbury Visitor's Pocket Companion or Salopian Guide*. Shrewsbury 1803.

N. Pevsner, *Pioneers of Modern Design*. New York 1949.

A. Raistrick, *Dynasty of Ironfounders*. London, 1953.

G. Rennie, « Account of experiments made on the strength of materials », *Phil Trans. Roy. Soc.* Part 1, pp. 118-136, 1818.

T. Telford, *Life of Thomas Telford*. London, 1838.

S. P. Timoshenko, *History of Strength of Materials*. New York 1953.

11
Thomas Tredgold (1788–1829): some aspects of his work (part 2: carpentry)

L.G. Booth

1. INTRODUCTION

When Thomas Tredgold was born in 1788, the Iron Bridge at Coalbrookdale was nine years old and the first hesitant steps in the new era of cast iron as a structural material had been taken. The place of timber as the most versatile structural material was threatened. When Tredgold began his apprenticeship as a carpenter (1802), Bage had completed the first multi-storey iron frame building, and by the time the apprenticeship was complete Telford and Jessop had built Pont Cysyllte Aqueduct. Whilst Tredgold was working in Scotland as a carpenter (1808-13), Wilson, Rennie and Telford established cast iron as the major bridge material. During the first five years of his work in London (1813-18) Tredgold doubtless watched the new bridges being built over the Thames at Vauxhall and Southwark, and heard of Telford's design for the suspension bridge over the Menai Straits. By 1820 cast iron was firmly established and timber was forced to reassess its role as a structural material. Undaunted, Tredgold drew on his eighteen years as a carpenter, his extensive reading of previous publications on both carpentry and the strength of materials, and wrote the first book on carpentry to be based on engineering principles. The book, *Elementary Principles of Carpentry*,[1] became the standard reference work on the subject for the next century; two editions were published during his lifetime and subsequently some twenty five editions from various publishers appeared, culminating in a final version in 1946, one hundred and seventeen years after his death![2]

2. ELEMENTARY PRINCIPLES OF CARPENTRY: 1820

2.1 The Art of Carpentry. Having served an apprenticeship as a carpenter and joiner, and having subsequently worked as a journeyman carpenter, it was natural that Tredgold's first major publication should deal with this subject. The title page of the first edition, which was published in London by John Taylor in 1820, clearly indicates that he intended to cover all aspects of the mechanical principles of the art of carpentry:

> Elementary principles of carpentry; a treatise on the pressure and equilibrium of beams and timber frames; the resistance of timber; and the construction of floors, roofs, centres, bridges etc. With practical rules and examples. To which is added, an essay on the nature and properties of timber, including the methods of seasoning, and the causes and prevention of decay, with descriptions of the kinds of wood used in building. Also numerous tables of the scantlings of timber for different purposes, the specific gravities of materials, etc.[9]

The title is virtually a contents list and it will obviously be impossible to discuss within this short paper all the topics covered by the book. The approach will be to present in general terms the philosophy of the book and to restrict the technical discussion to a few topics.

Tredgold begins by pointing out that *The Elementary Principles of Carpentry* would be a title which implied the inclusion of 'all that is essential to the art [and] therefore embraces a wider range than I have attempted to fill'[10] and so 'to avoid promising .... more than is performed,'[11] he omitted the definite article and made the title *Elementary Principles of Carpentry*. Having issued this disclaimer he proceeded to write a two hundred and fifteen page book which, in size and scope, surpassed all its predecessors. The only omission, and that a deliberate one, was any mention of descriptive geometry, which he considered 'had been so ably handled .... by Mr. P. Nicholson.'[12] Tredgold's book was the first British one devoted entirely, by title and contents, to carpentry, as distinct from Carpentry and Joinery.

Many treatises on carpentry had been written in the previous century but Tredgold criticises them on the grounds that 'in none of them is to be found any thing on the mechanical principles of the art, except it be a few rules for calculating the strength of timber; and these are founded on erroneous views of the subject; and therefore not to be relied upon.'[13]

THOMAS TREDGOLD (1788–1829): SOME ASPECTS OF HIS WORK

In the Preface, Tredgold clearly defines his philosophy of design. The following three quotations, written in 1820, would not appear out of place in any current engineering journal; they are timeless.[14,15,16]

> In the present state of society it is most desirable that stability and economy should characterize the different species of construction.[14]
>
> A just economy of materials should be one of the first objects of the builder's attention, and this desirable object is to be obtained only by judicious combinations of the materials to be used.[15]
>
> Science remains a dead letter unless it be applied 'to the business and affairs of men;' it informs him how .... to obtain maximum strength with a minimum of materials.[16]

2.2 Tredgold's Sources. There are two characteristics of the book that do not occur, or are not readily apparent, in previous books on carpentry: a wide knowledge of existing literature, and full acknowledgement of sources coupled with praise where it was merited.

Prior to Tredgold's book there had been many articles in encyclopaedias and books on carpentry. Although several made valuable contributions to the art they all had their deficiencies; frequently they emphasized architecture and joinery at the expense of carpentry; none concentrated on carpentry as did Tredgold's. Nevertheless he owed, and acknowledged, a debt to five major authors: Price, Smith, Nicholson, Robison and Young. Price[3] and Smith[4] had taken the first steps to show that Carpentry was an art in its own right and not merely a facet of architecture. Nicholson[5] was fascinated by the geometrical problems of construction ('finding the lines') and his books also contained much joinery; he did however introduce some elementary strength of materials. Robison contributed articles to the *Encyclopaedia Britannica* on the Strength of Materials, Carpentry, Roofs, and Centres for Bridges;[6] the articles gave a good indication of the state of the art in 1800 but it is doubtful if they had much influence on the average practitioner. Young's Royal Institution lectures,[7] and his later amendments to Robison's article on Carpentry,[8] probably had even less effect. It is generally acknowledged that Young's work on the strength of materials is difficult to understand and could only have been read with profit by able mathematicians; although his comments on carpentry are less obscure, it seems unlikely that there would have been many practising carpenters at the Royal Institution lectures.

When Tredgold discusses the strength of timber he quotes the well-known test data obtained by Emerson, Anderson[18] and Barlow;[19] the theoretical behaviour in bending and compression is obtained from Robison[20] and Young[21]. These are merely some of the British authors quoted, but they give a good indication of the breadth of his reading.

When Tredgold quotes French sources it is difficult to know if he was making use of his own reading of primary source material or whether he relied on summaries of the work given in articles in the various *Encyclopaedias (Edinburgh, Britannica* and *Rees)* and in Barlow's *Essay on the Strength and Stress of Timber.*[22] Tredgold is reputed to have taught himself French and he certainly owned a copy of the *Encyclopēdie Mēthodique.*[23] Whatever the source he was able to make reference to French work on the strength testing of timber (Parent,[24] Belidor,[25] Buffon,[26] Duhamel,[27] Lamandé,[28] Girard,[29] Gauthey[30] and Rondelet[31]), and its use in floors and roofs (Jousse,[32] Krafft[33] and Rondelet[34]). It was from these diverse sources that Tredgold drew his information to produce an engineering approach to the Art of Carpentry.

2.3 The Stiffness and Strength of Beams. If we now turn to Tredgold's treatment of theoretical topics we find that his arguments are difficult to follow even though they generally lead to the correct results. An example of this is his derivation of the deflection of a simply supported beam:[35]

> The extension of any part of the beam is directly as the force which produces it; and it is known by experiment, that the deflexion is as the weight; therefore, the deflexion is as the extension. Now the effect of a force to produce extension is as the leverage it acts with, and as the force itself; but the leverage is proportional to the length of the beam; therefore, the extension is as the length and weight directly; and inversely as the breadth, and square of the depth, as is shown by writers on the strength of materials. Also, the extension will be directly as the number of parts extended, that is, directly as the length; and as the quantity of angular motion, which will be as the length directly, and depth inversely. Uniting these proportions, we find the extension will be as the weight, and cube of the length, directly, and as the breadth, and cube of the depth, inversely;

CARPENTRY

consequently, the deflexion will be in the same proportion, that is, making L = the length of bearing in feet, W = the weight in pounds, B = the breadth in inches, and D = the depth in inches,

$$\frac{L^3 \times W}{B \times D^3} \text{ is as the deflexion}$$ [36]

Tredgold proceeds to use this (correct) answer to determine the required cross-sections of beams. He introduces (for the first time?) the concept that the permissible deflection should be proportional to the length and hence gives the expression $BD^3/aL^2 = W$, where a is a constant determined from experiment. (If the deflection $\delta = kL$, and B,D,L and W are defined as above, we would nowadays have $a = 36/kE$, where E is the modulus of elasticity.) Using load-deflection data provided by various experimenters (including himself) Tredgold tabulates the values of the constant a for oak and fir (from a number of different sources) together with values for another twenty species. We saw above that a depends on the chosen value of k, which Tredgold takes to be 1/480 (i.e. one-fortieth of an inch per foot). Although the formula for the deflection is given, Tredgold works all his examples through a set of rules. For example:[37]

> RULE. [When the breadth is given.] Multiply the square of the length in feet by the weight in pounds, and this product by the value of *a* [for] the kind of wood. Divide the product by the breadth in inches and the cube root of the quotient will be the depth required in inches.

Tredgold also shows how the constant a must be modified for fixed-ended and cantilever beams of both rectangular and circular cross-section under central point and uniformly diffused (i.e. distributed) loads.

To determine the strength of beams Tredgold quotes the correct formula and refers the reader to Emerson,[38] Gregory [39] or Young[40] for its derivation. For a simply supported beam of rectangular cross-section he determines the maximum load (W) from the expression $BDc = LW$, where c is a constant to be found by experiment: values of c for various species are then tabulated. He considers the formula will give 'the greatest weight a beam will bear without fracture .... sufficiently near for practical purposes.'[41] The expression appears to have been meant for checking purposes only since he does not discuss the application of a factor of safety.

2.4 The Design of Floors. The main use of timber in Tredgold's time was for floors and consequently he devotes a complete section to this topic.[42]

The sizes of single joists are found from the following rule:[43]

> Divide the square of the length in feet by the breadth in inches; and the cube root of the quotient multiplied by 2.2 for fir, or 2.3 for oak, will give the depth in inches.

The form of the rule shows that the floor was designed to limit the deflection (to a proportion of the span) and not the bending stress. The constants (2.2 and 2.3) depend on the deflection limit, the modulus of elasticity and the applied loads. Tredgold, however, makes no attempt to calculate them theoretically but states:

> The constant numbers .... in all the rules for flooring and roofing, are derived from the scantlings of timbers that were found to be sufficiently strong; this I considered to be the best method of obtaining those numbers, because it is difficult to calculate the weight that a floor has to support; yet it is easy to ascertain whether a floor be sufficiently stiff or not after it is executed. These comparisons have not been made from single observations, but from various ones on bearings of very different lengths.

The required depths of yellow fir joists at 12 inch centres are then tabulated for spans varying from 5 to 25 feet and breadths from 1.5 to 5 inches.[45]

Similar tables are given for girders (at 10 feet centres), binding joists (at 6 feet centres) and ceiling joists (at 12 inch centres).[46] Although various authors had previously given similar tables it is thought that Tredgold's were the first to be based on the combination of practical experience and theory. Certainly no previous author had given a formula for the derivation of his tabulated sizes.

Tredgold's practical comments on floors are also based on the combination of practice and theory. He emphasises the advantage of using bridging joists in considerable lengths so that they extend over several binding joists and have increased strength and stiffness; to substantiate this increase he quotes theoretical work on fixed-ended beams by Belidor[47] and Robison.[48] He points out that to make a 'strong floor with a small quantity of timber, the joist should be thin and deep; but a certain degree of thickness is necessary, for the purpose of nailing the boards, and two inches is perhaps as thin as the joists ought to be made'.[49]

The rule that the depth of a floor joist may be found in inches by dividing the span in feet by two and adding two, is often associated with Tredgold. The origin of this rule has not been found, but it does not appear to have been devised by Tredgold. He does however suggest that the size of ceiling joists may be found by making the breadth 2 inches and the depth half the span in feet,

2.5 The Design of Columns. Tredgold devotes several pages to the design of columns with rectangular, square and circular cross-sections.[50] He begins by noting that it 'is certain, from experience, that there is a certain force [W] that will bend a piece of timber when acting in direction of its length.'[51] Using a similar argument to that previously used to find the deflection of beams he shows (correctly), for rectangular columns, that $W = BD^3/eL^2$, where *e* is a constant to be determined from tests on different species. (We would now recognise this expression as the Euler buckling load, where $e = 12/\pi^2E$.) Tredgold calculated the values of *e* from tests on oak columns reported by Lamandé and Girard and finds that *e* varies from 0.00015 to 0.00129. He points out that a column should not be loaded until the deflection becomes 'sensible' and therefore the calculated values of *e* are too low for use. He considers that *e*

> should be taken of such a value as may give all the security that is necessary, without employing more timber than is consistent with just economy .... If the number [*e*] be taken 0.0015 the result will always be sufficiently strong for use, as in good timber it would seldom be loaded with above one-fourth of the weight that would produce sensible flexure, if its strength were determined by the rule.

Tredgold attributes the wide variation in the values of *e* to the likely lack of straightness of the timber and to the difficulty in applying the load in the direction of the axis of the column. An examination of Girard's test apparatus and results would confirm his speculations.[53]

Girard's tests were restricted to oak columns and Tredgold needs to give values of *e* for other species. He does this by quoting Young who had shown that the load 'which would hold a beam [i.e. column] in equilibrium, under a small degree of flexure'[54] is proportional to the modulus of elasticity of the timber.' From his knowledge of the modulus of elasticity Tredgold is then able to tabulate the values of *e* for a range of species. It is interesting to note that although Tredgold makes a brief reference to Euler in the Preface, no mention of him is made in connection with the behaviour of columns.

For a short column Tredgold states that 'it cannot be bent by any pressure less than that which would crush it.'[55] To find the load that can be applied 'without risk of [the timber] being crushed'[56] he finds the product of the area and the 'weight that has been found capable of crushing a square inch of the same kind of wood. Then one-fourth of the product will give the load .... that the piece would bear with safety.'[57] (Tredgold used the same factor of safety of 4 for the determination of a working stress in tension[58] and shear.[59])

The section on column strength is completed by comparing the rule for the resistance to flexure with that for crushing. For oak columns he shows that if the length is greater than ten times its thickness, its strength must be determined by use of the constant *e* (i.e. from the reduced Euler load); when it is less than ten times the thickness, the reduced crushing strength must be used. From this description of column design it will be seen that what subsequently became known as Tredgold's column formula was not developed for timber.

Tredgold does not apply his two rules to develop design tables for timber posts (columns) to the same extent that he did for floor joists. Although he gives four tables for the sizes of posts (for 2, 3, 4 and 5 storey buildings) he considers they are of limited value because the required size depends on the nature of the building.[60] (The tables do not specify the species of timber, but it may be assumed that the posts are either oak or fir; although oak is stronger than fir, Tredgold considers it is liable to warp in small sections, and is seldom so straight grained or so free from knots as fir.) He therefore recommends that the load on each post should be estimated and the size calculated.[61]

2.6 Remaining Topics. Only a brief mention will be possible of the remaining sections of the book.

CARPENTRY

After floors the most frequent use of timber was, and still is, in small span roofs. Tredgold gives rules for the determination of the cross-sections of members in King and Queen post trusses. These rules are limited in that they do not take into account the different spacing of the trusses, different weights of roofing or different slopes. Tabulated values of the sizes required are given for several configurations assuming that the timber is yellow fir, and that the trusses are 10 feet apart and covered with slates. As with floors, the constants used in the rules are determined from a combination of practical experience and theory.[62]

The section on Wooden Bridges is particularly interesting and is more detailed than any previous British book on carpentry; frequent reference is made to European bridges.[63]

The crux of carpentry is forming the joints, particularly in trusses and built-up beams, and Tredgold rightly devotes a complete section to Joints and Straps. When it is remembered that the theory for shear was even less well understood than bending and compression, it is not surprising that the section is devoid of theory and concentrates on giving practical advice on the forms of joints that had proved successful in the past.[64]

The last section occupies one-quarter of the book and is devoted to the physical properties of various species.[65] This last discussion is probably more detailed than is required in a book on carpentry but it is clear that Tredgold, like all who have worked with timber, was enthralled by it. The comparable chapter in a modern book on Timber Engineering would contain rules specifying the maximum size of defects (knots etc) allowed by various grades. How pleasant it would be if the modern Timber Engineer could write, as Tredgold did, 'timbers that are exposed to considerable strains ought to be of a good quality .... Good timber is that which is perfectly sound, straight grained, free from large knots or other defects, particularly near the strained points, and seasoned.'[66] Halcyon days!

3. LATER EDITIONS OF ELEMENTARY PRINCIPLES OF CARPENTRY

Two editions of Tredgold's book were published in his life time. The 1828 second edition is virtually the same as the first edition; printing errors are corrected, references are brought up to date and several minor articles are added.

This is not the place for a full examination of the later editions of his *Carpentry*.[67] The dates of the various editions are given in a separate paper, but we may briefly note that Peter Barlow edited three editions (between 1840 and 1870) before handing over to E.W. Tarn for the sixth and seventh editions (1885 and 1886). In parallel with those last two editions, Crosby Lockwood also published a version in Weale's rudimentary series; there is some confusion between editions and impressions but it is impressive to note that a 15th impression was published in 1946. The book also appeared in two American editions in 1837 and 1847.

The selling power of Tredgold's name is indicated by the fact that Spon also thought it worth while to publish their own version, edited by J.T. Hurst, in 1871; this line ran until 1919 with twelve editions.

The multiplicity of editions is firm evidence of Tredgold's contribution to the literature on Carpentry. Sadly, one suspects that they were a steady source of income, not to Tredgold's poverty stricken family, but to several publishers.

4. TREDGOLD'S BUILDINGS

Having considered Tredgold's book it is natural to proceed to a description of his major timber structures. Astonishingly, we know of no timber structure that can definitely be attributed to him. Neither of the two editions of *Elementary Principles of Carpentry* that were published in his life time (1820 and 1828) makes any reference to his own work; none of the subsequent editions remedies this omission (assuming that it was a remediable omission). Doubtless he was responsible for the carpentry in the houses designed by Atkinson's practice and presumably he undertook work in connection with Atkinson's appointment as Architect to the Board of Ordnance, but no details have yet been found. The only structures that we can attribute to Tredgold are the floor and roof of the Riding School of the Horse Bazaar, King Street, Portman Square. Tredgold covered the 46 ft span with trussed girders. Sadly, from the point of view of this paper on Tredgold's carpentry achievements, it must be recorded that the trusses were of cast iron and that timber was only used for the secondary joists and purlins.[68]

5. EPILOGUE

The final words must be with Tredgold himself; in writing these words little did he realise that his book would become a standard text for the next hundred years.[69]

> As the mechanical principles of Carpentry have never been published in a separate form, I have attempted .... to supply that defect; but, sensible of the extensive acquirements that are necessary to do justice to such a work, I should not have undertaken the task, had I had a hope of seeing it executed by an abler hand .... But though I feel much diffidence in submitting the result of my labours to the public, I also am encouraged by the hope that they will be useful; and that if these labours have not been directed by the profound views of superior science, nor elucidated by the elegant and perspicuous language of a scholar, at least they have arisen out of a sincere love of knowledge, and have been guided by a zealous desire to extend and improve the art of building.

NOTES FOR PART 2

1. Tredgold, Thomas. *Elementary principles of carpentry.* J. Taylor, London. 1820.
2. Booth, L.G. *Thomas Tredgold (1788-1829) - Some aspects of his work. Part 5. Tredgold's publications.* pp. 86-92, this volume.
3. Price, Francis. *A treatise on carpentry.* London. 1733.
4. Smith, James. *The carpenter's companion.* London. 1733.
5. Nicholson, Peter. *The carpenter's new guide.* London. 1793. Nicholson, Peter. *The carpenter's and joiner's assistant.* London. 1797.
6. Robison, J. *Enc. Brit.* 3rd ed. Edinburgh 1797. *Enc. Brit.* Supplement 3rd ed. Edinburgh. 1801.
7. Young, Thomas. *A course of lectures on natural philosophy and the mechanical arts.* London. 1807.
8. *Enc. Brit.* Supplement 4th, 5th, 6th ed. Edinburgh. 1824.
9. Tredgold. *op. cit.* (1).
10. Tredgold. *op. cit.* (1), x.
11. *Ibid.*
12. Tredgold. *op. cit.* (1) vii.
13. *Ibid.*
14. Tredgold. *op. cit.* (1), viii.
15. *Ibid.*
16. Tredgold. *op. cit.* (1), ix.
17. Emerson, W. *The principles of mechanics.* 4th ed. London. 1794.
18. Anderson. *Edinburgh Enc.* Article on carpentry.
19. Barlow, Peter. *An essay on the strength and stress of timber.* London. 1817.
20. Robison. *op. cit.* (6).
21. Young. *op. cit.* (7).
22. Barlow. *op. cit.* (19).
23. Inst. Civ. Eng. Min. Council. 2 June 1829 records that the Institution bought Tredgold's copy after his death.
24. Parent, M. *Les résistances des poûtres .... (The strength of beams ....).* Mém. de l'Acad. Roy. des Sci. 17-31. Paris. 1708.
25. Belidor, B.F. de. *La science des ingénieurs dans la conduite des travaux de fortification et d'architecture civile (The science of engineers in the construction of military works and civil architecture).* Paris. 1729.
26. Buffon, Comte de. *Expériences sur la force du bois (Tests on the strength of wood).* Mém. de l'Acad. Roy. des Sci., 292-334. Paris. 1741.
27. Duhamel, H.L. *Du transport, de la conservation, et de la force des bois (On the transport, conservation and strength of wood).* Paris. 1767.
    Duhamel, H.L. *Expériences pour connaître la force des bois (Tests to find the strength of wood).* Mém. de l'Acad. Roy. des Sci. 534-537. 1768.
28. Lamandé's test data is quoted by Gauthey. See Gauthey *op. cit. (30).*
29. Girard, P.S. *Traité de la résistance des solides et des solides d'égale résistance (Treatise on the strength of solids and solids of uniform strength).* Paris. 1798.
30. Gauthey, E.M. *Traité de la construction des ponts (Treatise on the construction of bridges).* Paris. 1809-13.
31. Rondelet, J. *Traité théorique et pratique de l'art de bâtir (Theoretical and practical treatise on the art of building).* Paris. 1805-10.
32. Jousse, M. *L'art de charpenterie (The art of carpentry).* Paris. 1624.
33. Krafft, J.C. *Plans, coupes et élévations de diverses productions de l'art de charpente (Plans, sections and elevations of various works of the art of carpentry).* Paris. 1805.

CARPENTRY

34. Rondelet. *op. cit.* (31).
35. Tredgold. *op. cit.* (1), 30-42.
36. Tredgold. *op. cit.* (1), 31.
37. Tredgold. *op. cit.* (1), 37.
38. Emerson. *op. cit.* (17).
39. Gregory. O. *Treatise on mechanics, theoretical, practical, and descriptive.* 3rd. ed., London, 1815.
40. Young. *op. cit.* (7).
41. Tredgold. *op. cit.* (1), 42.
42. Tredgold. *op. cit.* (1), 61-71.
43. Tredgold. *op. cit.* (1), 62.
44. *Ibid.*
45. Tredgold. *op.cit.* (1), Table III.
46. Tredgold. *op. cit.* (1), Tables I, II and IV.
47. Belidor. *op. cit.* (25).
48. Robison. *op. cit.* (6).
49. Tredgold. *op. cit.* (1), 62.
50. Tredgold. *op. cit.* (1), 47-59.
51. Tredgold. *op. cit.* (1), 48.
52. Tredgold. *op. cit.* (1), 53.
53. Girard. *op. cit.* (29). See also Booth, L.G. 'The strength testing of timber during the 17th and 18th centuries.' *Jnl. Inst. Wood. Sci.* no 13. Nov. 1964, 24-8.
54. Tredgold. *op. cit.* (1), 53.
55. Tredgold. *op. cit.* (1), 58.
56. Tredgold. *op. cit.* (1), 59.
57. *Ibid.*
58. Tredgold. *op. cit.* (1), 28.
59. Tredgold. *op. cit.* (1), 47.
60. Tredgold. *op. cit.* (1), Tables XVI-XIX.
61. Tredgold. *op. cit.* (1), 58.
62. Tredgold. *op. cit.* (1), 72-87.
63. Tredgold. *op. cit.* (1), 106-35.
64. Tredgold. *op cit.* (1), 136-48.
65. Tredgold. *op. cit.* (1), 149-218.
66. Tredgold. *op. cit.* (1), 26.
67. Booth. *op. cit.* (2).
68. [Nicholson, P.] *Practical carpentry, joinery and cabinet-making.* London. 1826 , 24.
69. Tredgold. *op. cit.* (1), viii.

12
Thomas Tredgold (1788–1829): some aspects of his work (part 3: cast iron)

R.J.M. Sutherland

1. INTRODUCTION

This paper attempts to analyse Thomas Tredgold's writings on cast iron mainly from the angle of a practising user of the 1820s. The bulk of his knowledge of cast iron, whether published elsewhere or not, is embodied in his book *Practical Essay on Cast Iron.*[1] This book provides the framework of the present paper.

Having stated in his title that the essay is 'practical' Tredgold says that it is 'for the assistance of Engineers, iron masters, architects, millwrights, founders, smiths and others engaged in the construction of machines, buildings, etc.' The question is whether, for cast iron, he was writing the equivalent of Charles Reynolds *Reinforced concrete designers' handbook* or the prototype of the modern steel section book or a code of practice or a series of state of the art reports. The answer must be a little of each.

Tredgold's book is an uneven assemblage of load tables, test results, mathematical expressions and rules for the design of beams, columns, tie rods, machinery shafts, etc., with guidance on impulsive loading and an extensive table of data on the properties of materials. Part of this book is intensely practical, part is theoretical and some of it seems just speculative. It doubled in size between the 1st edition of 1822 and the 2nd edition of 1824, with almost every section expanded and its range extended to include some, although limited, information on wrought ('malleable') iron, brass and gunmetal.

Before discussing Tredgold's advice, it is worth looking briefly at two broader topics:

(a) The state of the structural cast iron industry at the time when he was writing.

(b) The broad framework of Tredgold's thinking on structures.

THOMAS TREDGOLD (1788–1829): SOME ASPECTS OF HIS WORK

2. STATE OF THE STRUCTURAL CAST IRON INDUSTRY IN THE 1820s.

When Tredgold's Practical Essay was first published in 1822 cast iron was a new structural material as can be seen from Fig. 1. Even ignoring random examples from its pre-history the use of

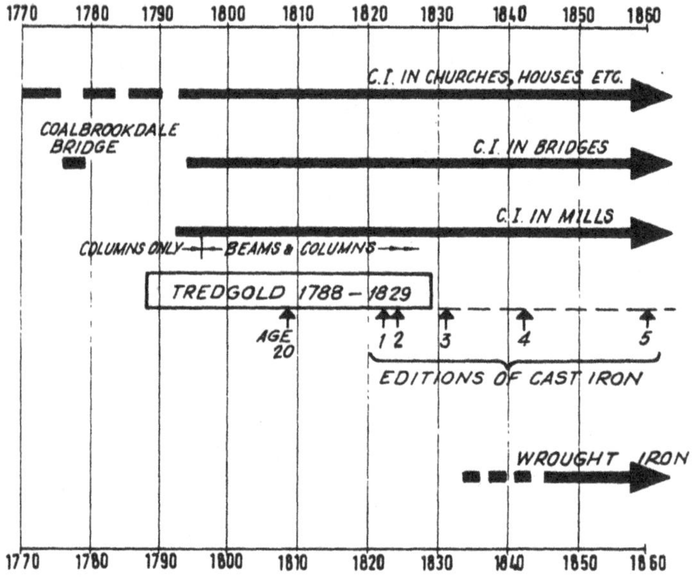

Fig. 1. State of the structural cast iron industry in Tredgold's day.

cast iron for columns in churches started fifty years earlier and for the full inner structures of mills twenty years or more before the publication.[2] Further by the 1820s scores of cast iron bridges were in use. Cast iron was very much the advanced and exciting material for structural work at this time although it was being used widely with very mixed understanding.

3. TREDGOLD'S STRUCTURAL THINKING.

It is easy to misinterpret the criteria on which the engineers of 150 years ago based their decisions. The statements in the following paragraphs should be accepted at least in part as conjectural and may need to be challenged.

Tredgold did not envisage the complete cast iron structure or, if he did, he did not see this as concerning him. He saw cast iron as being very good for some components in buildings, bridges or machines and all his thinking was about components and their individual performance. Beams, columns, shafts for engines, tie rods and wheels were the items in his mental catalogue. Overall stability of structures did not come into his thinking–on cast iron anyway–and even techniques of jointing cast iron passed virtually unmentioned, although they formed a large part of his writing on timber. Similarly structural continuity, accepted as advantageous with timber is virtually unmentioned.

Tredgold's design process was direct. A beam had to carry a certain load over a certain span. His thinking went straight from load and span to section size without mentioning bending and shear. The same applied to columns and machinery shafts. Safety was scarcely considered, anyway not in our present sense of having a factor of safety related to rupture. For this reason there is comparatively little about ultimate strength in Tredgold's writings and many of the tests quoted were not taken up to rupture.

CAST IRON

There is one exception to Tredgold's rejection of the ultimate condition in his book and that is in the Introduction where he briefly mentions factors of safety. After stating that 'something more than the mere resistance to fracture should be calculated' he goes on to say 'where the parts are short and bulky it may do very well to employ the rules for resistance to fracture and make the parts strong enough to sustain four times the load'. Then he virtually dismisses factors of safety[3] with a warning that 'where long pieces are loaded to one fourth of their strength we may expect much flexure [i.e. deflection] vibration and instability.' Having dismissed factors of safety Tredgold discusses the dangers of exceeding the elastic limit for any material. He says 'I find, that while the elastic force, or power of restoration, remains perfect, the extension is always directly proportional to the extending force, and that the deflexion does not increase after the load has been on for a second or two; but when the strain [stress] exceeds the elastic force, the extension or deflexion becomes irregular, and increases with time.' In a second edition footnote he adds 'A permanent alteration is a partial failure and hence the proper limit of strength.' These extracts imply that Tredgold was moving from the 'rules of fracture' and safety factors to a new and better basis for design. What basis?

The short answer in modern terms is 'servicability' or more precisely the limit of the material's elastic behaviour for all types of member, with the addition of a deflection limit for beams. In numerical terms Tredgold's structural design criteria for cast iron were:

a. **Elastic Limit,** based on a stress (f) = 15,300 lb/in^2 (6.83 ton/in^2) before a permanent set takes place. He established this figure for bending from beam tests, but he also used it as the limit for direct compression and tension. The logic of this is discussed below.
b. **Deflection.** A limit of 1/40 of an inch per foot of span for a simply supported beam when stressed at up to the elastic limit. (i.e. permissible deflection not more than 1/480 span). This is a repeat of his limit established earlier for timber.

One point which needs emphasis is that to Tredgold the stress at the elastic limit was a working or safe stress, or more precisely the upper limit of the working range. Tredgold seemed happy with this because his analysis of tests on small beams convinced him that for bending it provided a factor of safety of between 2.6 and 3.7. He then assumed, or so it seems because he says very little directly about it, that using the same stress in tension for cast iron, he would have the same factor of safety as in bending. He assumed further that using this stress again in compression the factor of safety would be at least as high. Thus all one needed to do was to design to the elastic limit and all would be well.

The error in deducing the tensile strength from the modulus of rupture of small test beams was pointed out by Eaton Hodgkinson in his companion volume to the 4th (1842) edition of Tredgold's essay[4] after Tredgold's death. The error which is not easy to comprehend, and on which Hodgkinson himself was not wholly right, is discussed in Sections 5, 6 & 7 which deal with Tredgold's design studies for beams, columns and ties. Before embarking on these studies, it is worth looking quickly at Tredgold's load tables. These were calculated from the formulae discussed later and were apparently intended for the practical man who had neither the time nor probably the ability to follow Tredgold's mathematics.

4. TREDGOLD'S BEAM AND COLUMN TABLES*

Table I gives the required depths for square-sectioned beams based on a limiting deflection of 1/40 inch for each foot of span. He considered only a central point load.

A footnote says that the load supported must include the weight of the beam and one wonders why he did not allow for self-weight in calculating the data in the table. He certainly considered this and suggested that the beam's weight might be allowed for by increasing its breadth in the same proportion as its weight increased the load—a singularly inefficient solution.

*The tables referred to in this section are the tables in Tredgold's *Practical Essay on cast iron.*

At one end of the table one finds a beam 3.7 inches deep supporting 0.05 ton over a span of 40 feet. A span/depth ratio of 130 for a beam is surprising enough in any material. Adding width to allow for the weight of even so slender a beam (0.75 ton) reveals that this part of the table is nonsense. Effectively the beam must be weightless. At the other end of the table is a formidable beam 36.7 inches square in section carrying a load of 500 tons over 40 feet. This is a casting weighing some 75 tons even without increasing its width to hold itself up–not practical even if not quite nonsense.

These are the limits of the table. Many of the sections and spans given are reasonable but equally much of the table is not. Even forgetting the problem of the dead weight, why did Tredgold base the table on square sections when he knew that rectangular, or better, I sections were more efficient?

Table II is much more sensible. Unlike Table I (based on deflection), Table II deals primarily not with ultimate strength but the elastic limit already discussed. This time he based it on rectangular sections per inch of width but regretably the dead weight still needs to be added and the section re-checked. However, the table does include a deflection for each entry. Thus having sized the beam on strength it is easy to check whether the deflection is less than his limit of 1 in 480 or not.

Given his more useful Table II it is hard to see what Table I has to offer and even more why it remained inviolate through all the editions. Tredgold seemed unable to edit his own work.

Table III (not in the 1st edition) relates height, diameter and load for solid circular cast iron columns. Like Table II it is sensible and useful in format.

From the tables it is logical to follow him, as in his book, into the main subject, the behaviour of cast iron beams.

5. ANALYSIS OF BEAMS.

5.1. General In his writings Tredgold went quite deeply into the method of calculating the deflection and strength of cast iron beams. His component approach was not concerned with the performance of complete structures but only with the intensity and distribution of load which could be resisted by single beams of different forms and support conditions.

The conditions were always determinate - cantilevers from walls, balanced cantilevers (as in beam engines) or simply supported beams. However, he considered numerous forms of beam. Given for the first time a material to be cast rather than cut from the solid, refining the shape became a worthwhile economy.

5.2 Deflection In his companion paper on timber Dr. Booth shows that before the first publication of his book on cast iron Tredgold had already recognized the correct relationship of beam deflection to member size, shape, span, fixity and loading. The constant governing the magnitude of deflection for each material was derived by experiment, but it was not until the second edition of his book on cast iron that Tredgold positively identified this as Thomas Young's modulus of elasticity. He gives the modulus as 18×10^6 lb/in^2 for cast iron compared with an average of about 1.5×10^6 lb/in^2 for timber thus showing cast iron's great superiority in stiffness.

5.3 Shear Before discussing bending strength (where Tredgold started to go astray) it is worth noting how he by-passed the problem of shear. Shear stresses in cast iron beams were not considered either by Tredgold in the 1820s nor, it seems, by Eaton Hodgkinson in the 1830s and 1840s. Coulomb and Vicat were aware of the problem of shear before these times but, according to Timoshenko, a rigorous treatment had to wait for Saint-Venant and Jourawski well after Tredgold's death.[5]

While not recognising shear Tredgold certainly understood the tied arch and for open web beams he illustrated a monolithic bowstring girder with the centre line of the bow and the tie meeting at the point of support. He also advocated beams shaped to suit the change of bending moment along their lengths. It may be that he envisaged every beam acting as some form of bowstring girder or it may be that the idea of shear strength never arose because all beams tested failed in bending near midspan. Whatever the truth Tredgold's readers were only invited to look at deflection and bending.

CAST IRON

5.4 Bending strength Although neither bending moment (M) nor section Modulus (Z) is mentioned one can readily find them in most of Tredgold's formulae for bending strength. For instance in the case of a cylindrical cantilever of length L and diameter D loaded at its extremity with a weight W, Tredgold's formula is $W = 0.7854fD^3/8L$ where f is the bending stress; this is the same as $WL = \pi D^3 f/32$ or $M = Zf$. The first impression given is that Tredgold's understanding of beam action in the elastic range was little different from ours today. For symmetrical cross-sections (rectangular, I, circular, open web, etc.) he made 'the neutral axis divide the section into identical figures'. Also when he analysed these sections he always seemed to arrive at the correct elastic solution for the relationship between load and stress.

It was only when he looked at unsymmetrical cross-sections that his expressions started to go wrong, or so it seems at first. For these irregular sections he says 'to find the place of the neutral axis we must compare the strength of the lower side with the upper one'; calculating from this premise with the same maximum fibre stress top and bottom he fails to get the neutral axis at the centroid. Having noted this error one doubts whether in the case of the symmetrical cross-sections he was right for the right reason.

Though some of Tredgold's formulae are not elastically correct they are not necessarily far off and they are readily useable. By taking the working stress (f) in the formulae as that at the elastic limit (i.e. 6.83 ton/in^2) and relating this to the particular beam section, span and type of loading, Tredgold's readers could calculate working loads for much more complex sections than

Table 1: Bending Tests on Cast Iron Beams

Beam section		Modulus of rupture ton/in^2	Factor of safety in relation to 6.83 ton/in^2	Author and notes on tests.
Square	1" x 1"	18 - 26	2.6 - 3.8	Tredgold (pub. 1822-4)
Rectangular depth x breadth	1" x 1" 3" x 1" 3" x 2" 3" x 3"	20.5 15.4 16.2 12.9	3.0 2.3 2.4 1.9	Capt. James quoted by Edwin Clark 1850[6]
Small I Sections	4" to 5" deep	9.6 - 13	1.4 - 1.9	Typical results by Eaton Hodgkinson Pub. 1842[7]
Large I beams various flange proportions	17" deep	7.9	1.2	Typical result by T. Cubitt pub. 1842[8]
	12" * deep 21" * deep 32" * deep	7.5 - 8.7 6.9 - 8.7 6.5	1.1 - 1.3 1.0 - 1.3 0.95	Tests by Chettoe, Davey and Mitchell in 1940s[9]
	* Small sections cut from these beams were shown to have a modulus of rupture of 16 - 20 ton/in^2			

those covered by his tables. For instance they could see that even quite a stocky I section carries appreciably more load for a given weight of iron than a rectangular one. The worrying point to the modern reader is not the correctness or otherwise of Tredgold's formulae for strength in bending but the level of the working stress. Something over six tons per square inch would be unthinkable for foundry cast iron today. However all the tests he quoted gave a modulus of rupture of 18 to 26 tons/in^2 indicating a mean factor of safety of over three in relation to his 6.83 ton/in^2. These tests were all on very small sections, mostly one inch square. Other tests, mostly later, on larger beams give much less reassuring results as can be seen from Table 1.

It is notable that as the beams get bigger the modulus of rupture reduces towards the tensile strength of the material. Even without the results of bending tests on large sections Tredgold might have sensed a danger from the results of tensile tests known to him (see Section 7). He chose to dismiss these by discrediting the experimental methods. In Tredgold's defence it must be said that the difference between the modulus of rupture for large and small sections of the same material would have been difficult to see from the available evidence without accepting the tensile tests. Eaton Hodgkinson's later explanation based on the rise of the neutral axis as rupture approaches[10] is certainly not the full story. Even in the 1940s Chettoe, Davey and Mitchell went no further than to talk of scale effects.[11] Looked at practically it must be admitted that Tredgold's advice on the bending strength of cast iron was dangerous. For any practical size the factor of safety would have been far too low.

6. COLUMNS IN CAST IRON

6.1 General When dealing with columns, Tredgold's logic is rather less clear than in the case of beams. He makes no mention of Euler's 18th century work, but he recognises the problem of buckling, although not by name, and offers two basic sets of formulae one for short and the other for long columns.

6.2 Short columns For short columns (length up to 10 or 12 times the least dimension of the section) he says that given truly axial loading (Fig. 2A) the working load is equal to the area of cross-section multiplied by his stress of 6.83 ton/in^2 at the elastic limit.

He dismisses axial loading as 'altogether impossible to arrive at in practice' saying that the force may be on one edge only as indicated in Fig. 2C. For this loading condition he states correctly, on an elastic basis, that in a rectangular section the strength is one quarter of that of an equivalent axially loaded one and that the load must be reduced accordingly. He recommends that this eccentricity should always be assumed in practice. When dealing with short cylindrical columns his logic—or clarity of expression—starts to waver but as indicated in Fig. 2 this uncertainty is not serious. He does not consider hollow columns at all.

6.3 Long columns With long columns Tredgold adds a further load reduction factor related to the length and size of section to prevent buckling which he saw clearly as being caused by deflection due to the compressive force increasing an initial eccentricity. His proposed load formulae for long columns of rectangular or circular section allow for both slenderness and for varying eccentricities, all as shown as on Fig. 2.

For general use he recommends the design formulae given in Column C of Fig. 2 which are concise and easy to use even though the derivation is not wholly clear. These two formulae can equally well be used for short columns with little loss of economy and may be looked on as universal. In fact the table of column loads in the 2nd edition of Tredgold's book were calculated from the formula for long cylindrical columns in Fig. 2.

It seems that these column formulae were widely used and were without effective precedent. Timoshenko[12] credits Tredgold with being 'first to introduce a formula for calculating safe stresses for columns'. One curious feature of all his column formulae is that, unlike those for beams, they were derived entirely from theory without the help of tests. With a crushing strength of 40 to 65 ton/in^2 a working stress of 6.83 ton/in^2 at a supposed elastic limit certainly gives a more than ample safety margin for short columns even without Tredgold's recommended allowance for eccentricity. For long columns the same stress with a reduction for slenderness comparable to that used today, as shown in Fig. 3, is equally conservative.

CAST IRON

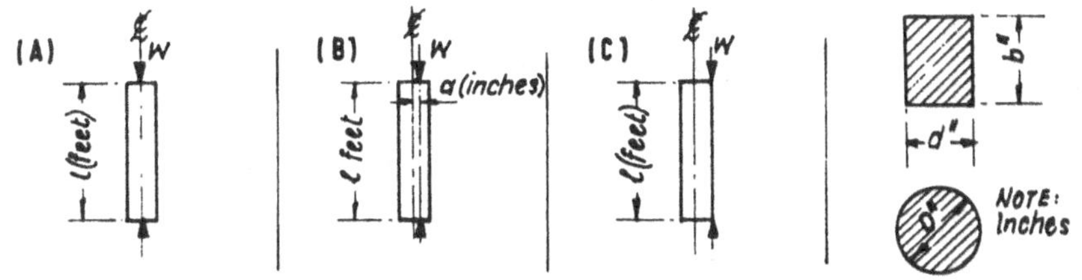

SHORT COLUMNS	(l/b or l/d or l/D ≮ 10 or 12)		
$W = f.b.d.$	Not mentioned	* $W = \frac{f.b.d.}{4}$	Rectangular Section
$W = f \times area$	Tredgold seems to be saying that a circular column is $\frac{1}{1.6}$ as strong as a rectangular one.		Circular Section
		See footnote to expression for long columns	

LONG COLUMNS			
$W = \frac{f.b.d^3}{d^2 + 0.18l^2}$	$W = \frac{f.b.d^3}{d^2 + 6da + 0.18l^2}$	$W = \frac{f.b.d^3}{4d^2 + 0.18l^2}$	Rectangular Section
+ $W = \frac{fD^4}{1.6(D^2 + 0.18l^2)}$	$W = \frac{fD^4}{1.6(D^2 + 6Da + 0.18l^2)}$	⁂ $W = \frac{fD^4}{1.6(4D^2 + 0.18l^2)}$	Circular Section

+ If l is negligible logically this should reduce to

$$W = \frac{f.\pi D^2}{4} = \frac{fD^2}{1.27}$$

instead of Tredgold's

$$W = \frac{fD^2}{1.6}$$

* This is correct elastically provided that the load is applied so that the stress (f) is distributed →

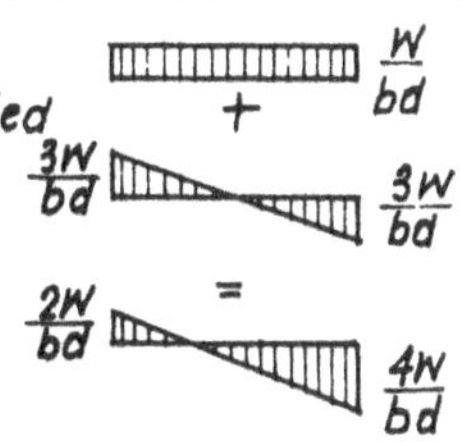

⁂ If l is negligible this should reduce to

$$W = f.\frac{1}{5}.\frac{\pi D^2}{4} = \frac{f.D^2}{6.37}$$

Which is virtually same as Tredgold's

$$W = \frac{f.D^2}{1.6 \times 4} = \frac{f.D^2}{6.4}$$

Fig. 2. Tredgold's column formulae based on a stress (f) of 6.83 ton/in² at elastic limit.

THOMAS TREDGOLD (1788–1829): SOME ASPECTS OF HIS WORK

7. CAST IRON IN TENSION (TIE BARS & RODS)

7.1 General When comparing ties with the columns in the opening paragraph of his section on tension[13], Tredgold makes the following surprising statement about tensile members, 'Instead of the strength being diminished by flexure [deflection] it either has not effect, or has a directly contrary effect'. He then goes on to divide them, as columns, into long and short ties.

7.2 Short Ties These he treated in the same way as short columns, with eccentricity allowed for as in the top part of Fig. 3, and the working tensile stress (f) taken at his universal figure of 6.83 ton/in^2 for the elastic limit. The recommendation for columns, that in practice, the eccentricity should always be assumed to be half the width or diameter of the section still applies, at least by implication.

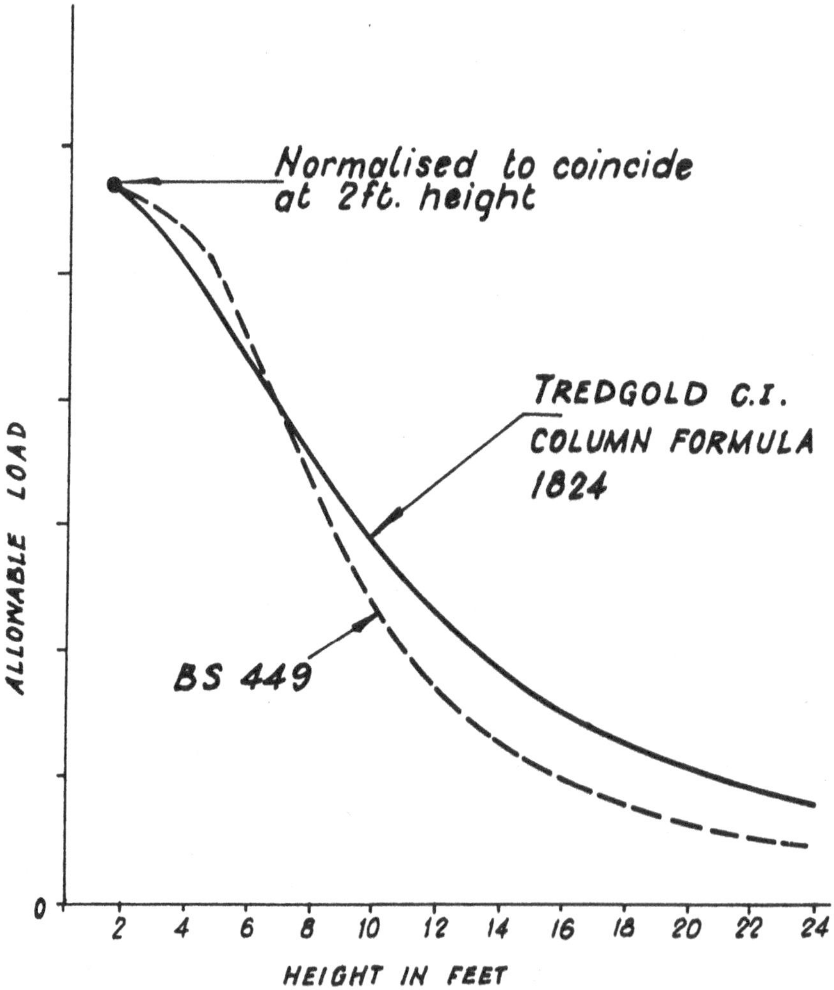

Fig. 3. Tredgold's column formula compared with 20th century equivalent for 4in diameter column with axial load.

CAST IRON

7.3 Long Ties Tredgold treated long ties as being similar to long columns but with one important difference. For long ties he states 'But when it [i.e. the tie] is long, and the bar is either curved or the force is not in the direction of the axis, then the effect of flexure may be considered.' Here he seems to be saying that, due to something equivalent to buckling in reverse, long ties are strengthened in the same way that long columns are weakened. This is confirmed by the formulae which follow and which are shown in the lower part of Fig. 4; the 0.18 l^2 factor is subtracted this time, rather than added as in the case of the columns.

With eccentric loading as in B and C on Fig. 4 one can begin to see what may have been Tredgold's thinking; the ties will deflect under the eccentric load so that the force in the middle of their length will tend to centralise. With short ties this deflection will be negligible but when infinitely long the force in the middle will become truly axial. If this was his thinking he did not seem to see that at the ends the eccentricity was unchanged.

Looking at these long tie formulae in practical conditions at least two snags emerge. If in (B) on Fig. 4 the eccentricity reduces until it is effectively zero then there is still a possible benefit from subtracting the 0.18 l^2 factor. This is at variance with his formula in (A). Where is the limit intended to be? If 0.18 l^2 becomes as great as (d^2 + 6ad), which is possible, the denominator becomes zero and the load carried infinite! What happens when the length is still greater?

After baffling any simple-minded reader with this dubious mathematics Tredgold goes on to say 'It was desirable to show what constituted the advantage of tensile strain [stress] but I do not intend to adopt these equations in practical rules .... ' His stated reason is that the equations are not so easy to apply as those for short ties, but one cannot help thinking that he might have been having doubts. He then gives worked examples based on the short tie formulae.

The reason for dwelling on what might be considered a small lapse on Tredgold's part is that it emphasises an apparent contradiction in attitude and intentions. If just writing a useful handbook he would have done better to omit all speculation about the advantages of tension. It is likely that many practical designers would give up reading long before they got to the point where they were told to use the simple formulae for all cases.

7.4 Adequacy of tie formulae Accepting the simple short tie formulae rather than those for long ties, the question of their adequacy remains. How do they compare with tests? Is f = 6.83 ton/in^2 reasonable as a working tensile strength for cast iron? The answer to the latter must definitely be 'no'. Tredgold had few tensile tests on cast iron to work from. He quotes four results from three sources; Musschenbroeck 28.3 ton/in^2 (basis uncertain), Capt. S. Brown 7.26 ton/in^2, J. Rennie 8.32 ton/in^2 and 8.70 ton/in^2.

He says Peter Barlow criticised the apparatus used by Captain Brown and he criticises that used by Rennie thinking their results too low.

His idea of tensile strength appears to be derived from an erroneous interpretation of his own ultimate flexural tests. From these he deduced 'cohesions' (ultimate tensile strengths) of around three times his design figure of 6.83 ton/in^2. Later pure tensile tests, notably those carefully conducted by Eaton Hodgkinson[14], showed the ultimate tensile strength of cast iron to be around 7.0 to 7.5 ton/in^2 and French tests quoted by Hodgkinson from Navier average 7.19 ton/in^2 with some only just over 5 ton/in^2. Tredgold was unwittingly using the ultimate tensile strength as a working stress. Admittedly by assuming eccentric loading as in Fig. 4C for a situation where, as in one of his examples, the actual load was sensibly axial he would have got a theoretical factor of 4 based on his calculations. Thus this first worked example for tension is plausible because it is based on the eccentricity in Fig. 4C.

Another of Tredgold's examples is worrying and even more alarming are some of his statements about this. He takes a suspension bridge of 370 ft span with cables sized *without any allowance for eccentricity* using the simple expression in Fig. 4A.

Thus accepting a working stress virtually at the ultimate he extolled the virtues of cast iron in such a situation. This he maintains would be less expensive, more durable and less subject to vibration than the 'very slight and temporary' bridges with wrought iron cables of which he knew.

There can be no doubt that Tredgold's ideas on cast iron in tension were suspect. What is so surprising is that in the 4th edition Eaton Hodgkinson added no doubting footnote as he did over Tredgold's other errors.

THOMAS TREDGOLD (1788–1829): SOME ASPECTS OF HIS WORK

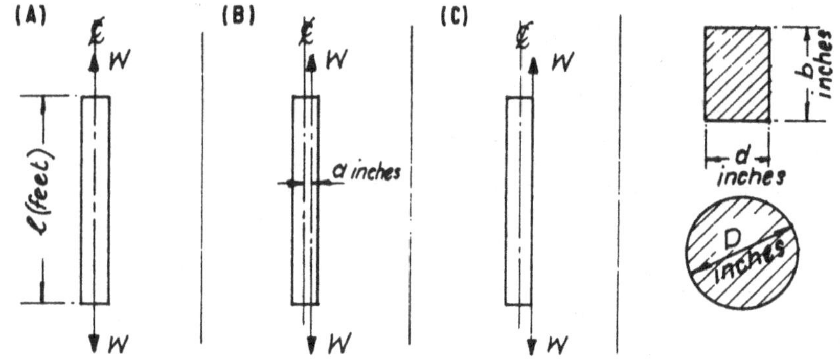

SHORT TIE BARS			
$W = f.b.d.$	Not mentioned specifically	$W = \frac{f.b.d.}{4}$	Rectangular Section
Not specifically covered	(see note on figure 2 for columns)		Circular Section

LONG TIE BARS			
$W = f.b.d.$	$W = \frac{f.b.d^3}{d^2 + 6ad - 0.18l^2}$	$W = \frac{f.b.d.^3}{4d^2 - 0.18l^2}$	Rectangular Section
	$W = \frac{f.D^4}{1.6(D^2 + 6aD - 0.18l^2)}$	$W = \frac{f.D^4}{1.6(4D^2 - 0.18l^2)}$	Circular Section

NOTE:
At the end of his article on cast iron in tension Tredgold advocates for practical work the use of the short tie formulae for long ties.

Fig. 4. Tredgold's formulae for tie bars in cast iron, based on a stress (f) at the elastic limit of 6.83 ton/in^2

CAST IRON

8. TORSION AND IMPULSIVE LOADING

Having dealt at length with beams, columns and ties Tredgold then considers more shortly the effect of torsion on shafts and of impulsive loads such as when a vehicle goes over a step in the surface of a bridge. Here as with tie bars he was on uncertain ground but he did recognise the existence of twist and impact and gave guidance to those who were faced with real problems. Regrettably space has precluded assessment of Tregold's theories on these forms of loading.

9. TREDGOLD'S OVERALL UNDERSTANDING OF THE BEHAVIOUR OF CAST IRON

If confusion is the only result of this attempt to analyse Tredgold's design methods for cast iron some part of this is due to contradictions and confusions in his thinking.

Tredgold deserves considerable credit for his advocacy of practical experiment as the best means of finding out how structural members behave. He deserves further credit for trying to devise a consistent set of principles to link the behaviour of beams, columns, ties and shafts. However, he is open to criticism for not seeing the need to apply the same level of testing to all types of member to prove his principles; also for pursuing these even in the face of contrary evidence.

His awareness of the need to consider deflection as well as strength is impressive and has a distinctly modern ring about it. Further, in his assessment of deflection limits he is hard to fault.

Tredgold's philosophy of design for strength does not stand up so well as his design for stiffness. He was right in seeing that for many metals there is an elastic limit. Even though Eaton Hodgkinson said that this limit did not strictly exist for cast iron, in practice Tredgold's figure of 6.83 ton/in^2 for it was quite reasonable. However, the idea, discussed above, of taking this stress both as the elastic limit and the safe or design stress in bending, tension and compression, while having an appealing simplicity, gives very varied margins of safety. It takes too little account of the great strength of cast iron in pure compression and over-estimates it in bending and particularly in tension. If Tredgold had done more tests, not only on large beams but also on columns and ties, he might have seen the fallacy in some of his assumptions.

If instead of working with cast iron Tredgold had been using wrought iron he would virtually have established the safe or working stress design method for wrought iron and steel which remained standard throughout the world for over a hundred years and is still widely used. By present-day criteria his big mistakes were firstly in thinking that cast iron should be designed with the same maximum stress in tension and compression and, secondly, in making the tensile stress about three times too high. Tredgold had some odd gaps in his appreciation of the nature of the material he admired so much; perhaps his enthusiasm carried him further than his reason.

Cast iron seemed to him the great new wonder material likely to supersede timber for all purposes because of its greater 'security' in supporting heavy loads and in resisting fire, wear and rot. He advocated grey cast iron which, unlike the white variety, dented when struck with the edge of a hammer instead of chipping or breaking and 'yielded easily to the file'. This he looked on as a malleable material, although for all types of cast iron he did concede that when it fails it 'gives no warning of approaching fracture'. Today we may under-estimate the malleability of structural cast iron but in practice we are probably right in looking on it as a treacherously brittle structural material, especially when used in tension.

To relate Tredgold's understanding to that of his contemporaries and predecessors would need more research and more space than is possible here; clearly he owed much to Dr. Thomas Young and to Robison and the great theorists of the Continent.

Tredgold lived in a time of conflicting and rapidly changing views on structural matters both theoretical and practical as is exemplified by the swing in design theories for cast iron beams. Before 1809 Charles Bage worked out, but did not widely publish, a wholly practicable theory for cast iron beams based effectively on the discredited Gallilean assumption that the whole section acts in tension with the compression edge as a pivot[15]. In the 1820s Tredgold opted for equal tensile and compressive stresses and, although not precisely stated, a double triangle stress pattern. In the 1830s Eaton Hodgkinson showed that cast iron was stronger in compression than in tension and developed his famous Ideal Section of cast iron beam and a formula reverting to the compression edge as a pivot[16].

THOMAS TREDGOLD (1788–1829): SOME ASPECTS OF HIS WORK

One cannot help wondering to what extent practical designers used Tredgold's book on cast iron. The fact that in 1832 William Turnbull found it worthwhile to publish a greatly simplified version[17] with full credit to Tredgold suggests some demand. The further fact that Tredgold's own text was apparently worth reprinting even in the early sixties, over thirty years after his death, proves still more, especially in view of the advance in knowledge and growing preference for wrought in place of cast iron.

10. CONCLUSION: TREDGOLD AS A MAN

Tredgold's actual influence on later developments is hard to assess. His writing on cast iron has been referred to in many later publications, not always with kindness. Only one cast iron structure or component designed by him has come to light. This is a shallow 40 foot truss of cast and wrought iron, Fig. 5, which was found by Dr. Booth in a book on timber.[18] If he had lived longer would he have left many influential even if not monumental, structures as well as his books? It seems doubtful. One must conclude that Tredgold was not an active thrusting designer like Telford, Rennie and many others of his contemporaries but a man with a marvellously enquiring mind whose breadth of interest was uncontrollable and who just read, thought and wrote.

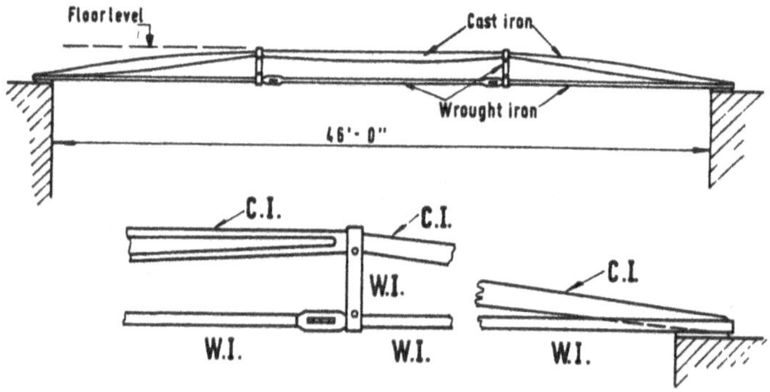

Fig.5. Truss in cast and wrought iron designed by Tredgold

NOTES FOR PART 3

1. First published 1822: 2nd ed. (enlarged) 1824, 3rd ed. 1831, 4th ed. 1842-6 (with notes by Eaton Hodgkinson), 5th ed. 1860-1.
2. Wendy Slemen, 'Liverpool's cast-iron churches'. *Foundry Trade Journal* 6 March, 1975, 307.

   A.W. Skempton & H.R. Johnson 'William Strutt's Cotton Mills 1793-1812' *TNS.* **XXX** and 'The first iron frames'. *Architectural Review* March 1962.
3. It is notable that a factor of safety (always 4) features much more widely in Tredgold's book *Elementary Principles of Carpentry*, first published in 1820.
4. Eaton Hodgkinson: *Experimental researches on the strength and other properties of cast iron* (published as the 2nd part of Tredgold's Essay) 1842, 384.
5. S.P. Timoshenko, *History of the strength of Materials 1953*, 141.
6. Edwin Clark, *The Britannia and Conway Tubular Bridges 1850*, 442.
7. Eaton Hodgkinson, *op. cit.* (4), 424.
8. Eaton Hodgkinson, *op. cit.*(4), 464.
9. C.S. Chettoe, N. Davey & G.B. Mitchell, 'The Strength of cast-iron girder bridges'. *Institution of Civil Engineers Paper No. 5418*, October 1944, 266.
10. Eaton Hodgkinson, *op. cit.* (4), 384.
11. Chettoe, Davey & Mitchell, *op. cit.* (9), 265.
12. Timoshenko, *op. cit.* (5), 100.
13. Tredgold, *Cast Iron*, 2nd ed. 240.
14. Eaton Hodgkinson, *op. cit.* (4), 308.
15. A.W. Skempton, 'The Origin of iron beams', reprint from *Actes du VIIIe Congress International d'Histoire des Sciences*, Florence 1956. 1029-39.
16. Eaton Hodgkinson, *op. cit.* (4), 444.
17. William Turnbull, *A treatise on the strength, flexure and stiffness of cast iron beams & columns*, Josiah Taylor 1832.
18. Peter Nicholson (revised by Thomas Tredgold) *Practical carpentry, joinery and cabinet-making:* 1868 ed. Kelly & Co., London, 24.

13
The birth of stress: a historical review

R.J.M. Sutherland

INTRODUCTION

Up to the late 18th century structures were designed essentially on the basis of proportion. To some extent this meant no more than deciding whether sizes looked right - that is, familiar - but in many, perhaps almost all, periods there were some rules or statements by authorities which were almost as firm as our codes of practice today. The difference is that they were not based on strength or stress but on shape and scale. Stress in the sense that we use the word in engineering today did not exist.

The structural materials were masonry and timber, and the criteria for acceptance were respectively stability and stiffness. Even the rules of proportion which governed these were not always written down and the records which survive tend to be fragmentary.

Throughout the 19th century the concepts of stress and strength gradually took over until today they dominate almost all structural thought. Starting from a time when structural redundancy was a virtue born of limited understanding we moved towards structures which were deliberately simplified to suit the limits of current knowledge.

Now with computers we can analyse the elastic behaviour of frameworks or solids of a complexity undreamed of 40 years ago and the dominance of stress is still greater. The advantages of this effortless accuracy are great. In addition to stress, deformations which used to be tedious to assess - and seldom were - are now a bonus given free with almost every packet of computed strength.. What is more, the scope for effective modelling of inelastic behaviour is growing. The gain in our capacity to check quickly is wholly desirable, but it is less clear that the thinking of the designer is advancing as fast as the tools he uses. I am using the word 'designer' deliberately here; I do not mean analyst.

In this paper I am trying to see how the change took place and how different the design engineer is today from his counterpart of, say, 1784 or 1834 or 1884. This is not a history of the development of structural theory, a subject which has been well covered by others (1). Here I am concerned with the reaction of engineers, whether simple or sophisticated, to the changing state of knowledge, to fresh demands and to new materials.

Progess in construction practice in the last 200 hundred years has been out of all proportion to that in several millennia before. This progress has been due in part to new materials, or effectively new ones, but particularly to the introduction of the concepts of stress, strength and measurable levels of safety. Before looking at these revolutionary years it would be worth looking briefly at structural thinking before, say, 1800.

MASONRY AND TIMBER BEFORE 1800

Perhaps the earliest surviving equivalents to our structural code clauses are those in Vitruvius's 'de Architectura' which was written before AD 27 (2). Although often stated with rigid authority these are nearer to the broad functional requirements planned for the new Building Regulations than to precise rules. For instance in the case of retaining walls 'the thickness of the walling must answer to the amount of earth' and 'supporting walls and buttresses are to be carried up at the same time'. Excellent sense but there is nothing on the dimensions which had been found satisfactory and no hint at how to arrive at such dimensions. Possibly if the illustrations to 'de Architectura' had survived they would have given some indication.

The difficulty with most writing on masonry before the 19th century is the lack of distinction between beauty, economy and structural sufficiency. The discipline of the 'orders' of architecture although basically aesthetic led to proportions which largely controlled stability, while also ensuring that stresses were never critical. If the origins of the orders were in part structural these origins had been lost even before the time of Vitruvius.

Visual and structural criteria were equally confused in the age of the Gothic cathedral. Safe rules on form and proportion were controlled by the masonic lodges and still used long after their origins had been lost. Jacques Heyman's view (3) that the rules were good and safe but could not be reconstructed outside living memory is convincing. The same difficulty persists today when one tries to reconstruct the basis of some clauses in existing codes.

In the cathedral era, as in Roman times, stress in masonry was almost never significant. Stability was becoming more and more critical, but the basic principles for ensuring it remained wrapped up in the rules for the northern Gothic architecture, which were neither universally believed nor followed. For instance, in the dispute in 1400 over the size of buttresses at Milan cathedral the Italians maintained that pointed arches did not exert a horizontal thrust whereas the French consultant, Jean Mignot, knew that they did (3), but not how much.

The apparent ignorance of Gothic thinking in Italy in the 14th and 15th centuries is more understandable when one considers that their tradition was to use timber roofs, or tie rods mainly with round arches, whereas the Northern Goths were buttress men, even if they did not quite know how the buttresses worked. The 'line of thrust' analyses of the cathedrals were all made in the 19th century well after the work on arches by Couplet, Coulomb and others. The important point is that at the time of its construction Gothic architecture, which looks so extremely assured to us today, was based on rules of proportion in which aesthetics were bound up with stability without a thought for stress.

In the Renaissance there was a return to Vitruvius as the master. Writing in the 15th century Alberti (4), one of the most influential writers on architecture, extolled the virtues of the 'ancients', both aesthetically and structurally, with no apparent yearning for innovation. His writings dwell at length on such mundane matters as the selection of timber for beams but without a word on how to assess their sizes.

Similarly in the 16th century Palladio (5), another convinced Vitruvian and probably more influential in Britain than Alberti, gave no advice on wall thicknesses or beam sizes in his 'Quattro Libri' of 1570 but became more specific on foundations which he suggested should normally extend into the ground one sixth of the height of the building and be twice as wide as the walls they support. In bad ground or at 'the Waterside', he advocated improving the soil under all walls, not just the front ones, with closely spaced piles, because the expense of these is 'less to be feared than the falling of the Building'. His plans seldom show walls thinner than 750-1000mm for buildings some 20 to 25m high with two to four storeys, and thus neither stability nor strength were likely to be critical and foundations automatically reasonably wide.

Even though the diameter of columns and the thickness of walls may have been established by the architectural orders it is far from clear who between the 15th and 19th centuries fixed the sizes of timbers or the connections in trusses. Probably, it was the carpenters working on experience, observation and possibly copy books of details. In spite of growing knowledge of the strength and stiffness of different materials, this unscientific approach sufficed for the majority of construction until well into the 19th century - at least in Britain, but perhaps less so on the Continent.

STATE OF STRUCTURAL KNOWLEDGE IN BRITAIN IN THE EARLY 19th CENTURY

I have said that this is no place to follow the development of structural theory, especially in relation to bending, except to the extent that this affected the ordinary designer.

The fact that some of the theories were incorrect was of no importance provided that like was being compared with like. For instance, having established that for a rectangular beam the bending strength was proportional to:

(bd^2) x (a constant depending on the material)

(where b and d are breadth and depth of section, respectively) it did not matter whether you used Galileo's or Mariotte's incorrect theories of the 17th century or Parent's elastically correct one of the 18th, provided that the constant was derived from bending tests and used in comparable circumstances for the assessment of the bending strength of other cross-sections. In 1803, Charles Bage developed a perfectly valid method of designing cast iron beams on the basis of tests and Galileo's bending theory (6).

Among the earliest mathematical design handbooks in Britain, if not actually the first, were Peter Barlow's book on timber, originally issued in 1817 (7), and Thomas Tredgold's books on timber and cast iron, first issued in 1820 (8) and 1822 (10), respectively. It is worth looking quickly at the methods advocated in these books to get some idea of how an engineer could have tackled the problems of fixing the sizes of structural members in the 1820s. The extent to which these handbooks were actually used is uncertain.

DESIGN OF TIMBER IN THE EARLY 19th CENTURY

In the simple case of direct tension, Barlow used the word 'cohesion' which is 'proportional to the number of fibres or to the area of section'. He tabulated 'cohesion on a square inch', as did Tredgold, both basing their values on their own experiments or those by Musschenbroek, Emerson, Rondolet and others. Thus for direct force the concept of stress was there in all but the name.

For timber Barlow stated in relation to 'absolute strength' that 'practical men assert that not more than one fourth of this ought to be employed' but implied that so large a reduction was not necessary. Neither the effect of knots and other defects nor the concept of an overall factor of safety to cover all variables seemed to come into his thinking. Tredgold merely accepted a factor of safety of 4 on the ultimate strength of timber as disclosed by tests.

With timber there was little need to consider beams of anything other than rectangular section. Barlow and Tredgold gave practical rules both for strength and deflection. For instance, for a rectangular beam of length L with a load of (W) Tredgold's rule for strength amounted to

$$W = \frac{C.bd^2}{L}$$

where the constant C allowed for the strength of the material, the loading conditions and different units for length and cross-section. There was no reference to bending moments or section moduli. All was direct, the tabulated values of C being derived from tests on small sections of comparable timber loaded in the same way.

It is notable that both Barlow and Tredgold devoted as much space to deflection as to strength, a clear follow-on from the time when sagging was the first - and probably the only - indication of inadequacy.

Tredgold suggested 1 in 480 as a reasonable limit for deflection in relation to span.

When considering floor joists, Tredgold's emphasis on deflection was particularly strong. He gave a rule, again controlled by a mysterious constant, which rightly relates the span, spacing and breadth of the joists to the cube of their depth but curiously is independent of the load. He explained that the constant was based on scantlings 'found to be sufficiently strong' whereas 'it is difficult to calculate the weight that a floor has to support'. Thus in this field anyway the dominance of strength rather than proportion was not yet complete (9).

DESIGN OF CAST IRON IN THE EARLY 19th CENTURY

For cast iron Tredgold, who certainly produced the first real calculator's guide to the material (10), moved closer to modern thinking than in his book on timber but in some respects went very wrong, although pardonably so.

Again, he advocated a deflection limit of 1 in 480 but also what we would call a safe working stress (f)* of the frighteningly high value of 106N/mm^2 (6.8 tonf/in^2); this he considered to be the elastic limit in bending (based on tests on 25 x 25mm bars of cast iron). He found the 'absolute strength of cast iron bars to resist a cross-strain' (modulus of rupture) of these small bars to be 280-400 N/mm^2 and thus thought he had what amounted to a factor of safety of 2.6 to 3.8.

He then assumed, or so it seems because he said very little directly about it, that using the same working stress (f) in direct tension he would have a similar margin of safety as in bending. He assumed further and with more justification that using this stress (f) again in compression, the safety margin would be at least as high. Thus all one needed to do was to design to the elastic limit as a working stress and all would be well.

In the case of direct tension Tredgold discounted the testing techniques which had given ultimate tensile strengths of around 110-120 N/mm^2 and had no reason to know that later bending tests on larger beam castings were to show a modulus of rupture of as low as 110 N/mm^2 for comparable iron. The last of these errors was specially understandable because the variation in the modulus of rupture with size of casting has still not been fully explained. Nevertheless, his thinking led to a potentially dangerous set of assumptions. He even suggested cast iron links at his universal working stress of 106 N/mm^2 as more robust than wrought iron ones for suspension bridges.

It must not be implied that Tredgold got it all wrong. His method of calculating deflection appears to be generally correct. Further, with cast iron, there was a demand for cross-sections other than rectangular, and Tredgold went into the properties of these at some length getting the right answer with the symmetrical ones, but possibly not for the right reason, and going only slightly astray on the position of the neutral axis with T-sections and similar shapes.

On cast iron columns, Tredgold's recommendations were generally very sound, but for ties he got into a tangle once more on the effect of length. He thought some long ties to be stronger than short ones, visualizing them as being subject to something like buckling in reverse which increased their strength with length. (11)

The sad point about Tredgold's safe working stress, apart from his curious error on tension which had only a limited effect, is that if it had been applied to wrought iron, it would have been almost universally sound if not always absolutely correct. Also it would have been well ahead of any other practical guidance of the time.

The thinking behind some of Tredgold's methods is not always easy to understand today, and it is doubtful whether many of his readers succeeded or even tried to follow this in detail. It is even more doubtful how many engineers read or understood the writings of men like Thomas Young or John Robison or the works of the vast galaxy of Continental theorists. Some certainly tried but probably very few, and the level of success would be hard to measure today.

Tredgold's book on cast iron was translated into French and German and ran into five editions, with the same errors perpetuated, the last being issued in 1860. However, from the 1830s onwards his practical advice was challenged by Eaton Hodgkinson's advocacy of his 'ideal section' for cast iron beams and his simple formula related to this. (12)

Eaton Hodgkinson showed by test that cast iron was about six times as strong in compression as in tension and proportioned his beam accordingly. His simple formula (Fig 1) has been repeated in engineering handbooks until well into this century. All was derived from tests and would be equivalent to saying:

$$\text{Ultimate resistance moment} = N.D.A.t$$

where t is the ultimate tensile strength of cast iron. If N = say, 0.9 the value of t derived from his formula would be 6.7 to 7.2 tonf/in^2 which is a very plausible range. The significant point is that even Eaton Hodgkinson was not thinking in terms of stress but of a constant relating tests under one set of conditions to practical use in the same form.

DESIGN OF WROUGHT IRON IN THE EARLY 19th CENTURY

Until towards the middle of the 19th century, wrought iron was used almost exclusively in tension for such applications as chains, straps, tie rods and boiler plates. In this limited sphere its properties were much better understood than those of cast iron. In Britain, wrought iron beams date from as late as the 1840s and even the French all wrought floor beams of the late 18th and early 19th centuries (Fig 2) were really trusses.

Although Tredgold, Barlow and others touched on the properties of wrought iron in their writings, the best

* Tredgold did not use the word 'stress' but 'strain' defining (f) as 'the strain in lbs. a square inch of cast iron would bear without permanent alteration'.

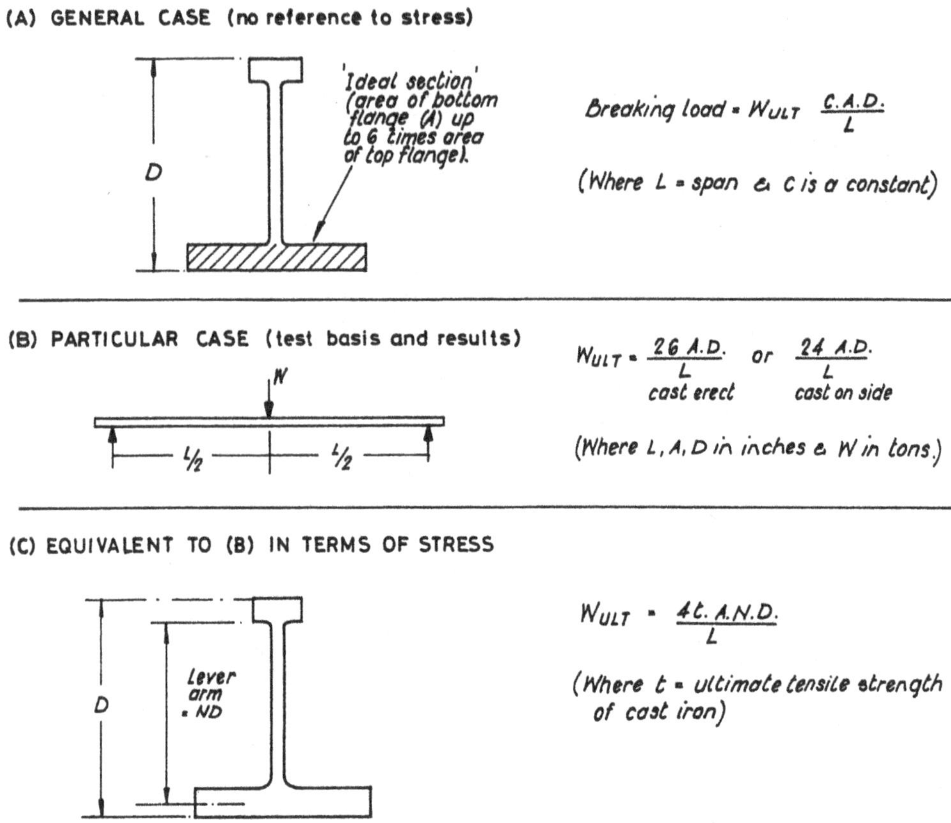

Figure 1 : EATON HODGKINSON'S 'IDEAL' CAST IRON BEAM & ASSOCIATED DESIGN FORMULA

impression of the understandings of this material in the early 19th century can probably be found in Provis's book on the Menai Suspension Bridge (13). Based on tensile tests and a 'factor of safety' of 5, the origins of which are uncertain, Telford established the following criteria for the design and acceptance of the chains for this bridge:

Ultimate strength:	417 N/mm^2 (27 tonf/in^2)
Elastic limit:	209 N/mm^2 (13.5 tonf/in^2)
Strain to which links proved:	170 N/mm^2 (11 tonf/in^2)
Working strain:	83 N/mm^2 (5.4 tonf/in^2)

With every link proved to more than twice its working load but only to 80% of the load which would deform it permanently, it would be difficult to find a sounder basis for acceptance.

TERMINOLOGY: STRAIN,STRESS,COHESION,ETC.

This may be the point where a short diversion on terminology is appropriate. In the first half of the 19th century the word 'stress' virtually did not exist in engineering. What we refer to as stress was called strain or sometimes, if tensile, cohesion, but 'strain' also seems to have been used to denote a force (e.g. a strain of 10 tonf). There was some uncertainty in the use of these terms.

The relationship which really meant something was the proportional one between member size and load. If in Tredgold's words, 'the strain in lbs. a square inch of any material would bear' was x then four square inches would bear 4x. That was all right for direct tension and compression but with bending the explanations are less clear.

According to Timoshenko the concept of 'stress on an infinitesimal plane' was due to Augustin Cauchy and published in 1822. Cauchy also developed the valuable concept of principal stress, but again according to

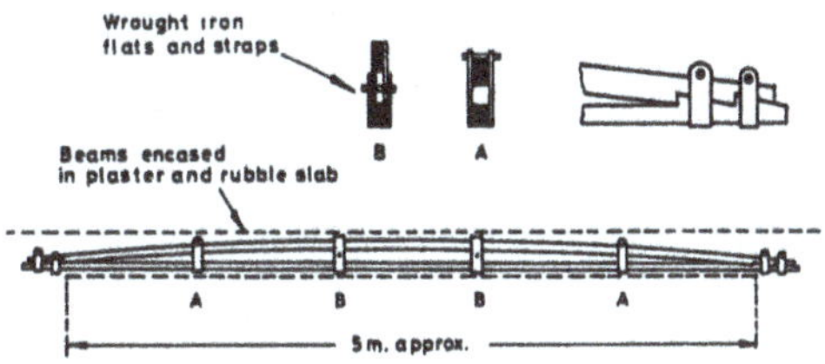

Figure 2 : FRENCH WROUGHT IRON FLOOR BEAM (e.g. ANGO : 1782)
(EFFECTIVELY FILLER JOIST CONSTRUCTION)

Timoshenko, it was St. Venant who first defined stress in its final form in 1845. Both Todhunter and Pearson, Timoshenko (1) and others give W. J. M. Rankine the credit for being the first to provide rigorous definitions of stress, strain, working stress, proof strength, factor of safety and other phrases which are now commonplace in engineering.

THE YEARS OF INNOVATION AND TESTING: 1820 to 1850

The period from about 1820 to 1850 was one of great creativity and unmatched opportunity for engineers. It encompassed the maturity and finally the virtual eclipse of cast iron, the development of splendidly inventive combinations of wrought iron, cast iron and timber, the 'invention' of reinforced concrete by Wilkinson in 1845 (14), and by 1850 the establishment of wrought iron as the pre-eminent structural material. The opportunities were mainly, but not exclusively, provided by the railways, some 6500 miles being completed in the first 25 years up to 1850 with accompanying structural challenges almost all met by the use of iron. This was in fact an iron age in the most literal sense with pig iron production in Britain rising at a rapidly accelerating rate. (Fig 3).

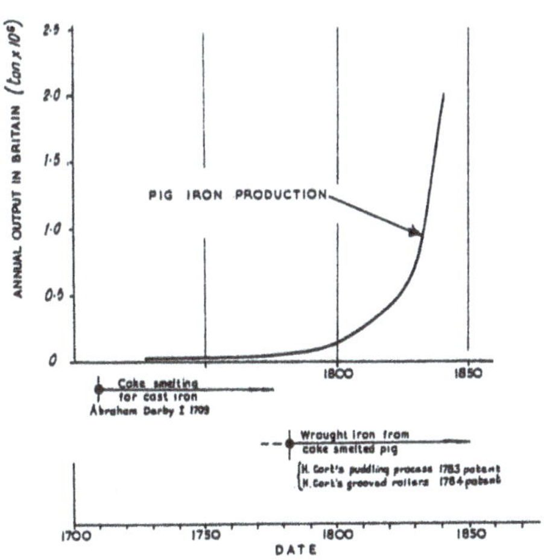

Figure 3 : PRODUCTION OF PIG IRON IN BRITAIN 1700-1850

Like other creative periods it was an age not so much of theory and calculation as of speculation and practical testing. Engineers wanted to understand how materials behaved and observed fractures and deformations closely. (Fig 4).

On some projects such as the Great Exhibition Building of 1851 each cast iron girder, or truss, was proof-loaded; weighed on arrival, slipped into a hydraulic testing frame, proved to twice its design load, taken out of the frame and stacked all, one is told, in the space of 4 minutes (15). On other contracts, as for example the Albert Dock warehouses in Liverpool, each beam had a strip about 1m long cast on in such a way that it could readily be broken off and then tested in bending. The proportion of failures in tests on iron castings is uncertain but in the case of the wrought iron links for the chains of the Menai suspension bridge the failure rate was as follows (13):

Number of main links submitted to proof:		10476
Visually rejected:	249	(2.48%)
Cracked under proof:	100	(0.95%)
Broke under proof:	47	(0.45%)
Total rejected:	396	(3.8%)

Even for multi-link chains proved to twice their working load the considerable number of rejects makes the case for testing.

Not only were full-size components such as beams and columns tested but also small sections of different materials to establish their properties. Further the development of new forms depended almost entirely on tests. Effectively the tubes for the Menai and Conway bridges were designed by experiment. Starting from the concept that wrought iron was just a less brittle form of cast iron initial calculations were based on Eaton Hodgkinson's formula for cast iron beams. Tests then showed that unlike cast iron, wrought iron was apparently weaker in compression than in tension. Further tests proved that this was not a property of the material but due to buckling. Other tests proved that for tubular beams a rectangular shape was more efficient structurally than was a circular or elliptical one provided that its top and sides were stiff enough.

All this led to the seemingly perfect form of a continuous tube with cellular top and bottom flanges, web stiffners on its sides and trains running through the middle. It would perhaps be unfair to speculate on the amount of iron which might have been saved if the sides of the tubes had been open and triangulated. Such trusses could not then have been analysed, but nor, when work started, could riveted wrought iron box or I-beams.

There is no space here to go into all the advances in understanding which accrued from the two year development programme for this seemingly impossible structure nor to try to disentangle the disputed contributions of Stephenson, Fairbairn and Eaton Hodgkinson (16). The more one looks at this stupendous achievement the clearer it becomes that it was the testing which came first and showed 'how' and the theory which followed up and explained 'why'.

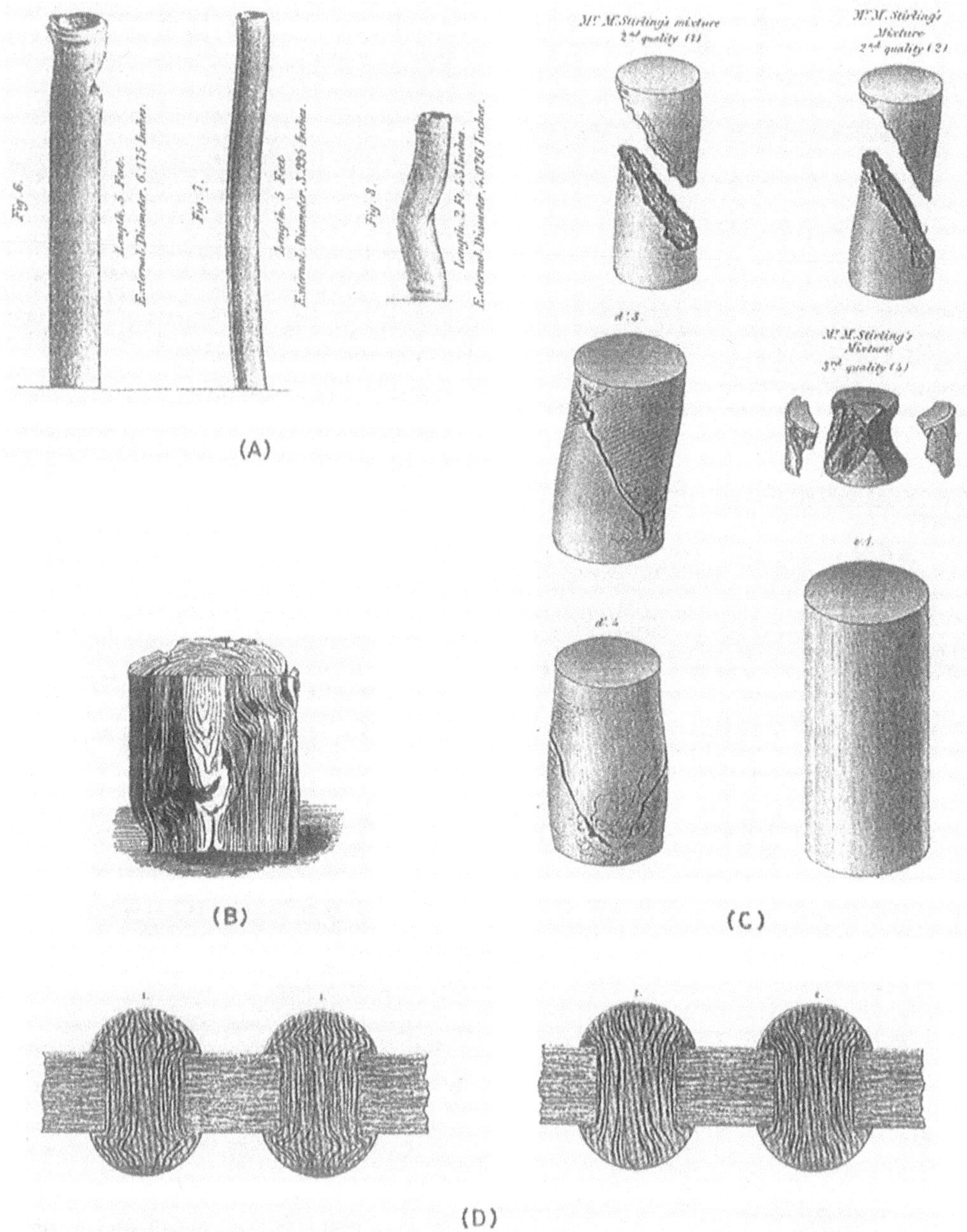

Figure 4 : EXAMPLES OF CLOSE OBSERVATION OF FORMS OF FAILURE IN THE 19TH CENTURY AND OF ALIGNMENT OF FIBRES.

COMPRESSION FAILURES OF WROUGHT IRON TUBES (A), TIMBER (B) & CAST IRON (C).

COMPRESSION OF FIBRES WITH MACHINE RIVETTING (LFET) & HAND RIVETTING (RIGHT)

Engineers in Britain throughout the 19th century were frightened of mathematics. Largely they still are.

It is notable that in the official account (17) published in 1850 after the completion of the tubular bridges (a work far from free of mathematics) Edwin Clark still felt constrained to say of 'transverse strain':

'The complete theory of a beam, in the present state of mechanical science, is involved in difficulties. The comparative amount of strain at the centre of the beam where the strain is greatest, or at any other section, is easily achieved but the exact nature of the resistance of any given material almost defies mathematical investigation.'

Because of the magnitude of the achievement we may be over-estimating the understanding of those responsible.

STRUCTURAL DESIGN BETWEEN 1850 AND 1890

While there is a danger of over-elevating the Menai Bridge designers today, there is an even greater risk of assuming that their new-found understanding was immediately absorbed by all other engineers. It was not, but there was a very great change in attitude mainly in the years between 1850 and 1860. This was the period when ordinary engineers learnt to calculate the sufficiency of most simple structural forms, beams in particular, and to believe in their calculations -even for major structures - without testing.

1850-60 was also the period when it became possible to analyse the forces in trusses with certainty. Several researchers contributed to the understanding of the forces in complex but determinate trusses, but it seems that, although perfectly practicable, all these methods proved tedious until R. H. Bow introduced his famous notation in 1873 (18). This was exactly the sort of systematic and almost foolproof graphical method to appeal to engineers. It has retained its popularity through many generations and has been superseded only recently for speed by the computer.

In spite of growing confidence testing was not yet dead in the mid-19th century. Large scale tests were still being used although possibly as much to satisfy clients as to reassure designers. In the late 1840s three of the crescent trusses of 47m span for the first Lime Street Station in Liverpool (19) were erected as a unit in Turner's works in Dublin and tested first for a uniform case of almost $2kN/m^2$ and then for eccentric loading. This was a record span, and the need for assurance was understandable.

The proving of the 65m trusses for New Street Station in Birmingham (another record span completed in 1854) was even more elaborate (20) as shown in Fig 5. Apart from testing the performance of a complete section of the roof, each tie member was proved to 139 N/mm^2 before incorporation.

After about 1860 confidence in wrought iron had grown enough for testing even of major structures* to be played down.

* This was certainly true for building structures. Bridges continued to be tested and sometimes they still are.

Provision was made for testing in the contract for St. Pancras (completed 1868) but it was never used. The Albert Hall roof (1867-71), Figure 6, was erected in Fairbairn's works in Manchester to make sure it fitted together but was not load tested. These are just examples. One could cite others to illustrate the change from intuition and physical verification to the calculation of sizes with confidence.

One reason for this change was of course the displacement of cast by wrought iron.

Wrought iron was soon recognized as a wholly reliable material and, with rivets of definable strength, it could be built up into structures virtually limitless in scale in spite of restrictions on the sizes of plate and angle which could be rolled. Further, and most important of all, by 1850 it had become a calculable material not just for ties and struts but also for beams.

While it was the triumvirate of Stephenson, Fairbairn and Hodgkinson who established the riveted wrought iron beam it was the elasticians of the mid-century like Rankine (22) who showed engineers how the material really behaved and how to design with it.

Fig 7 shows a comparison of the structural behaviour of cast iron and wrought iron in tension and compression and also the $77N/mm^2$ (5 $tonf/in^2$) Board of Trade stress limit which was dominant in the second half of the 19th century even in areas where it was not imposed. It can be seen that this working stress - still generally referred to as a strain in spite of Rankine - was well within the elastic limit of wrought iron and incorporated a very adequate factor of safety in both tension and compression.

The reality of the known behaviour of wrought iron was limited to the range of stress within which the theorists were thinking. With a working stress generally not exceeding $77N/mm^2$ there is no doubt that wrought iron behaved elastically and that the theory of elasticity, which became the gospel for engineers in the third-quarter of the 19th century, was wholly relevant.

Hooke's law held. Young's modulus was a constant. There was no need to think about factors of safety. You had a working stress to control your design, even though you might still have been calling it a strain, and you had every reason to feel confident.

Stress as we understand it had not only been born but by now was the controlling factor in almost all structural design, at least with iron, and iron was becoming increasingly dominant where a high level of performance was needed.

Looking back in, say, 1860 the ingenious combinations of wood, cast iron and wrought iron of the 1840s and the sometimes ill-judged attempts to increase the span of cast iron beams with wrought iron tie rods must have seemed far away.

The period from 1860 onwards was an age of confidence with blinkers on. It is not surprising that in this atmosphere Wilkinson's reinforced concrete did not take hold. There was no real need for alternatives to wrought iron.

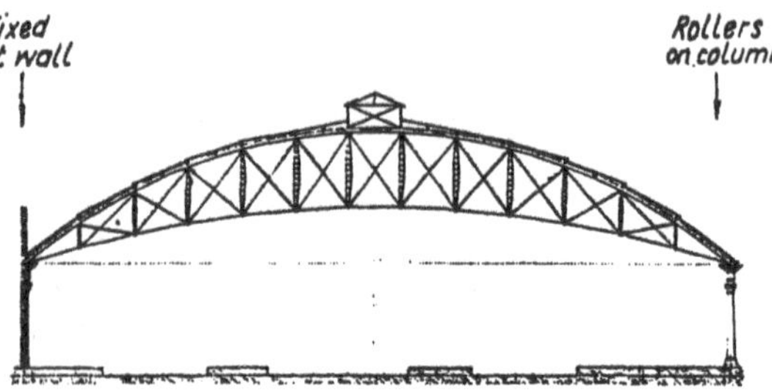

Max. span of truss: 64.6m. (212ft)
Spacing of trusses: 7.3m. (24ft.)
Approx. weight of cladding & purlins (part of test load) 0.25 kN/m²
If design live load taken as 0.75 kN/m²
Max. test load = $\frac{2.53}{1.53}$ = 1.65 x design load.

Self weight of truss	25 tonne (0.53kN/m²) U.D.	25 tonne (UD)	25 tonne (UD)
Test load	45 tonne (1kN/m²) U.D. 67 tonne (1.5kN/m²) U.D. 90 tonne (2.0kN/m²) U.D.	45 tonne (2kN/m²) UD — 22½ tonne (1kN/m²) UD	45 tonne (2kN/m²) UD — 11 tonne (0.5kN/m²) UD
Deflection under test load	84mm (max.) (1 in 768) 64mm. (at roller)	76mm.	75mm
Recovery	Complete	Complete	Complete

Figure 5 : LOADING TEST ON 65 METRE TRUSS FOR NEW STREET STATION, BIRMINGHAM (COMBINED WITH AN EARLIER COMPRESSION TEST ON AN ARCH MEMBER AND PROOF LOADING OF ALL BOTTOM TIE MEMBERS).

STABILITY

In spite of the grip of elastic theory and the dominance of riveted wrought iron there was little calculation of stability by frame action before almost the end of the 19th century. For the most part stability continued to be provided by masonry, that is by walls, often with few or no returns, which relied on their thickness to resist lateral forces. A good example is Cannon Street Station (1863-66) where the arched walls in brickwork (Fig 8) remain even after the demise of Hawkshaw's 58m crescent trusses.

Where columns were needed, cast iron curiously held its popularity until the end of the 19th century and even beyond. These columns were sometimes braced by decorative brackets or haunched cast iron beams which often carried no vertical load, but appearances point to this bracing being assessed on no more than a judgment of what would be stiff enough. The one possible example of an early 'moment-frame' is the Boat Store (23) at Sheerness (1858-60). Although devoid of braced bays or bracing walls this has been carefully detailed to ensure rigidity at the joints within the depth of the beams. However, there is no evidence that this action was calculated.

19th century textbooks devote chapter after chapter to the design of spanning structures - bridges, floors and roofs - but have scarcely a word to say on stability.

Figure 6 : TRIAL ERECTION OF THE ROOF STRUCTURE FOR THE ALBERT HALL IN FAIRBAIRN'S WORKS IN MANCHESTER.

POSTSCRIPT

Logically this paper should stop at about 1890 but the temptation to look on, however briefly, is irresistible. By that date the principles of elasticity and the practice of riveting wrought iron had been established for more than 40 years. The Forth Bridge, exploiting both elastic design and riveted connections on a breath-taking scale, was just being completed but with one new factor. Its 58 000 tons of shipbuilding practice was being built, not in wrought iron, but in steel. Structural steel, with higher working stresses than wrought iron, but riveted as before and still obeying Hooke's Law, was clearly seen at the time as the material of the future.

The challenge of reinforced concrete had not yet arrived but was just round the corner. What is more its behaviour was not, and is not, elastic, although it took some time for engineers to realise this. Next, with welding, came the plastic hinge in steelwork and ultimate load design in concrete and steel. In spite of these advances the analysis of actual structures, rather than structural members, has remained largely elastic with only minor adjustments as a sop to non-linear behaviour.

In the 1950s and early 1960s there was a return to fundamental thinking, testing and innovation, mainly associated with prestressing, prefabrication and new materials, but this has largely faded away or been codified. Now that the structural codes are enshrined in easy-access computer programs the blinkers of elastic thinking fit even more tightly. This thinking has not helped with any of the much-publicised structural problems of the last 25 years. These have been due to structural instability, chemical attack (in the broadest sense) and moisture movements, yet about two-thirds of engineering education and the bulk of practical design thinking are still locked into the 19th century ideas of elastic behaviour. Strength and stress are not the foremost problems today. For

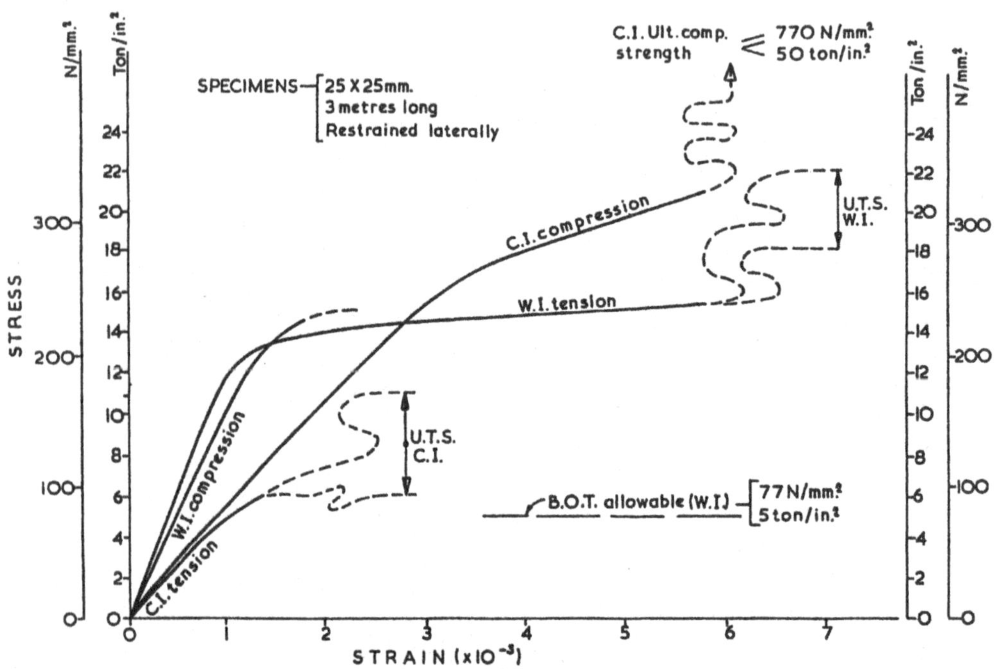

Figure 7 : STRESS/STRAIN CURVES FOR WROUGHT AND CAST IRON IN TENSION/COMPRESSION.

Figure 8 : CANNON STREET STATION TODAY, SHOWING THICK MASONRY WALL WHICH GAVE STABILITY TO THE FORMER 58 METRE CRESCENT TRUSS ROOF.

instance, we need to know whether our concrete will crack and how noticeably and what effect this will have on its deflection and durability.

Perhaps we need a fairly fundamental shift in our education. Perhaps we need to return more to the emphasis on test and speculation of the early 19th century. To see things bend and break and observe their fracture (Fig 4) is far more revealing than days of lectures or reams of text. Also perhaps we need to learn more about stability and chemistry.

One thing which is quite certain is that we need to monitor our structures much more closely - all structures that is and not just the ones we are worried about.

REFERENCES

1. **Todhunter, I and Pearson K: A history of The Theory of Elasticity and the Strength of Materials Cambridge University Press 1886–1893**

   Timoshenko, S P: History of Strength of Materials, McGraw-Hill Book Company Inc, New York 1953

   Charlton, T M: A History of the theory of Structures in the nineteenth century Cambridge University Press 1982

2. Vitruvius on Architecture (translated Frank Granger, Loeb Classical Library, Heinemann, London 1933).

3. Heyman, J: Beauvais Cathedral, Trans-Newcomen Society XL, 1967-68

4. Alberti, L B: Ten Books on Architecture (translated Leoni 1726), reprinted by Alec Tiranti London 1965

5. Palladio, A: The Architecture of A Palladio in Four Books (translated Leoni 2nd edition, John Darby London 1721)

6. Skempton, A W: The origin of iron beams Actes du VIIIe Congres Interntional d'Histoire des Sciences: Florence 1956

7. Barlow, P: Essay on the Strength and Stress of Timber 1817, John Weale, London (Note: the use of the word 'stress' in the title is curious because the word 'strain' seems to be used exclusively in the text)

8. Tredgold, T: Elementary principles of carpentry, J Taylor, London 1820

9. Booth, L G: 'Thomas Tredgold 1788-1829: Some aspect of his work', Trans Newcomen Society 51. 1979-80. (Dr Booth reviews 'His life' in Part 1, his 'Carpentry' in Part 2 and his 'Publications' in Part 5).

10. Tredgold, T: A practical essay on the strength of cast iron, J Taylor, London 1822

11. Sutherland, R J M: 'Thomas Tredgold (1788-1829): Some aspects of his work : Part 3 : Cast iron'. Trans. Newcomen Society 51, 1979-80

12. Hodgkinson, E: 'Theoretical and experimental researches to ascertain the strength and best forms of iron beams'. Memoirs of Literary and Philosophical Society of Manchester, 2nd Series, V 1831

13. Provis, W: The Menai & Convey Suspension Bridges, London 1828

14. Brown, J M: 'W B Wilkinson (1819-1902) and his place in the history of reinforced concrete', Trans Newcomen Society, XXXIX, 1966-67

15. Wyatt, M D: 'On the construction of the building for the exhibition of the works of industry of all nations in 1881', Proc. ICE, X, 1950-1951

16. Sutherland, R J M: 'The introduction of structural wrought iron'. Trans Newcomen Society, XXXVI, 1963-64

17. Clark, E: The Britannia and Conway tubular bridges, Day and Son and John Weale, London 1850

18. Bow, R H: Economics of construction in relation to framed structures, London 1873

19. Turner, R: 'Description of the iron roof over the railway station, Lime Street, Liverpool', Proc ICE, IX, 1849-50

20. Phillips, J: 'Description of the iron roof, in one span over the joint railway station, New Street, Birmingham', Proc. ICE, XIV, 1854-85

21. Barlow, W.H.: 'Description of the St. Pancras' Statilon and Roof, Midland Railway. Proc. ICE, XXX, 1869-70

22. Rankine, W.J.M. 'A manual of applied mechanics, 1st edition, London 1858; many editions subsequently

23. Skempton, A.W.: ' The Boat Store, Sheerness 1858-60 and its place in structural history', Trans Newcomen Society, XXXII, 1959-60

Where practicable original sources of information have been quoted. However further details and illustrations of structures referred to in this paper, notably the Menai Suspension Bridge, the Britannia Bridge, the Sheerness Boat Store, the pioneer railway roofs and the Forth Bridge can be found in the Institution's anniversary volume Structural Engineering - two centuries of British achievement, Tarot Print Ltd., 1983

14
The genesis of engineering: theory and practice

A. Rupert Hall

If I were to begin speaking of the origins of engineering in the most obvious and literal sense, I should have to speak of the building of the great monuments that are still available for inspection in Egypt and elsewhere, and of others known to us only from literature or legend. The Pyramids of Egypt, the dams and irrigation canals of Mesopotamia, the ships that sailed to Punt, the tunnel of Samos and the Corinth boat-lift, the water-raising wheel and the Archimedean screw, the water-wheel, are all over a period of two millennia the diverse achievements of the first engineers. But in fact we know virtually nothing of the methods by which these monuments, ships, or machines were planned and executed. We know some details; we know something of the way in which the ancient Egyptians cut and prepared stones in their quarries because partially completed blocks remain there; we know from pictures that their rather primitive ships were strengthened by a longitudinal hogging-truss; we may be certain (though we have not the exact method) that some geometrical technique was required before the carpenter could assemble the wooden pieces of an Archimedean screw. But there is far more of which we are ignorant; we know nothing of the oldest methods of surveying and levelling; little of the transport and hoisting of blocks of stone; of the diverse invention of two types of water-wheel. Much of the first history of engineering, therefore, can amount to no more than a description of what we discover to have been made, with some speculations about the methods that were probably employed. However, we can be quite certain that no great sophistication was involved either in the planning or in the execution; great works were accomplished by great expenditure of time and muscle-power. Theoretical analyses and computations were of course quite absent, except as relating to quantities. Exercise-book problems of the type " If one man can move a ton of earth 100 yards in one day, how many will be required to move 1000 tons a mile in five days? " do exist in Egyptian mathematics, for example; the earliest " theory " of machines—based on the principle of the lever—was formulated by the Greeks in the fourth century B.C., but this had little or no relevance for actual machines. Certain ancient works must, in all probability, have involved the preparation of simple plans before the site was actually marked out; but of these things, too, we know nothing. So far as we can tell, apart from the facts that stone is harder to work than wood, and that the masses to be moved were far greater, the methods used in building a Pyramid were no more sophisticated than those required for making a wooden hut; and an irrigation canal was simply a field-ditch on a gigantic scale.

I really wish to discuss the beginnings of engineering in the modern sense, that is, engineering founded on design and analytical procedures. I can claim no specialised knowledge of any one of the branches of engineering through which this enquiry might be pursued in depth, but I shall try to elicit their common character in historical development.

Mathematical studies of problems relevant to the engineer begin in the 17th century, and I come to them later. When does design begin? It seems to me that there are at least two sides to this question, for we can consider design in the more literal sense as the preparation of a drawing—from the crudest sketch upwards—of what the craftsman is supposed to realise in three dimensions, either as a model or at full scale; or we can consider design as the procedure by which the drawing and other instructions given to the craftsman are formulated. As regards the former sense of the word there is, I think, no great problem; at least the historian can produce a somewhat thin series of drawings extending to a remote antiquity. I show here a few examples from the pre-Renaissance period; we have excellent reasons for believing that Hero's *Mechanics* of the first century A.D. was illustrated from the first by sketches of the devices he described; here are some well-known studies by Villard de Honnecourt in the 13th century; and finally some mechanical drawings illustrating his new ideas by an inventor of 1335.[1] As a technical drawing each of these is, no doubt, very imperfect, indeed the interpretation of drawings of this sort can be a principal problem for an historian of technology. But since they exist they do at least give us some notion of the mediation between the concepts of an inventor, or a design engineer or architect, and the minds of the craftsmen who will realise these concepts, if that may be done, with the aid of their experience and the skill of their hands.

The history of design in the larger sense is far more complex. Effectively we can only begin to study this as an historical problem from the Renaissance, because we have so little information about the preceding period and, as I have indicated, good reason to believe that there is little to know. We can be sure that the observance of due proportion was the foundation of design; but the definition of what due proportion was, on aesthetic or other grounds, remains pretty mysterious.

The Roman architect and engineer Vitruvius, writing in the time of the Emperor Augustus, illustrates this. Starting from the principle that the architect must be both theoretician and craftsman, trained in philosophy and mathematics as well as building practice, he declares that Order is one of its foundations, by Order meaning " the balanced adjustement of the details of the work separately . . . This is made up of Dimension [which] is the taking of modules from the parts of the work." To illustrate: he tells us that in laying out a temple, its proposed length is to be divided into 11½, 18, or 24½ parts according to the style adopted; one of these parts is the module, which is the thickness of a column. The columns are variously spaced 2¼ or 3 modules, and the height is another ratio dealt with at length in another section. If you should suspect that these proportions are matters of aesthetics alone, consider that Vitruvius also uses modules in describing how to make his artillery, catapults. One type is designed round a module which is one-ninth of the length of the arrow shot off; this is the diameter of the skein of twisted cords providing the elastic force. The members containing these cords are of one by one-and-a-half modules thick; the size of the tenons joining them are stated in the same measure, with many other dimensions all in proportion.

When the catapult throws a stone, the module is calculated from its weight, varying from 5 digits if the projectile weighs 2 pounds, to 1½ feet if the weight is 360 pounds. The curve relating weight to module is non-linear.

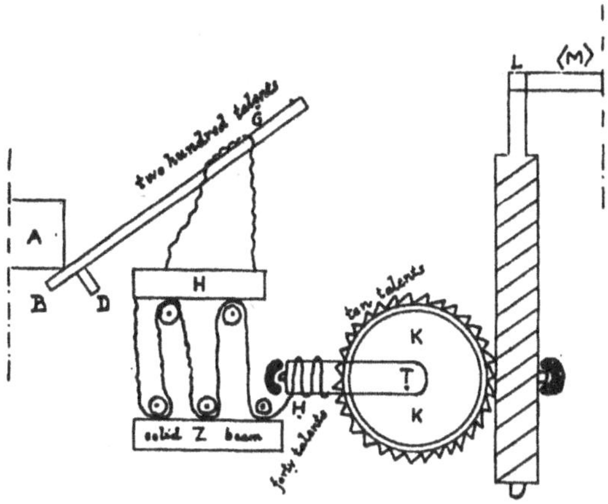

Hero's sketch of a powerful mechanical system: the handle (*top right*) rotates a screw, driving a gear with a windlass on the same shaft; this raises the weight by means of multiple pulleys and a lever. This is pure fantasy First century A.D.

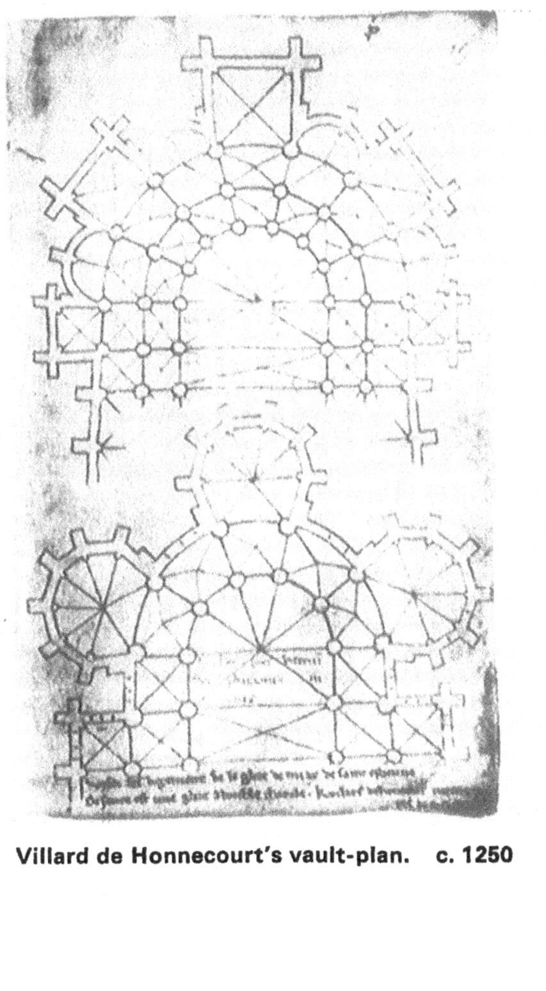

Villard de Honnecourt's vault-plan. c. 1250

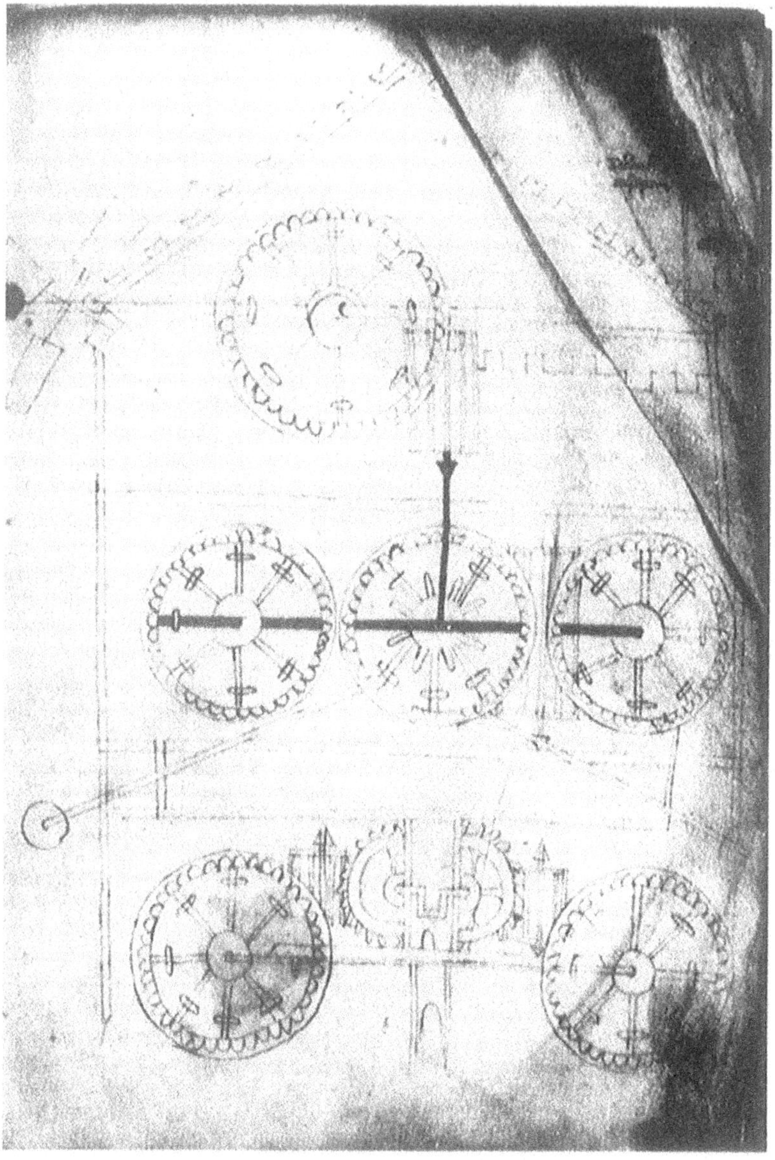

Guido da Vigevano's design for a wind-driven military vehicle: the sails are top right. Note the " prop-shaft " and system of gearing. The top housing carrying the wind-shaft is mounted on wheels or rollers, so that the sails can be rotated into the wind. Note the compound crank for front-wheel drive. 1335.

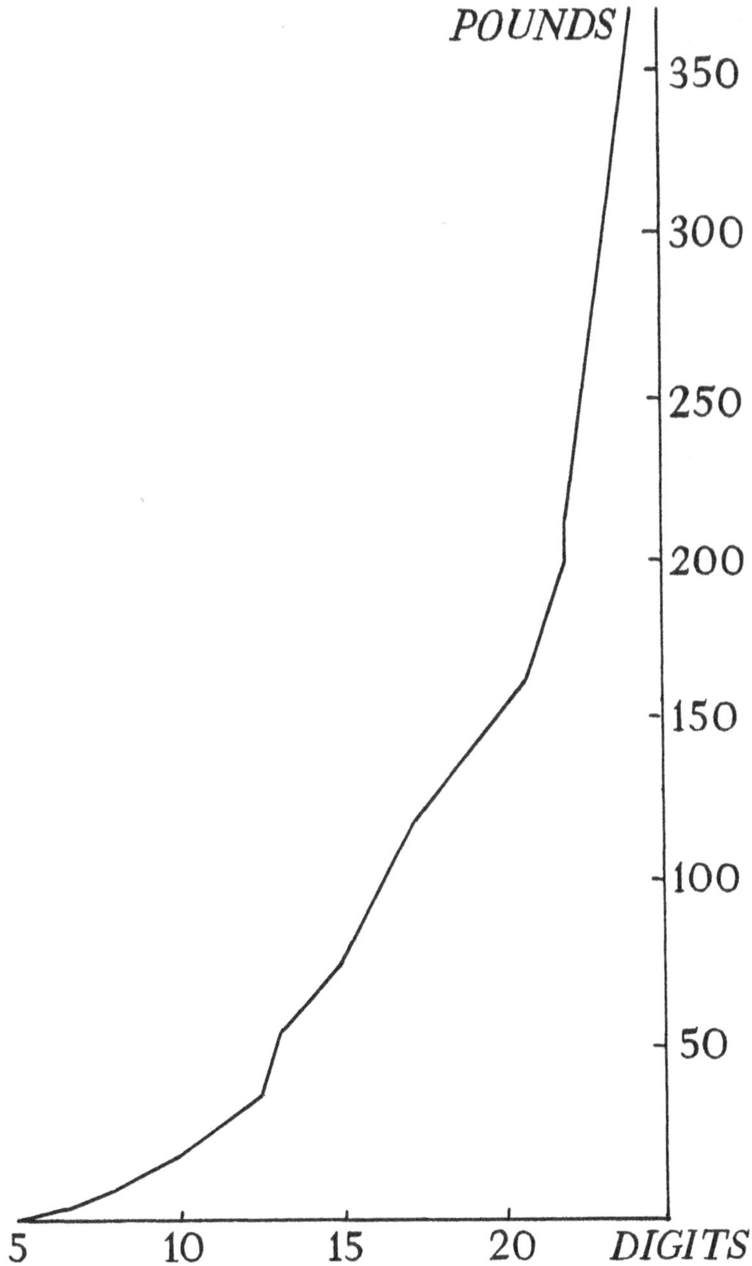

Vitruvius's rule relating the modulus (digits) of a stone-throwing catapult to the weight of the projectile (pounds).

In post-classical times it is more difficult, perhaps impossible, to find examples of *mechanical* design based on proportionality, though there are hints of it in Guido da Vigevano's treatise of 1335 describing his mechanical inventions. To run further ahead for a moment, the many machines described by Agricola (1556) are measured by dimensions, not by proportions. In architecture however the story is very different, as Mr. Mainstone has demonstrated, partly because of the aesthetic factor that is irrelevant to machines. Proportion was all-important to the Gothic architect, at least to those of the later period when the revival of the sciences from the 12th century onwards had restored some knowledge of geometry. In Mr. Mainstone's words, " the main lines and form of a structure—such as the heights of vaults and the proportion of piers and buttresses—[had] to conform to a certain predetermined geometric grid . . . the need for some control of this kind on the design was generally accepted and was strikingly acknowledged by the calling in of a professional mathematician in 1391 [this was at Milan, in connection with the cathedral] to compute the irrational height of the equilateral triangle then favoured as the control on the cross-section ". Prescriptions based on proportionality are commonly found in early writers on architecture and are very familiar in the writings of Leonardo, for instance; Mr. Mainstone gives the example of the complete derivation of all the dimensions in plan and elevation for a small pinnacle from the double or half square of a single dimension of the ground plan.[2]

The very simplicity of such schema, whether of Vitruvius or a Gothic architect, proves that they are artificial, that is to say not based on analysis of fundamental mechanical principles. What their origin was we do not know; one can only guess that each schema was worked out to fit the known dimensions of successful constructions, elaborated and improved in process of time. Though intellectually more appealing, such schema based on proportions were similar in their manner of use to modern " codes of practice " which indicate to the practical engineer the magnitudes he needs, whether he be choosing a column size or the sizes of resistance and capacity in a particular electronics circuit. There is the difference, however, that the former had no backing but experience, while the latter are based on analysis.

To go on with architecture for a moment, there were of course other approaches to design, at least from the Renaissance onwards. In general the history of statics, concerned as it largely was with producing a rational explanation for the law of the lever and applying this to explain a number of other phenomena, contributed little to engineering, but it is again well known that Leonardo was not singular among his contemporaries, though he was the most imaginative, in applying on paper statical methods to architectural and engineering problems. One of Leonardo's sketches shows how the horizontal outward thrust at the springs of an arch could, in imagination, be counter-balanced by cords internally passing over pulleys, and hanging weights. But so far as is known Christopher Wren was the first architect to make even qualitative use of conceptions drawn from statical considerations. More attention was paid to what one might call the negative theory of design, that is to say the elucidation from the study of failing structures of the faults to be avoided in building successful ones. Obviously even if the limits at which failure occurred were completely known—which they could not be—this would only have furnished one boundary to the designer's freedom ! Still, the study of failures would teach lessons about the need to supply lateral support in the form of buttresses, struts in wood roofs, or tension bars in masonry, as well as teaching the need for care in preparaing foundations, mixing mortar, or bonding brickwork. Such negative considerations can induce Alberti to remark: " That the entire [semicircular] arch is strongest of all, appears not only from experience but reason: for I do not see how it can possibly disunite itself unless one voussoir thrusts out another, a thing they are so far from doing that they assist and support one another". The merely negative character of such an argument is, however, demonstrated by the fact that architects both in Islam and in Christendom had long experience of the successful use of various forms of pointed arch.

The contributions of the architectural historian to the history of theory and practice in engineering are particularly important because of the richness of the physical remains available for examination and because there is relatively ample documentary evidence for the early period also. But one must not forget that design, on whatever basis, features in many other crafts. Machines will not work effectively if they are not duly proportioned; ships will not sail well unless the dimensions of the hull, of the sails, and the rudder are in a certain range of relations one to another; the civil engineer has corresponding problems in dealing with dams, or the dimensions of watercourses, or indeed in making the accurate surveys that must usually precede his labours. For most of such aspects of engineering we have little or no information about design before the Renaissance or even afterwards; we have to assume that the forms of things were determined by craft experience in practice. We know, for instance, that the New River here in London (completed in 1613) was accurately surveyed, but we do not know how the survey was effected. However, we do have with the Renaissance, and for the first time since antiquity, an engineering literature of some size and detail, a literature that begins to enter at least into the practice of design, if not its theory. It would lead me far afield to discuss this, but one can at least conclude that it teaches how the engineer should proceed *by example*, as indeed craft practice has also. Let me take two cases: machines and ships.

The literature of mechanics begins with Aristotle and his school. Their treatise on *Mechanical Problems* explains, among other things, the multiplication of effort furnished by the lever, windlass (or capstan), pulley and wedge, the basic machine elements to which later centuries added gears and the screw. The explanation is derived from a very rudimentary apprehension of the principle of virtual velocities. The practical literature of mechanical engineering begins with other Greeks, Ctesibios and Hero among them, who described war machines, automata, screw-presses, pneumatic devices and the like. We learn more about ancient machines from Vitruvius, whom I mentioned before.

Then follows a long silence, although as we know from other sources, there occurred in the high Middle Ages great developments in mechanical engineering involving the use of water- and wind-power for many industrial purposes—fulling cloth, reeling silk, sawing wood, blowing iron furnaces and working their hammers, stamping rags for paper and many other things.[3] This great extension of machine technology is slightly reflected in technical manuscripts of the 14th and 15th centuries, of which the work of Leonardo da Vinci is by far the best known, but is only fully documented in the "machine-books" of the 16th, whether those of Biringuccio and Agricola that describe real machines, or the more numerous ones depicting, often, complex and improbable devices.[4] The inventors' brazeness, no less than their ingenuity, proves that the machine had arrived in a big way.

Now all this happened—and we are practically on the verge of steam—without a single addition to the theory of machines ! To us it might well seem that natural power demanded the addition of a new dimension to that theory. It seems self-evident to a modern man that if an ox, a donkey or a man is doing work, no machine, however complex, will enable him to do more work without greater muscular effort. The best the machine can do is to make the work more convenient, or spread the action of the force over a longer period of time. But I should add that even this elementary point was by no means clear, in the theory of machines, in the 17th century, or even in the 18th. Many inventors still supposed that the work done by a given muscular effort could be indefinitely increased by suitable machines; Leibniz could describe a perpetual motion. Even Galileo, who wrote on this subject in the late 16th century, did not really add much to knowledge except clarity, though indeed Galileo did emphasize (perhaps for the first time in the history of mechanics) that machines do not " perform work ", but only modify the manner of its application. In his own words: " at the end of the action we will find that the only profit we have gained from the machine is to have trasnported the given weight in one piece with the given force to the given end; which weight, divided into [many small] pieces, would have been transported without any machine by the same force, in the same time, through the same distance ".[5] However, the action of water or wind is harder to understand, since their force is more difficult to measure; even the concept of there being a fixed quantity to measure is difficult to grasp. Since, taking a given stream, experience shows that by improving the design of a water-wheel worked by it, it is possible to obtain more and more power, it is by no means easy to perceive that there is a limit to the power obtainable, or to reach (therefore) a concept of relative efficiency.

Galileo is better known for two other applications of science to engineering on which, because of their familiarity, I need not dwell; these are the *Two New Sciences* of the book he published in 1638. The first of these analyses the strength of materials; in this Galileo demonstrated that the strength of a beam is directly proportional to its thickness, and inversely proportional to twice its length; demonstrated, I mean, according to his own assumptions which placed the neutral axis at the base of the beam. His development of this theory is no less elegant than his physical conceptions are ingenious, if mistaken; and it is, be it noted, a theory about proportions, not absolute strengths. Unfortunately, as we know now, this theory should not have been applied, and indeed it was not. The other new science described by Galileo was kinematics, which led him on to ballistics. In the 17th century ballistics placed excessive demands on mechanical engineering; the methods available at this time for casting and boring guns, and fabricating projectiles, were quite unable to produce the range and accuracy in shooting which alone make a mathematical treatment of the trajectory worthwhile. So Galileo's parabolic theory had some possible application in mortar-shooting, but nothing more.

As these examples suggest, Galileo was most emphatic in his belief that " mathematics " and " philosophy " are one; in his own words again:

" I expect a terrible rebuke from one of my adversaries, and I can almost hear him shouting in my ears that it is one thing to deal with matters physically and quite another to do so mathematically, and that geometers should stick to their fantasies and not get involved in philosophical matters where the conclusions are different from those in mathematics. As if the truth could ever be more than one; as if geometry in our day was an obstacle to the acquisition of true philosophy; as if it were impossible to be a geometer as well as a philospher ! So that we must infer as a necessary consequence that anyone who knows geometry cannot know physics !"

Moreover, mathematics and philosophy together could be practically useful; Galileo effected the great change in his career that took him from Padua back to Florence by pointing out the use of the telescope at sea. In the book I have been speaking of, he described the isochronism of the pendulum, the principle of improved timekeepers, and drew attention to the limitation of Nature's resistance to a vacuum. From this observation sprang the science of pneumatics, and from that, deviously, the atmospheric steam-engine.

But now I should turn back to ships. We know little about the methods of construction used for ancient shipping, so I shall begin with the method of design employed by the shipwrights of the Arsenal at Venice in the late Middle Ages.[6] The hull of a ship is a most complex form, dicatated as you know by the frames attached at intervals to the keel; each frame is a complex curve—especially in the older sailing ships—and the frames on either side of the broadest or midships section (often placed forward of the centre of the keel), towards bow and stern, vary progressively in shape and size. The Venetians based their dimensions, as Vitruvius did, on a module which was normally the beam of the proposed galley; this was multiplied in a fixed proportion to give the deck-length, and a fraction of this in turn gave the length of the keel. The outline of the midships section was plotted by taking stated proportions of the beam at each foot depth; through the points a freehand curve was traced. The other frames were derived from the midships section by similar methods of proportionality. As in architecture, we need see no more in this than an appreciation of symmetry combined with a desire to codify experience so that vessels of the same type but different dimensions would have approximately the same desirable qualities. As Lane wrote:

" It would be superfluous to suspect a Platonic passion for geometric proportions. As a practical means of making experience valuable it was desirable to simplify the number of different measurements used in determining the lines of the ship. The more simply these determinants could be stated the greater was the possibility of finding widely applicable ratios between the various measures ".[7]

It is true that the schema used by shipwrights did leave the individual master much latitude for the exercise of his skill, and that apparently similar vessels did correspondingly differ considerably in their sailing qualities. Nevertheless, it is clear that the measurements thus defined by the various schema were quite precise enough to determine the typology of vessels, known by distinct names. As in architecture, the schema had no rational or analytic foundation; they are another example of the marvellous way in which, for purposes of design, experience can be clothed in a numerical pattern which is in itself quite irrational, though serviceable. Experience, as a basis for design, can show what forms of construction are *good*, but only analysis can determine the optimum.

Reflecting, by the way, on the shipwrights' belief that design was largely a matter of scale-adjustment, we can understand better the necessity for another piece of Galileo's argument in the *Two New Sciences* of 1638, which I mentioned just now, a book that opens with some comments on the practical skill of the men working in the Arsenal. For Galileo there shows that if a large member or structure is merely a scaled-up copy of small one, it must be weaker. Shipwrights and others must often have been puzzled to know why large ships had to be so much more massively constructed than small ones.

The Italians naturally preceded the English in writing and printing books on the shipwright's art; English manuscripts on this subject date from the 16th century; the earliest printed work is Robert Dudley's *Arcano del Mare* (1646), published in Italian as the title suggests.[8] The English practice seems to have differed somewhat from the Italian in that the schema gave rules for choosing the centres and radii of curves forming the side of a ship. However, it seems clear that in England as at Venice the shape of the hull was by no means left to the unaided intuition or whim of the " mystery " of shipwrights. In the 17th century the subject began to assume a more rational form; both the hull, and the action of the sails in propelling the ship, began to receive theoretical attention. In 1671 Nicholas Witsen published a splendid folio entitled *Scheep's Bouw en Bestier*, or *Naval Architecture*, which not only reviewed the history of shipbuilding but gave an elaborate account of contemporary design and constructional practice. Witsen also attempted to answer some more abstract questions: I quote from the account of his book in the *Philosophical Transactions*:

> [He explains] and gives reasons for the several sizes and shapes of the parts of a ship; as why the masts ought to be just of such a bulk and height? Why some of them must incline backward, some stand upright? Why a small rudder can turn a great ship, and a little anchor stay it? What maketh ships not feel the rudder? Why vessels too broad are weak and prove inconvenient in high winds? Why long and moderately narrow ships endure the sea better, than short and broad ones? How the keel ought to be placed? Why a ship is to be broader before, than abaft?[9].

A similar, more mathematical treatise was that of Paul l'Hoste, *Art des Armées Navales* published in 1967, which contained a *Théorie de la construction des Vaisseaux*. Far more truly abstract, and scientific, were the studies in hydrodynamics made by Christiaan Huygens in Paris, and Isaac Newton, the latter being described in Book II of his *Mathematical Principles of Natural Philosophy* (1687). Huygens investigated the hypothesis that the resistance of a fluid to a moving body increases as the square of its velocity by making experiments in both air and water; Newton initiated the mathematical study of the problem: what shape must a body have to encounter least resistance in traversing a fluid? I think I need not dwell on these researches, nor on the further extension of rational fluid mechanics in the 18th century, notably by Leonhard Euler (1707–83). But I must note the publication by the French mathematician and physicist Pierre Bouger (1698–1758) of his *Traité du Navire*, generally regarded as giving the theoretical foundation for naval architecture as it has developed in modern times. The fruits of these studies were slow to ripen; only in the second half of last century did analytical and experimental studies assume serious importance, and begin to affect actual practice. The great factors here were, of course, practical ones; the change to all-metal construction, increased speeds and hence greatly increased engine-power, enforced abandonment of traditional design processes that had produced the clipper ship, the apotheosis of sail, which had so large a share of world-wide trade from 1849 to about 1875, from the yards of Aberdeen, the Clyde, and the United States. At least so far as I am aware the design of the clipper's hull, giving her great speed at the cost of great " wetness " and discomfort, owed nothing to science or analysis.

This, I think, gives a hint of an historical generalisation of some interest. When the historian attempts to consider the manner in which modern procedures first began to supplant traditional ones in the period between 1750 and 1850—which is also the period of the industrial revolution—he may at first be inclined to suppose that the change was brought about solely by the increasing power and flexibility of scientific knowledge, which was exploited by a more enlightened class of manufacturer who was willing to be taught—by the *entrepreneur* or innovator who

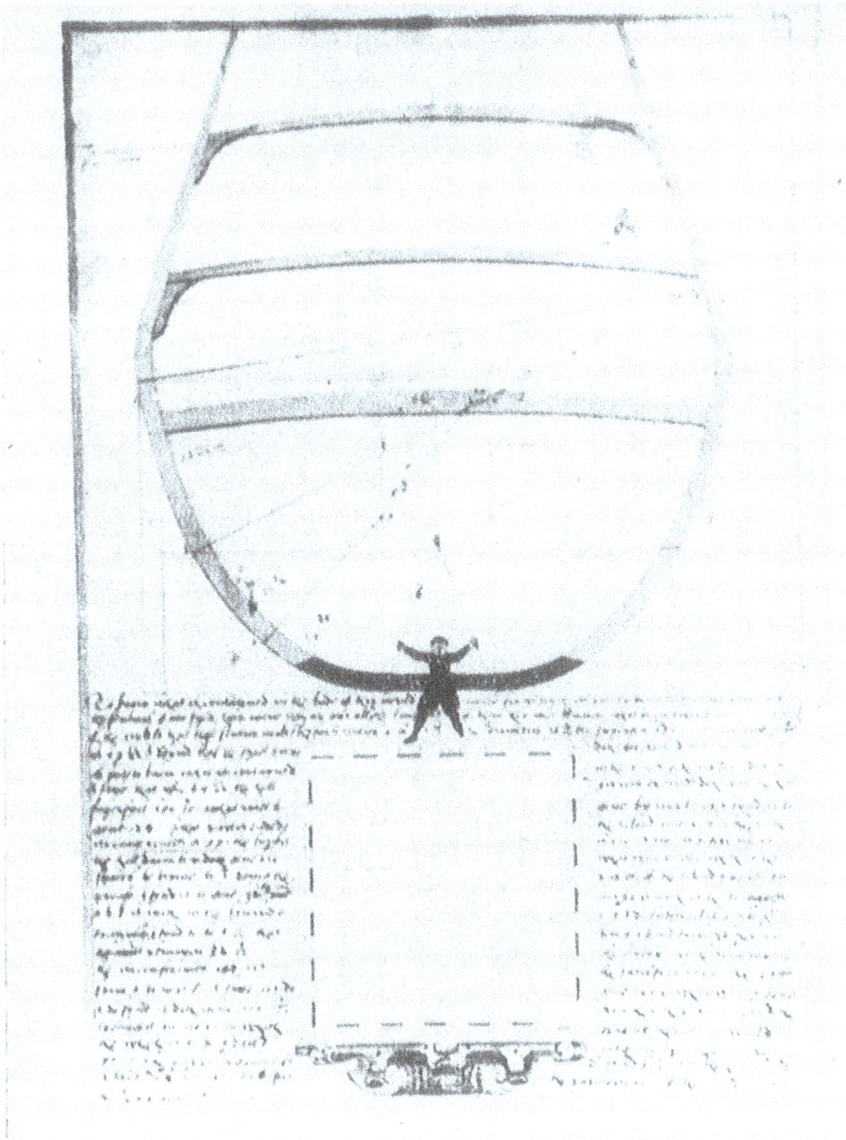

Mathew Baker's sketch of the procedure for striking the curves of the midships section of a vessel. 1586.

would actually seek advice from the chemist, mathematician, or natural philosopher. It seems to me that this is too simple a view and that other circumstances—alterations in the whole context and subject of design—had much to do with the change. Design procedures had to be altered because non-traditional materials were coming into use, and non-traditional problems had to be faced.

Let me revert to building construction. Here we see the development of a considerable body of mathematical theory and experimental measurements that was of slight immediate effect, but that bore fruit later in a different technological context. I have mentioned Wren's interest in statical analysis, a subject considerably advanced (so far as the problem of the arch is concerned) in the *Traité de la Mécanique* of Philippe de la Hire (1695). La Hire's smooth-voussoir theory was obviously inapplicable in practice, where friction imparts much strength to the arch, and was above the heads of engineers. During the 18th century sounder notions were gruadually worked out by Gautier, Couplet, Danisy, Perronet, Coulomb and others—this was largely a French field. But at the end of a century which was one of very great advance in applied mathematics the problem had still not been given a useful solution; it was still necessary to base engineering practice on rules found by means of experiment, as modified by experience. In other words, the best way of ensuring that a bridge would stay up was to copy bridges that had passed this test of time; for although theoreticians were by 1800 agreed that arches failed by a hinging at the joints, they could not define the conditions under which this failure would occur.

During the 18th century also extensive experiments on the strength of conventional building materials—stone, brick, lime-mortar, timber—were made by such natural philosophers as Musschenbroek, Buffon, Coulomb, Gauthey, Girard and others. Again, from the time of the military engineer Vauban to that of Coulomb, who produced a very elegant mathematical analysis, the action of earth-pressure upon a retaining wall had been carefully investigated experimentally and mathematically. All this work, however—confined, as I have remarked, to one nation—seems to have had little effect upon conventional building. Consider the case of Charles Auguste Coulomb (1736–1806), for example. Although Coulomb later won great distinction as an experimental physicist, he was trained as an engineer in the *Ponts et Chausées*, and he held the post of Inspector of Hydraulic Works at the time of the Revolution. While serving as an engineer in Martinique he completed in 1773 the work I have several times alluded to, entitled *Essai sur l'application des règles de maximis et minimis à quelques problèmes de statique relatifs a l'architecture*. This great work on the theory of structures remained virtually unknown after its presentation to the Academie des Sciences in Paris; it was Thomas Young's praise of this essay some 40 years later that called attention to it, at least in England.

A transition did take place when a new material, cast iron, became available for construction (the Iron Bridge at Coalbrookdale, you will recall, was completed in 1779), and when new problems arose, such as the great fire-risk in the vast mills and warehouses created by the industrial revolution. Not very long afterwards wrought iron also entered into structural use, presenting further problems in design since its merit lay in its tensile strength, as in suspension bridges. Structures relying heavily on the strength of metals could not be designed by following traditional rules: it was necessary to guess, to experiment, to attempt analysis. As one can see from the Iron Bridge itself, some strange expedients were adopted both in the design, and in the actual fabrication. Some notion of the state of the art in England about 1800 may be gained from the Parliamentary enquiry in that year into Telford's proposal to throw a single-span cast-iron bridge of 600 feet across the Thames to replace London Bridge It is clear both that the properties of a cast-iron arch bridge were little known in 1800 and the expert witnesses were eager to calculate the stresses involved, though of course their attempts to do so were inadequate. A year or two later (1805) John Robison, writing the article on " Strength of Materials " for the *Encyclopedia Britannica* was ignorant or contemptuous of what the Continental mathematicians and experimenters had achieved; this was an evil consequence of the long insularity of British mathematics. As late as 1879 the failure of the first Tay Bridge seems to reveal defective utilisation of theoretical investigation in desigining a major structure; but the story of this disaster is complicated.

With respect to prime-movers, to take another example, momentous changes in the design process occurred only after the whole technique of generating power had been revolutionised, first by the introduction of the steam-engine, then by that of the water-turbine. Yet it would be a mistake to suppose that there had been no interest in the analysis of the older prime-movers depending on wind and water. Not to mention the geometrical studies of the shape of the gear-teeth vital to mill-work made in the 17th century by Huygens, Roemer and La Hire, from the early 18th century there was very great interest in machines, and in such problems as deriving the maximum power from a given head and volume of water by one or other of the various forms of water-wheel. With regard both to theory and to practice the most important early work published in France was that of B. F. de Belidor, *Architecture Hydraulique* (1737–53), a work dealing with all aspects of hydraulic engineering. His contemporary counterpart in England was John Theophilus Desaguliers, author of *A Course of Experimental Philosophy* (London, 1734, 1744). Desaguliers was much interested in the more striking working machines of his day, including the steam-engine, and obtained many good descriptions from a practical Midlands engineer, Henry Beighton. He also perceived that the theory of machines was important, particularly with regard to the relative efficiency of various types. Then as now there were projectors eager to deceive the ignorant public. There was no doubt in Desagulier's mind that the amount of useful effort to be got out of, say, a given weight of water flowing down a given height in a given time was limited, and accordingly he understood that the mechanical problem was simply that of harnessing as much as possible of this potential effort. Of course, as with everyone else at this time, his abstract handling of this question was vitiated by inability to decide whether " motion " was more truly measured by momentum (MV) or *vis viva* (MV^2), the problem being that *vis viva* is not conserved when imperfectly elastic bodies collide. To take the specific case of overshot and undershot water-wheels, Desaguliers records that he had:

> " generally found that a well-made Over-shot Mill ground as much Corn in the same time as an Undershot Mill with ten times less water, supposing the Fall of Water at the Over-shot to be 20 Feet, and at the Undershot to be six or seven Feet. I generally observed [he goes on], that the Wheel of the Overshot Mill was of 15 or 16 Feet Diameter, with an Head of Water of four or five Feet, to drive the Water into the Buckets with some momentum ".[10]

Which leads him to enquire whether or not this dropping of the water to impact on the wheel is worthwhile. Of course it is obvious that Desaguliers is not comparing like with like. And if we should turn to another authority, Belidor, we should find exactly the opposite view, that is that the *undershot* wheel is preferable.[11]

Desaguliers' exhortation:

> "He that would meddle with water-works, should know so much of Mathematicks, as to understand mechanical Principles; and be so much a Philosopher, as to be skilled in Hydrostaticks and Pneumatics . . ."[12]

was very proper; but so long as authorities differed so markedly, their practical influence must be diminished. Here, we may say, the debate was ended by the classical enquiries of John Smeaton (1759) which definitely established the superiority of the overshot wheel, and gave the optimal relation between the speeds of wheel and water.

Smeaton's investigations—which included the windmill as well as the water-wheel—possibly did contribute to the improvement of these machines, though it may be that the use of iron in the construction of mills contributed more. But as everyone knows, the really great discoveries about energy and its harnessing did not come from this source at all, but partly from science proper and partly—perhaps one should say largely—from the new prime-mover, the steam-engine. (When I say great discoveries I mean the laws of thermodynamics.) Comparing falling water with rising weight, as Smeaton did, was one thing; but how could you compare rising water with bushels of coal? Watt did this empirically. But the question remained—what was the *maximum* weight that a bushel of coal would raise through a given height in a given time? I need not go over Carnot's ideal reversible heat-engine and other steps in the story. In brief, after 1850, but not before, rational sense could be made of the ways in which machines and other devices either release energy, or convert it from one form to another, with varying degrees of efficiency. But again it was long before thermodynamics counted in engineering design; around 1850 the direct methods available to the working engineer for comparing two machines were all that he required.

As I have spoken of the water-wheel, I should say a word about its equivalent in modern engineering, the water-turbine. Vertical shaft water-wheels having curved blades and sometimes a curved profile also, the so-called tub-wheels, are known from at least the 16th century, and were made by Persian craftsmen into modern times. In these machines the water strikes the wheel in a single jet, and it seems likely that they worked rather from this impulse than by reaction. The true reaction or pressure turbine was undoubtedly invented by Benoit Fourneyron (1802–67) in 1827, this being an outward-flow machine with fixed central guides. Fourneyron's turbine was succeeded by the axial-flow type of Jonval (1843) and the radial inward-flow machine of James B. Francis (1840 onwards), which is the ancestor of most modern forms. With low heads and at low speeds the water-turbine was, and still is, used for ordinary mechanical purposes such as grinding corn, but its great opportunity for development was, of course, furnished by the parallel evolution of a quite different technology, that of electric power generation. The water-turbine is of little direct importance as a new invention replacing the thousands of vertical water-wheels that once functioned over northwestern Europe—and to a declining extent do still function in the more primitive lands of the south-east—this work of destruction was performed by the internal combustion engine and electric motor; it is important as the basis of a whole new technology that took shape more than 50 years after Fourneyron's invention. And I would emphasise that that was a theoretical invention; Fourneyron's design was not based on experience or intuition, but on a careful study of fluid-flow and its ability to do work.

This has been only a hasty sketch of the changing scene of engineering design in early modern times. It would be historically interesting to amplify and develop every aspect of it, for we are still pretty ignorant about the methods by which forms and dimensions were arrived at. What I have tried to show, from the examples I have briefly reviewed, is that the advent of modern, analytical design was by no means a simple and obvious result of the investigations made by mathematicians and physicists, nor, even when new methods became available, were they quickly applied to familiar materials and problems. Rather it was when novel situations had to be contended with that new approaches to design commended themselves. In this way the progress of technology is highly interdependent, progress in one branch constantly cross-fertilising another. We can never be sure, or at least the historian from his evidence cannot be sure, that the most important changes in any one particular branch will be effected simply by greater and greater advances in design within that one branch; something from outside, changing the whole context, may be far more importnat. Engineering, after all, is a social discipline that can never be completely dominated by analytical techniques.

NOTES

1. The slides shown at this point were taken from A. G. Drachman, *The Mechanical Technology of Greek and Roman Antiquity* (Copenhagen, 1963); the *Album de Villard de Honnecourt* (Bibliothèque Nationale, Paris, n.d.) and the *Texaurus* of Guido da Vigevano (Bibliothèque Nationale, Paris, MS. Latin 11015).
2. Rowland J. Mainstone, " Structural theory and design before 1742 ", *The Architectural Review*, April 1968, pp. 303–310.
3. See Bertrand Gille, *The Renaissance Engineers* (London, 1966).
4. G. Agricola's *De re metallica* (1556) is available in an English translation by H. and L. Hoover, reprinted by Dover Books; Biringuccio's *Pirotechnia* has been translated by Cyril Stanley Smith and Martha Teach Gnudi, the book re-issued by Basic Books, New York 1959. On the machine-books generally see A. G. Keller, *A Theatre of Machines* (London, 1964).
5. I. E. Drabkin and Stillman Drake (trans.), *Galileo Galilei on Motion and on Mechanics* (Madison, 1960), p. 149.
6. Here I am indebted to F. C. Lane, *Venetian Ships and Shipbuilders of the Renaissance* (Baltimore, 1934). A French revised edition was issued in 1965.
7. *op. cit.*, p. 90.
8. See Sir Westcott Abell, *The Shipwright's Trade* (Cambridge, 1948).
9. *Philosophical Transactions*, VI, No. 77, pp. 3006–3012.
10. J. T. Desaguliers, *Course of Experimental Philosophy*, Vol. II (London, 1744), p. 532.
11. *Architecture Hydraulique*, Vol. I (Paris, 1737), p. 286.
12. Desaguliers, *loc. cit.*, p. 414.

15
The functions of engineering research in the university

A.J.S. Pippard

WHEN I received the invitation of the Council of the Institution to deliver this Lecture, I felt that its acceptance was almost a pious duty to the memory of my distinguished predecessor, who from 1884 until 1904 occupied the Chair of Civil and Mechanical Engineering at the Central Technical College of the City and Guilds Institute. The praising of famous men, to which we are exhorted, requires much skill, knowledge, and tact, and my pleasure at the honour implied by the invitation was largely tempered by vivid awareness of my inadequacy for the task. It would be, I think, a little presumptuous and certainly superfluous to attempt a lengthy eulogy of William Cawthorne Unwin. His work stands for all time as his true memorial, and I think that Miss Unwin was wise when, in her bequest to the Institution, she stipulated that this Memorial Lecture should deal with engineering research. Engineers, however distinguished, must pass, but the influence of their work on future generations can provide an enduring fame which no form of words, however eloquent, will achieve.

It seems proper, however, that the early Lectures in this series should be connected with some subject which held special appeal for Unwin. Such a subject is not difficult to discover ; even a glance at the titles of his published works will indicate the breadth of his interests. His vision embraced a large part of the field of engineering science and, in his long and active career, there can have been very few of its problems which did not interest him. In pursuance of this idea, I propose to deal this evening with the story of research into the stresses in masonry dams, a subject which caused considerable controversy during the latter part of the last century and the first decade of the present one. I cannot hope to do full justice to the subject in the time available, and I am perhaps inviting trouble for myself, since some of those who played leading parts in the controversy are, happily, still with us and may dispute my interpretation.

A simple recapitulation of the facts, interesting as they are, would scarcely be justifiable on such an occasion as this unless something were to be gained from the re-telling. I hope, however, that the story may be of value in helping us to appreciate the spirit which animated engineering research at that time, to understand something of the differences between that period and our own in this connexion, and, above all, to assess the functions of engineering research in the Universities at the present time. I have chosen this topic principally because of the important part played by Unwin, but there is another reason, a mention of which will, perhaps, not be quite out of place. In 1911, a very recent graduate in engineering was introduced by a much admired teacher to a Paper by L. W. Atcherley and Karl Pearson, which had been published a few years previously and to which reference will be made again very shortly. The young man became sufficiently interested to follow the subject through the paths which he proposes to re-tread this evening. It was his first contact with Unwin's work and, without the inspiration gained from the study he would, in all probability, not be giving this Lecture to-night.

The earliest statements of the principles underlying the design of masonry dams of the gravity type were made by distinguished French civil engineers in the early part of the nineteenth century, and were very simple in essence; the design had to be such that the dam would not turn over by pivoting about the toe under the pressure of the impounded water, and the resistance to shearing had to be sufficient to prevent sliding along any horizontal section. At a later date these simple criteria were judged to be insufficient and a third was added, namely, a restriction of the compressive stresses at the toe and heel of the dam, when the reservoir was full and empty respectively, to a predetermined safe value.

In 1870, Professor W. M. Rankine was consulted about the proposed design for the Periyar dam in India, and his report [1] has become a classic in the literature on this subject. In it he rejected the views of the early French engineers, Graeff and Delocre, that the compressive stresses should be limited to the same intensity at heel and toe, pointing out that on the ordinary bases of calculation the stress was the vertical component only; the true stress was higher and was governed by the batter of the face or back of the profile at the particular section under consideration. It was, therefore, necessary to keep the vertical component of stress at the toe lower than that at the heel. More important than that, however, Rankine introduced the "middle-third rule" to determine the widths of horizontal sections required to eliminate the possibility of tensile stresses occurring in the mortar joints and initiating cracks. In the course of the discussion which it is proposed to recount, the suggestion was made that Rankine had been anticipated by several years in the adoption of this criterion, but whether this is true or not I have been unable to ascertain. It is certainly true that prominence was first given to it in Rankine's report, and it became thereafter a generally accepted criterion in masonry design. As will be seen shortly, it was also the cause of much subsequent trouble and controversy.

In 1895 the Bouzey dam, near Epinal in France, failed. The cause of the disaster was naturally the subject of much discussion, and in 1895 Unwin wrote an article [2] for *Cassier's Magazine* in which he showed that the no-tension condition had been seriously violated at the section of the dam where failure began, and he definitely ascribed the collapse to this cause. About the same time, in a Discussion [3] at the Institution of Civil Engineers on the Manchester and Liverpool water-supply, Unwin commented upon a remark by Mr Charles Hawksley that the stability of the Vyrnwy dam had been calculated to have a factor of safety of 2. Unwin expressed

[1] Bibliographical references are given on p. 283, *post*.

gratification that the words " by calculation " had been introduced, since he considered that the assumptions made were so wide of the truth that the result was fictitious and misleading. In his opinion, if the real factor of safety against overturning could be calculated, it would be more nearly equal to 10 than to 2. The French engineers, he said, perceiving the erroneousness of the assumptions on which the ordinary theory of overturning had been made, had abandoned them and taken refuge in stress calculations; in his opinion, the Bouzey dam was the first and only case on record of a high dam failing by overturning.

These two early references are quoted to show Unwin's long-standing

Figs 1

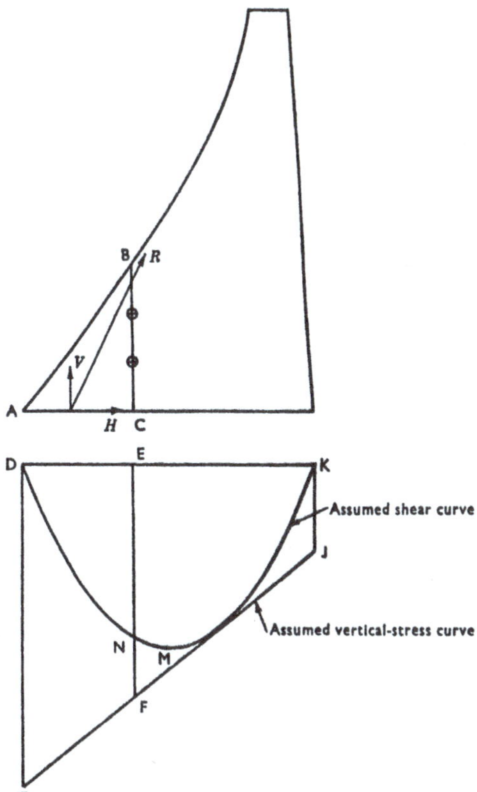

DEVELOPMENT OF TENSILE STRESS ON A VERTICAL SECTION

interest in this problem; it was no sudden fancy which made him take part in the historic dispute on this subject.

In 1903 the Drapers' Company of the City of London gave £1,000 to the University of London for research; that sum was allocated to the

Department of Applied Mathematics at University College, then in charge of the distinguished elastician, Professor Karl Pearson. In the words of Karl Pearson himself, "It seemed desirable to commemorate this—probably, first occasion on which a great City Company had directly endowed higher

Figs 2

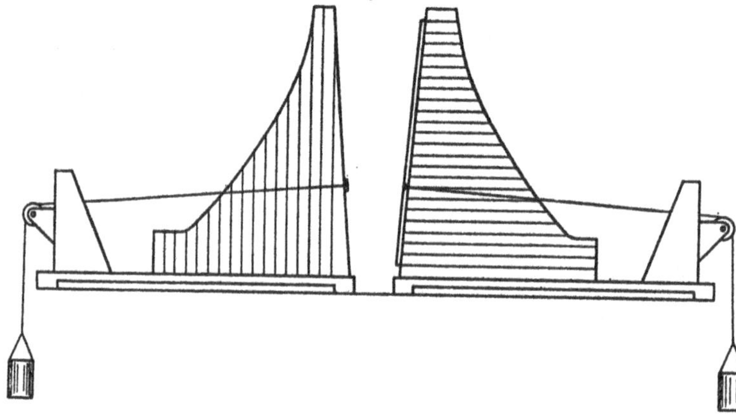

BEFORE APPLICATION OF WATER PRESSURE

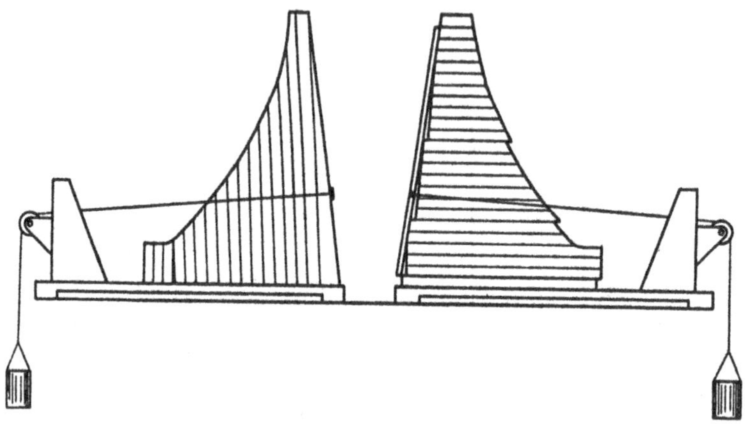

COLLAPSING UNDER WATER PRESSURE

MODEL DAMS OF ATCHERLEY AND KARL PEARSON

research work in mathematical science—by the issue of a special series of memoirs." These were known as the Drapers' Company Research Memoirs, and the second of the series,[4] published in 1904, was entitled "On Some Disregarded Points in the Stability of Masonry Dams," by

L. W. Atcherley with some assistance from Karl Pearson. This Paper directed attention to the fact that, in the past, only horizontal sections of a dam profile had been examined to ensure that no tension was developed, and then by mathematical argument showed that the assumption of linear distribution of normal stress on horizontal sections, which was always made *faute de mieux*, led with almost equal validity to a parabolic distribution of horizontal shearing across that section. If a vertical section were taken through the dam profile and the stability of the portion between that section and the toe were considered in a manner similar to that used for horizontal sections, the resultant action might fall outside the middle third of the vertical section, and tension would develop. This is shown in *Figs 1.* BC is a vertical section; DGJK is the assumed distribution curve of normal stress plotted in terms of masonry units; DMK is the assumed parabolic shearing-stress curve in the same units. If the section ABC of the dam be considered, it will be seen that there is a vertical resultant force, V, acting upwards, representing the difference between the areas DEFG and ABC, and a horizontal shearing force, H, represented by the area DEN. When these are compounded the resultant force indicated by R may pass outside the middle third of the section CB, causing tension at C. Model-dams were made of wooden strips arranged vertically and horizontally respectively, and pressure was applied to their backs to produce distortion (*Figs 2*). The Authors were satisfied that these tests confirmed the existence of the tension they had predicted, and they summarized the results of their research in the following terms:—

"(1) The current theory of the stability of dams is both theoretically and experimentally erroneous, because;—

(*a*) Theory shows that the vertical and not the horizontal sections are critical.

(*b*) Experiment shows that a dam first gives by tension of the vertical sections near the tail.

"(2) An accepted form of profile is shown to be stable for horizontal sections, but unstable for vertical sections.

"(3) The distribution of shear over the basis must be more nearly parabolic than uniform, and it is not unreasonable to assume that the former distribution will describe fairly closely the facts until we have greater knowledge.

"(4) In future we hold that, in the first place, masonry dams must be investigated for the stability of their vertical sections. If this be done, we feel pretty certain that most existing dams will be found to fail, if we accept the theory of stability usually adopted for their horizontal sections. This failure can be met in two ways:

(*a*) By a modification of the customary profile. We

think it probable that a profile like that of the Vyrnwy Dam would give better results than more usual forms.

(*b*) By a frank acceptance that masonry, if carefully built, may be trusted to stand a certain amount of tensile stress. It is perfectly idle to assert that it is absolutely necessary that the line of resistance shall lie in the middle third for a horizontal treatment, when it lies well outside the middle third for at least half the dam for a vertical treatment.

"Without being unduly alarmist, we should like to draw the attention of practical engineers to the point that the masonry dams they have designed to be, according to a defective theory, free from tensile stress are probably subjected not only to tensile but to shearing stresses far exceeding anything they have in the least imagined. We may even speculate whether this truth may not be to some extent the source of weakness in certain dams which have existed in the past."

These opinions of Atcherley and Pearson are quoted at length because they put the cat fairly among the pigeons. The first indication that the research had attracted attention from practising engineers appeared in the discussion on the Coolgardie water-supply [5] when Sir Benjamin Baker criticized the standard middle-third method of calculation on the ground that it was based upon an unsound theory. He considered that the problem was infinitely more complicated than was suggested by that method, and expressed pleasure at the serious attempts made by Atcherley and Pearson towards a better understanding of it. He said that he did not "care a pin" about the accuracy of the results, but it was the first fruits of an earnest attempt to carry the theory of an elastic solid a stage farther than it had been carried before. He had transmitted the memoir to the engineers of the Public Works Department in Egypt, with the comments that he considered the investigation deserving of every encouragement and assistance and that, although personally he did not agree with the conclusions, he believed that the mathematics was correct on the assumptions made. He had, however, told Professor Pearson that he thought that not all the conditions of the problem had been included. On the two assumptions of uniform shear-distribution and parabolic shear-distribution, there were tensions of 3 tons per square foot and 10 tons per square foot on a vertical plane of a typical dam, and the latter figure was very high.

Consideration was being given at this time to the problem of raising the dam at Aswan, and Sir Benjamin Baker, in his capacity of Consultant, made a number of stiff jelly models; the engineers in Egypt agreed, after 3 or 4 hours' experiment with these, that raising the dam was not so simple as it had appeared at first.

In March 1905 an article [6] appeared in *Engineering* announcing the postponement, if not the abandonment, of the scheme for raising the of the height Aswan dam by 20 feet. It was stated that, in view of the new light on the stability of dams afforded by Atcherley's Paper, it was considered imprudent to do so without much further consideration. Shortly afterwards, Unwin entered the field with a Note [7] to *Engineering*, in which he pointed out that the linear distribution of stress on horizontal sections was pure assumption and agreed that if the conclusions in Atcherley's Paper were true the position would be very serious. Referring to the pointed attention which had been drawn to this Paper by Sir Benjamin Baker, he emphasized that dams were monoliths having no plane joints, and there was no reason for supposing that they could not withstand a reasonable amount of tension. He agreed, however, that it was desirable that there should be no tension at the water face. On the much criticized assumption that normal stresses were distributed linearly on the base, he developed a method for obtaining the distribution of shear stress across a section of a triangular dam and showed that, on his assumptions, this distribution was also triangular, having zero value at the heel and the maximum value at the toe. This Note was elaborated 3 weeks later [8] when the normal and shearing stresses were completely worked out for a triangular profile.

Others now took part in the discussion. Mr E. P. Hill, in a letter to *Engineering*, pointed out what he felt to be the entirely erroneous nature of Atcherley's reasoning; a dam and its bed formed one body, and an arbitrary division into strips was unsound. He was particularly perturbed by the decision to postpone the raising of the Aswan dam on the evidence of this thesis, which had naturally caused anxiety among authorities in Britain, and he pointed out that, in the particular case of the Periyar dam in India, the application of the equations given would indicate that it had a tensile stress at the foot of the outer slope of nearly 19 tons per square foot, which was at least 50 per cent. above the strength of Portland-cement concrete placed under favourable conditions. He concluded with the pertinent query: " Is, therefore, the theory or the dam to give way ? " and the dogmatic opinion: " There is no reason for postponing the raising in consequence of anything in the thesis." A week later another contributor suggested that Professor Unwin had omitted an important point in that his solution did not satisfy the condition that there could be no shear stress on the face of the dam. He pointed out that Atcherley's work contained the same inconsistency and, after summarizing his views in the statement that " Atcherley's solution depends upon two concurrent inconsistent assumptions while Professor Unwin's does not satisfy an important boundary condition," he concluded that the assumption of uniformly varying stress over horizontal sections was therefore incorrect. This drew a prompt reply from Unwin, briefly indicating that he had not made the error attributed to him, and eliminating this particular contributor with the remark: " The problem is difficult enough without being complicated by hasty misrepresentations; your correspondent might have given me credit for not ignoring an elementary principle."

At the end of June 1905 Unwin published a third article [9] on the distribution of shearing stresses. In this, which was an extension of his original treatment for a triangular dam, he showed how the method could be applied to a dam of any shape, and concluded that, for dams with rectangular profiles, Atcherley's assumption of a parabolic distribution of shearing stress would be correct, but in no other cases. The heavy guns on the other side now came into action, and a fortnight later Karl Pearson dealt with the matter in an article [10] in *Engineering*. He dismissed contributors other than Unwin with the statement: "It is difficult to believe that the correspondents referred to can possibly have read the memoir which they profess to criticize," and then attacked Unwin on the grounds that he (Pearson) and Atcherley had already worked out in their memoir all that was included in Unwin's Notes. They had shown that the shear-stress distribution over the base would be parabolic; that the parabola would not be symmetrically placed with zero shears at front and flank of the dam; and that in the special case of a triangular dam the parabola reduced to a triangle. Unwin had also discovered these facts, the only differences between his treatment and that of Pearson and Atcherley being :—

(1) They had worked out the case of any form of dam whatever, and showed the triangle as a special case.
(2) Their treatment occupied nine lines only, as against ninety.
(3) They showed reasons why their treatment was not legitimate.

He felt that it was hard to see what Professor Unwin had contributed to the subject. The object of their memoir was to show that the usual treatment of a dam was fallacious and that the fallacy was really independent of the actual distribution of shear over the base. He concluded his article by remarking that it was certain that the hypothesis of linear distribution of pressures over horizontal sections was *wholly* erroneous.

In a postscript written after the appearance of Unwin's third article Karl Pearson gave his conclusions on the position: "That empirical forms of dams have been reached which are reasonably stable is undoubted; but I believe that this stability is purely empirical, and cannot be demonstrated by middle-third rules for the horizontal sections. The vertical sections are, I hold, weaker than the horizontal, and I conclude from our experiments that the substratum is weaker than either. The view I should take therefore, would be extremely cautious and tentative progress in any dam construction going much beyond any existing experience. Horizontal middle-third stability is not, from the theoretical standpoint, a safe criterion of what is or is not reasonable in dam construction."

Other correspondents intervened but most of them were promptly subdued. In August 1905, however, there appeared in *Engineering* an article [11] which was in the nature of a preliminary report by J. S. Wilson and W. Gore, describing a method they had devised for experimentally investigating the distribution of stress in dams. They subjected a rubber model, representing a complete cross-section cut out of the dam, to pressures causing distortions which were measured and analysed. An ingenious device was adopted to represent the effect of gravity loading of the dam; weights hung from cross-bars placed through the rubber profile increased the overall specific gravity of the dam to the value required to give measurable strains while retaining an accurate ratio of pressure to weight. In this article attention was restricted to a triangular dam, but much more was to be heard later of the development of this method. Response to the publication of this Paper was prompt, and only a week later Karl Pearson added a postscript to a letter to *Engineering* in which he rather summarily dismissed the method adopted in the experiments.

There seems to have been a lull in the controversy during the next 6 months but, in *Engineering* of the 22nd June, 1906, in a description of the Royal Society *Soirée*, which had been held on the 9th May, reference was made to a model-section of the Aswan dam exhibited by Karl Pearson and A. F. C. Pollard to illustrate an extremely delicate method of determining experimentally the distribution of shear stress. A full description of the work done by Karl Pearson and Pollard in this connexion was not published until the following year, when it appeared as No. 5 of the Drapers' Company Research Memoirs.[12] The model was made of a stiff jelly, the suggestion for this having apparently come from Sir Benjamin Baker, and the use of this material was, as will be seen, the cause of widespread criticism. The general type of shear distribution was found to be very peculiar, having no resemblance to the parabolic curve often assumed; no general law was deduced to fit the curves and it was recommended that models of projected dams should be investigated in this way before construction was put in hand. It will be remembered that Sir Benjamin Baker had actually done this in connexion with the proposed raising of the Aswan dam.

The experimental work contained in this memoir was the subject of an editorial review [13] in *Engineering* on the 13th September, 1907. No reference was made to Pollard, the work being ascribed to Pearson alone. The history of the previous 2 or 3 years following the publication of Atcherley's Paper was recalled, and it was suggested that Pearson, in his recently published memoir, had dealt with the various contributors to the discussion in somewhat drastic fashion.

The results obtained by Wilson and Gore were, according to Pearson, inapplicable to real dams because the experiments failed to take into account the horizontal bending which exists in practice. This point is of importance in connexion with the further discussion which ensued, and it must be explained that the real point of Pearson's criticism was that to take a slab from the centre of a dam and assume that it directly represented the same slab in position was to ignore the effects transmitted from the main structure on either side of the cut sections.

The Editor did not agree with this point of view, and questioned whether the inclusion of the missing terms would modify the results obtained by Wilson and Gore by more than a very small percentage. After paying tribute to the excellence of the mathematical portion of the memoir, he expressed great doubt about the actual value of the experimental results, since no evidence was given that the material obeyed Hooke's law, whilst there was positive evidence that it acted plastically. An attempt to deduce the distribution of the stresses within the elastic limit from such a model was, in his opinion, very likely to lead to totally erroneous conclusions, and for these reasons he felt that no confidence whatever could be placed in those results, even if considered as a first rough approximation. The memoir indicated that both the Aswan and Vyrnwy dams were safe structures, having no tensional stresses exceeding 5,500 lb. per square foot and no compressions exceeding 13,000 lb. per square foot. The Editor suggested that there the matter must rest, pending the execution of experiments on some material more suited to the measurement of elastic displacements than a plastic jelly could possibly be. The Editor concluded that it was well known to readers of Pearson's " History of Elasticity " that the author had no very exaggerated respect for the intellectual abilities of the engineer, but he was not quite prepared to find the intelligence of the practical man rated so lowly as in this Paper: Pearson actually believed him capable of erecting a masonry dam in such a position that the capacity of the reservoir behind it might be materially affected by the addition of a few cubic yards of concrete, to round out the re-entrant angle between the upstream face of the dam and the foundation.

The editorial brought a quick response, for, in the following week, *Engineering* published a letter from Pearson pointing out that the memoir reviewed was the work of himself and A. F. C. Pollard, assisted by C. W. Wheen and L. F. Richardson, who, he had no doubt, would be quite prepared to share with him the blame which the critic thought it deserved. In answer to the Editor's remarks about his views on engineers, he said that, on the contrary, he had enormous respect for ability, intellectual or practical, wherever he found it, and that his one regret was that Sir Benjamin Baker was no longer alive to give his opinion of the work. He claimed that Sir Benjamin, at least, would have seen that they were striving to reach the truth, and that " to slumber in the middle-third and reject any attempt to reach a logical theory in dam construction " was not an ideal position for a country which once led the engineering world.

Engineering, commenting on this letter, was quite unrepentant, and adhered to the point that the observations made gave little clue as to what the elastic distribution of stress would be in a dam.

A month later Pearson returned to the attack, having calculated the error involved in the use of a slab dam as a means of checking stress distribution. He claimed that the " small percentage " referred to in the review was rather more than 30 per cent. The Editor completely refused to accept this and said categorically: " We do not believe that the maximum stresses developed in Professor Pearson's experimental slab were 30 per cent. more or less than they would have been did the slab form part of a complete dam. In any case, the question still remains, what is the elastic limit of the jelly, if it does exist ? "

The issue was now reduced to these two points ; the elasticity of the jelly and the interpretation of results from a slab dam. In regard to the first, Pollard wrote that the small displacements used in their work never exceeded the limits for jelly mentioned by Lord Kelvin in his article on " Elasticity " in the *Encyclopaedia Britannica,* to which the Editor commented that Lord Kelvin's dictum was negatived in this particular case by Professor Pearson's statement that the permanent set in a typical observation amounted to 20 per cent. of the maximum strain measured. Very little more need be said on this point, except that in a later discussion at the Institution of Civil Engineers it seems to have been generally agreed that the jelly was a quite unsuitable and unreliable material.

The question of the 30-per-cent. discrepancy was raised by H. M. Martin in the columns of *Nature* and, after two or three interchanges, Pearson admitted that he was in error and that there was no such discrepancy.

On the 31st January, 1908, at an ordinary meeting of the Institution of Civil Engineers, three Papers were presented for discussion. The first of these,[14] by Sir John Ottley and A. W. Brightmore, described experiments they had made on plasticine models of dam sections. These were meant to be only qualitative but, of course, were open to the same criticism as those of Pearson and Pollard since plasticine displays scarcely any elasticity. The conclusions reached by Ottley and Brightmore were :—

(1) That a masonry dam, designed on the assumption that stresses on the base varied uniformly, would, in fact, have smaller stresses than those estimated.
(2) That there could be no tension near the outer toe.
(3) That there would be tension on certain planes other than the horizontal near the inner toe.

The second Paper,[15] by J. S. Wilson and W. Gore, described the extension of their work already mentioned to the case of a dam profile of a more practical section than the simple triangular one first considered. This work deservedly received very high praise from many speakers in the discussion that followed. The Authors stated that they had attempted to provide a solution to the mathematician's dam, and they felt confident that they had obtained very good approximate solutions to the problem, which was too complicated for a mathematician's analysis. This claim has been well substantiated by later work, as will be seen shortly.

The third Paper,[16] by E. P. Hill, was of an analytical character. In the discussion it was claimed by Unwin that Hill had done analytically only what he, Unwin, had already done graphically in *Engineering.* Little attention was directed to this Paper in the discussion, which showed clearly that the experimental work of Wilson and Gore was far more acceptable to engineers than either the results obtained by Pearson or the experimental values found by using plastic or semi-plastic models. One particular point of importance did arise ; all the Authors confirmed the fact that there was slight tension on some plane near the heel of the base section where it joined the foundation, but that nowhere else in the profile was tension possible. This tension was not considered to be of importance by any speaker.

On the 22nd May, 1908, *Engineering* summed up the position,[17] pointing out that 4 years had elapsed since engineers had been seriously disquieted by the Paper by Pearson and Atcherley. Meanwhile, experiments had shown that their conclusions were wrong, and that the perturbation which had followed the publication of the Paper was consequently unwarranted. The Editor held that, whilst Pearson's reply to critics of the mathematical work had been complete, Atcherley's experiments were very far from conclusive. The experimental material employed by Pearson and Pollard had proved wholly unsuitable, and their labour had been expended in vain, since an attempt to deduce the properties of an elastic material from jellies which gave a 20-per-cent. permanent set was absurd. Drawing attention to Pearson's criticism that Wilson and Gore had introduced errors of the order of 30 per cent. by their use of a slab dam, it was pointed out that *Engineering* had consistently disputed this and that Professor Pearson had now acknowledged that he was in error. The work of Wilson and Gore, in the Editor's view, demonstrated the absence of any material tensional strains, and had determined reasonable stress-distributions ; he considered their results to be of the highest importance, justifying in all essential points the present practice of engineers in designing gravity dams.

Thus ended the protracted controversy on the stresses in masonry dams. The engineers were vindicated and there was general agreement that mathematics was insufficiently powerful to provide anything in the nature of an accurate solution to the problem of stress distribution in an elastic solid of this shape. The experimental results of Wilson and Gore remained, therefore, as the best available solution.

There is, however, a postscript to be added to this. The development by Southwell of relaxation methods for the solution of problems in mathematical physics has provided a tool which makes it possible to obtain numerical answers to problems which have hitherto defied analysis. This method has, within recent years, been applied to the problem of stress distribution in gravity dams and it is appropriate that the work was done by a young civil-engineering research student, O. C. Zienkiewicz,[18] working in Unwin's old College. He took three profiles, one triangular, one rectangular, and one designed by the Wegmann method respectively, and made a complete analysis of the stresses on the assumption of perfect elasticity. This assumption, it will be remembered, corresponded with the claim made by Wilson and Gore as to the significance of their results. The analysis should be studied in detail by anyone who is interested in this matter, but attention may be drawn to certain general conclusions. Zienkiewicz finds that the maximum compressive stresses when the reservoir is full do not generally exceed those indicated by the classical method. The most severe conditions of compressive and shearing stress then occur on the downstream face of the dam, where the material is subjected to a state of uni-axial compression. The analysis also shows that, in all cases when the reservoir is full, tension develops in the portion below ground-level at the upstream side of the dam. The most dangerous region inside the dam is near the heel, where compliance with the middle-third rule does not always eliminate tension, but no other region of the dam tends to develop tension. In the upper portion of the dam, agreement between the classical and elastic theories is good. When the reservoir is empty, classical theories are adequate. It will be noticed that the general conclusions agree with those arrived at experimentally, and Zienkiewicz draws attention to the fact that the results for the Wegmann profile agree remarkably well with those of Wilson and Gore for a similar type of dam (*Figs 3*).†

As indicated at the beginning of this Lecture, the story of the controversy over the stresses in masonry dams has been told not only for its intrinsic interest, but also as an excuse to take stock of the position of engineering research in the Universities and to discuss its functions under present-day conditions.

The most casual student of this story must be struck by the differences between the years before the first world war and those after the second, with regard to the facilities existing for research and the attitude towards it. In the first place, the fact may be noted that the money which initiated not only this research, but also others on widely varying subjects, came from private sources, and I would take this opportunity of paying tribute to the far-sighted policy of the Drapers' Company which set a precedent from which Unwin's old College and others have since benefited so greatly. The City Companies have indeed nobly maintained the tradition begun by the Drapers. Secondly, the research begun at University College by Atcherley and Pearson stimulated workers in other places; Unwin, already retired from his Chair, at South Kensington; Ottley and Brightmore at Coopers Hill; Wilson and Gore in a private flat; and so on. The day of Research Committees and Government Laboratories had not yet arrived; otherwise it is probable that a Committee for the Investigation of Stresses in Masonry Dams would have been set up under the aegis of some Department of State and would have taken over the whole investigation. The principal protagonists would, presumably, have

† This illustration has been re-drawn, using masonry units of stress, in order to bring out more clearly the essential similarities between the results from the two methods.

Figs 3

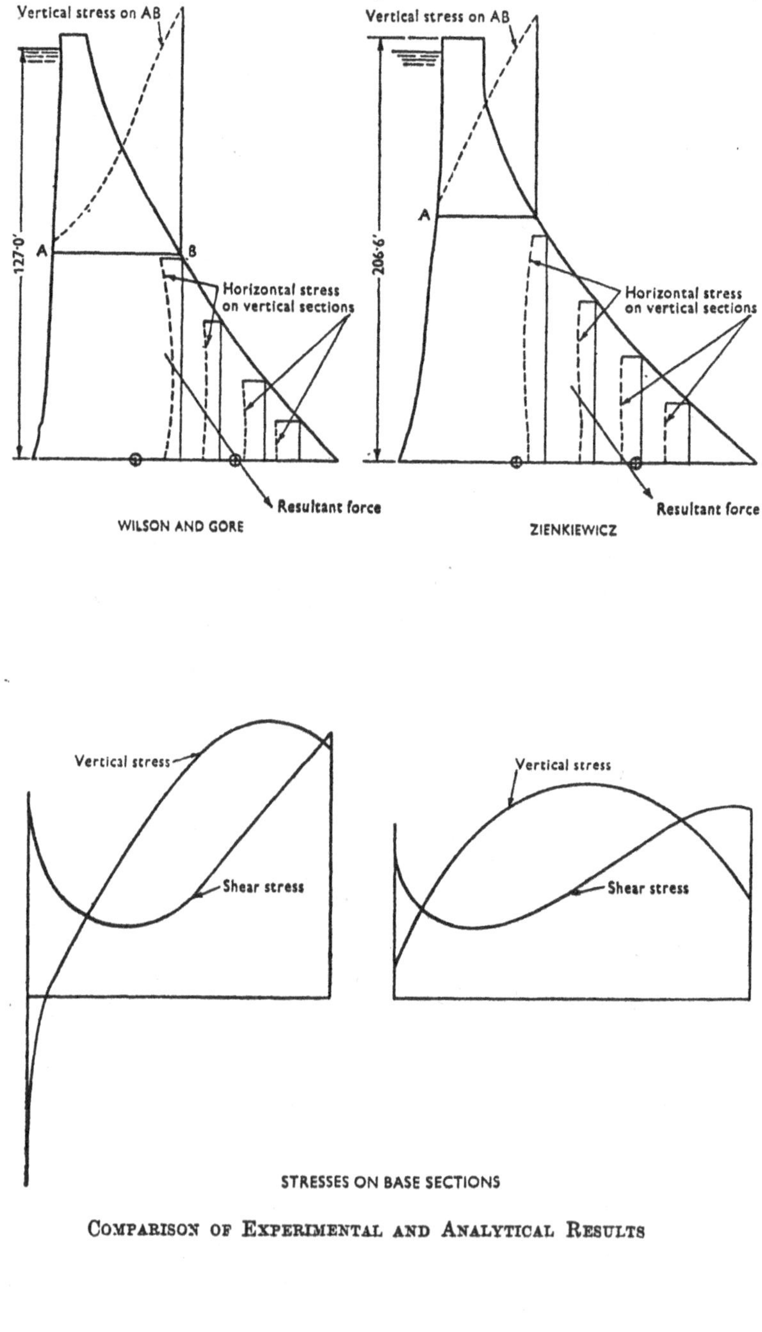

Comparison of Experimental and Analytical Results

been members of this committee and, instead of the lively and stimulating discussions carried on in the pages of *Engineering* and elsewhere, the interchange of ideas would have taken place across a committee table and the report which would ultimately have appeared would certainly have omitted all the truly human elements of the controversy and left engineering literature so much the poorer.

The third point of note is the way in which the engineering world responded to stimulus; some of the greatest of the practising engineers of the time were frankly sceptical of the value both of the mathematical analysis and of the model-tests, but that did not prevent them from paying tribute to the honest attempts of the researchers to discover the truth or from encouraging and helping them as far as lay within their power. Some, at any rate, disproved the jibe that the practical man was one who perpetuated the mistakes of his predecessors; here at least were men, hard-headed and cautious indeed, but willing to learn from sources which they obviously mistrusted, provided that the results did not too flagrantly violate their engineering common sense.

We have in many ways gained much since that time, but I venture to think that we have also lost something of value, the recovery of which would be to our advantage, were it possible.

References

1. W. M. Rankine, "Report on the Design and Construction of Masonry Dams." Miscellaneous Scientific Papers, No. xxxiv.
2. W. C. Unwin, "The Failure of the Masonry Dam at Bouzey, France." *Cassier's Magazine*, vol. xi (1896–97).
3. Discussion on the Manchester and the Liverpool Water-Supply. Min. Proc. Instn Civ. Engrs, vol. cxxvi (1895–6, Pt. IV), p. 89 (contribution by W. C. Unwin).
4. L. W. Atcherley and Karl Pearson, "On some Disregarded Points in the Stability of Masonry Dams." University of London, *Drapers' Company Research Memoirs*, Technical Series II, 1904.
5. C. S. R. Palmer, "Coolgardie Water-Supply." Min. Proc. Instn Civ. Engrs, vol. clxii (1904–5, Pt. IV), p. 50. *Discussion*, p. 120.
6. "The Stability of Masonry Dams." *Engineering*, vol. 79 (1905), p. 414 (31 March, 1905).
7. W. C. Unwin, "Note on the Theory of Unsymmetrical Masonry Dams." *Engineering*, vol. 79 (1905), p. 513 (21 April, 1905).
8. W. C. Unwin, "Further Note on the Theory of Unsymmetrical Masonry Dams." *Engineering*, vol. 79 (1905), p. 593 (12 May, 1905).
9. W. C. Unwin, "On the Distribution of Shearing Stress in Masonry Dams." *Engineering*, vol. 79 (1905), p. 825 (30 June, 1905).
10. Karl Pearson, "On the Stability of Masonry Dams." *Engineering*, vol. 80 (1905), p. 35 (14 July, 1905).

284 UNWIN MEMORIAL LECTURE

11. J. S. Wilson and W. Gore, " Stresses in Dams: an Experimental Method o Investigating the Distribution of Stress in Dams, etc." *Engineering*, vol. 80 (1905), p. 134 (4 August, 1905).
12. Karl Pearson and A. F. C. Pollard, " An Experimental Study of the Stresses in Masonry Dams." University of London, *Drapers' Company Research Memoirs*. Technical Series V, 1907.
13. " The Stresses on Masonry Dams." *Engineering*, vol. 84 (1907), p. 368 (13 September, 1907).
14. J. W. Ottley and A. W. Brightmore, " Experimental Investigations of the Stresses in Masonry Dams subjected to Water-Pressure." Min. Proc. Instn Civ. Engrs, vol. clxxii (1907–8, Pt. II), p. 89.
15. J. S. Wilson and W. Gore, " Stresses in Dams: an Experimental Investigation by means of India-Rubber Models." Min. Proc. Instn Civ. Engrs, vol. clxxii (1907–8, Pt. II), p. 107.
16. E. P. Hill, " Stresses in Masonry Dams." Min. Proc. Instn Civ. Engrs, vol. clxxii (1907–8, Pt. II), p. 134.
17. " The Stresses in Masonry Dams." *Engineering*, vol. 85 (1908), p. 692 (22 May, 1908).
18. O. C. Zienkiewicz, " The Stress-Distribution in Gravity Dams." J. Instn Civ. Engrs, vol. 27 (1946–47), p. 224 (January 1947).

16

The design of structural ironwork 1850–1890: education, theory and practice

Stanley Smith

The mid-nineteenth century was a period when a series of changes took place in the construction industry, especially in that part of the industry concerned with the design and erection of iron structures. Apart from the introduction of new materials, and the need for new building types, there were influential changes that occurred in the way designers envisaged the scope of their work in this field which led to an increasing specialisation of activity and the proliferation of professional organisations intended to serve the interests of specific groups. In addition, alternative views were developing about the provision of education and training for designers and what should be offered to new entrants to the engineering design professions. Another major area of activity was the attempt to apply the principles of scientific investigation to explain the performance of structures and to develop theories that could be used as the basis for design.

It has been suggested that changes in the practice of engineering generally had by 1890 led to the development of a 'scientific engineering', founded on the application of science and mathematical analysis.[1] By considering the writings and comments of some of the mid-century designers this paper attempts to establish if such a development took place in the practice of structural design in the second half of the nineteenth century. To do this, attention has been focused on four specific areas. First, on the specialisation that took place within the civil engineering profession, coupled with the growth of provincial engineering societies; second, on the published material intended to assist structural designers, its content and intended readership; third on the attitudes towards education and training that were expressed by designers in public and finally, on mid-nineteenth century views about the importance of practice in relation to theory.

Specialisation and Structural Design

Today a structural designer can be defined with general agreement as 'one who is competent to design stable and economical structures of different kinds to meet the requirements for which the structures are needed.' But it was not until 1908 that there was a specific professional institution to pursue the interests of the structural engineer.[2] In the nineteenth century, structural design of ironwork formed part of the wider activities of the civil engineer, whose needs were catered for by the Institution of Civil Engineers, founded in 1818. It was not until attempts were made to obtain a Royal Charter in 1828, that it was considered necessary to define what an engineer was or did. The Council requested Thomas Tredgold to prepare such a description and this was entered in the minutes on 4th January, 1828 and eventually, in an abridged form, included in the Charter. The work of the civil engineer was seen as concerned with, 'The art of directing the great sources of power in nature for the use and convenience of man', and as described by Tredgold, the improvement of the means of communication, the protection of property from natural forces and the control and manipulation of water.[3]

By 1866, when John Fowler gave his Presidential Address, his list of the activities of a civil engineer was much more specific and comprehensive than Tredgold's and listed seven categories of work, ranging from inland communications, works relating to public health, harbours and mining, to machinery and shipbuilding. His selection demonstrated how far the specialisation in engineering design had progressed. Fowler was also much more specific about what an engineer's duties involved, stated as 'design, prepare drawings, superintend and carrying out of the works', not as in Tredgold's description, 'the art of directing the forces of nature'.[4] The Fowler list defined much more clearly those fields where the design of spanning structures was likely to occur, a reflection of the increasing importance of this type of work for the engineer, especially in connection with new building types.

Another measure of the changes in the practice of design can be seen in the content and format of papers presented to the professional institutions over the years. The learned institutions had papers read on structural design matters, but it was the papers read at the Institution of Civil Engineers and at the Mechanicals, often reprinted at length in the technical press, that had the widest influence.[5]

A very typical presentation in the late forties was that by Turner, on the roof his firm had designed and erected at Lime Street Station, Liverpool.[6] This was a factual description of the structure, involving no theoretical explanation, and was followed by a discussion that concentrated on the economics of such a roof, the type of covering and the practicalities of this class of work. This approach was very common, and continued until the eighties, as can be seen in the description of the Clyde Bridges, given in 1880, where the author B.H. Blyth considered that all his audience were interested in was a description of the solution, one that could be the basis for a later scheme.[7] The Turner paper is also interesting for the light it throws on the approach to design of people such as William Fairbairn and Joseph Locke, who when asked to check the safety of the structure, gave answers that appear to have been based very much on past experience, not calculation.

By the early 1850s practising engineers were presenting papers that started to contain descriptions of the theoretical basis for their design, such as Barton's paper in 1854 on the stress in the web of beams as used at the Boyne Viaduct. It is interesting to note that Fairbairn's paper on tubular bridges in 1849 contains very little mathematical analysis, but the ensuing discussion, which continued over several subsequent meetings, included contributions from most of the theoreticians who were connected with the Institution of Civil Engineers.[8] In the sixties and seventies papers such as Calcott Reilly's on uniform stress and those by continental engineers like Gaudard contained some theoretical analysis.[9]

Papers of the Turner type continued to be published, but with the passage of time became more wide-ranging and reflected the continuing specialisation of interests about practical aspects of design. This was so in the case of the papers presented in 1863 and 1867 on the Charing Cross and Cannon Street bridges and roofs respectively (Fig. 1), and that on St Pancras station in 1870, all of which contained, in addition to practical information about construction, discussion of the economics of the methods chosen and the basis of the design adopted.[10] Later papers, like Baker's in 1879 on the strength of beams, contained a great deal of theoretical material, whilst still intended for the practical man, and in this instance reflected concern about how such factors as wear, daily use and exposure could affect design decisions[11] Other papers with a theoretical content attempted to summarise the state of knowledge in a subject, and present it in a way that made it possible to apply it in practical situations. Papers by Gaudard and Bender on wind forces in the 1890s are a good example of this type.[12] However the continuing publication of papers that concentrated on the practical aspects of the design and construction of ironwork suggest that for many this was the approach required. Theoretical descriptions, with the very criticial refutations of the theories stated in the following discussion, may have suggested that theories could not provide the answers to practical problems.

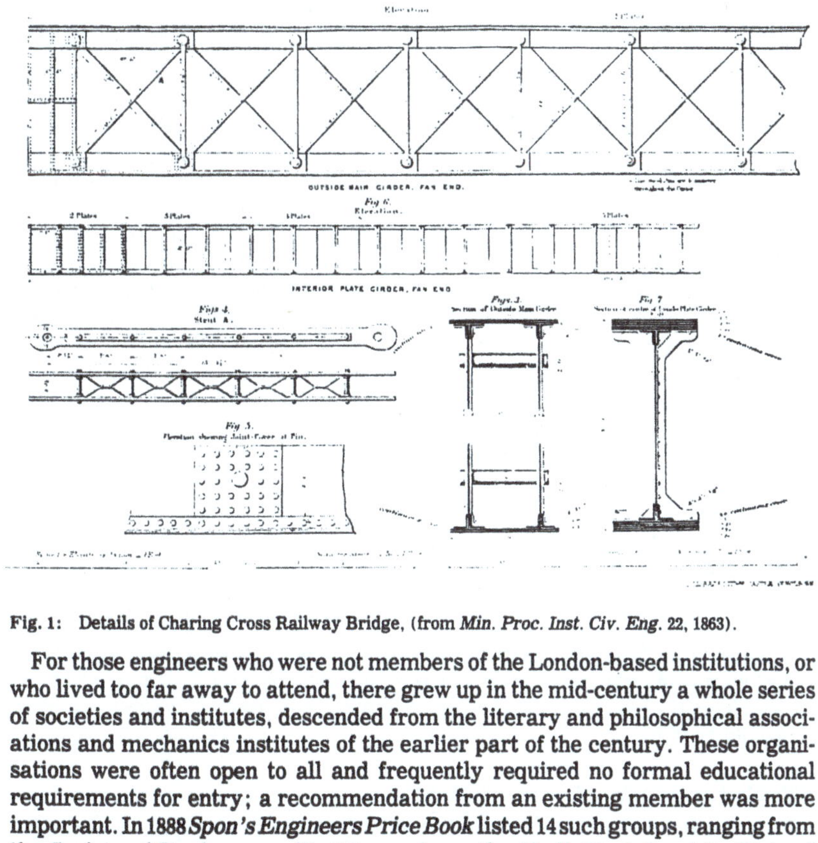

Fig. 1: Details of Charing Cross Railway Bridge, (from *Min. Proc. Inst. Civ. Eng.* 22, 1863).

For those engineers who were not members of the London-based institutions, or who lived too far away to attend, there grew up in the mid-century a whole series of societies and institutes, descended from the literary and philosophical associations and mechanics institutes of the earlier part of the century. These organisations were often open to all and frequently required no formal educational requirements for entry; a recommendation from an existing member was more important. In 1888 *Spon's Engineers Price Book* listed 14 such groups, ranging from the Society of Engineers, with 450 members, the North East Coast Institute of Engineers and Shipbuilders, with 600 members, to South Staffordshire and East Worcestershire Institute of Mining and Mechanical Engineers, with only 70. The total membership of all these societies was about 3500, at a time when the membership of the Institution of Civil Engineers was 7500.[13]

At the meetings of these societies the papers that were read had a very practical content compared to those given at the Institution of Civil Engineers. An interesting example was that given in 1877 to the Liverpool Society of Engineers by C. Graham Smith (an Associate Member of the Civils) entitled 'Wrought Iron Girder Work'. The first paragraph set the tone of the paper by saying, 'Strains in Ironwork. In treating this subject it is not here proposed to consider the abstract questions involved in the various methods of arriving at the theoretical strains on the different portions of girder work'. The paper discussed in some detail the technical problems of the engineer, on site and in the office, in attempting to cope with the problems of thermal movement, definitions of elasticity, the arrangement of members to meet strains, (used throughout to mean stress in present day terms), the differing qualities of iron and methods of testing.

The information about testing is full of detail about the preparation of samples, both for site and workshop tests, and a great deal of comment obviously the result of long experience of inspecting ironwork. The details of defects likely to originate in the rolling mill due to worn rolls, or inaccurate cutting of plates must have been extremely useful to inexperienced engineers.[14]

Another paper of a similar practical type was that given by Hamilton W. Pendred in October 1883 at the Society of Engineers. This Society, although London based, had been founded in 1854 as the Putney Club, intended for ex-students of the Putney College of Engineering. Its members originally were those who had not been articled to a member of the Institution of Civil Engineers and so could not easily join the Institute. The title of the paper, 'Designs, Specifications and Inspection of Ironwork', clearly describes its content. Plentiful information about the practicalities of ironwork detailing, riveting and fabrication was given, based largely on personal observation and experience. This paper was intended for those whose experience of preparing drawings, detailing and inspection were limited.[15]

The occupational tables in the census returns in the latter part of the century list very large numbers of people who described themselves as engineers, a number far exceeding the total membership of the recognised institutions.[16] Thus these people, plus those without educational qualifications in the provincial societies, formed a large group who would have very little opportunity of learning about and understanding the application of the more mathematical theories of structural design. It appears likely that in the 1880s and 1890s it was still common practice for design to be based on a set of standard rules of thumb, and it was very unlikely that structural design theory was widely used. The effect of the need for a wider use of theory, and the separation of functions that has been described, meant that the education of engineers had to respond to changing circumstances and to be divided into different streams, each dependant on the activity in which the student was to be engaged.

Published Information for the Engineer

Another measure of how far the new theoretical knowledge was applied in practice is to be found in the published material available to the student and practitioner which set out the result of new research in an easily accessible manner.

When the first volume of the Todhunter and Pearson, *History of the Theory of Elasticity and the Strength of Materials* was published it listed several thousand articles, papers, memoirs and books on the subject, written between 1650 and 1886.[17] Although many were not of direct use to the designer, a large number were.

What is difficult to establish is how many practitioners would have been able to see them, and in case of those originally published in a foreign language, the ability to translate them. Some of the English articles had appeared in the *Transactions of the Royal Society*, the *Proceedings* of the Institution of Civil Engineers and of the Mechanicals, and were reprinted in the technical press, but it would appear that the extent of their circulation would have been limited.

From the mid-century onwards there was a variety of printed material intended to supply the designer with the information he needed, and this can be divided into four main groups. There were text books, intended for both students and practitioners; books that contained detailed illustrations of structures, intended as examples to be copied; manufacturers' handbooks; and finally the encyclopaedic pocket books.

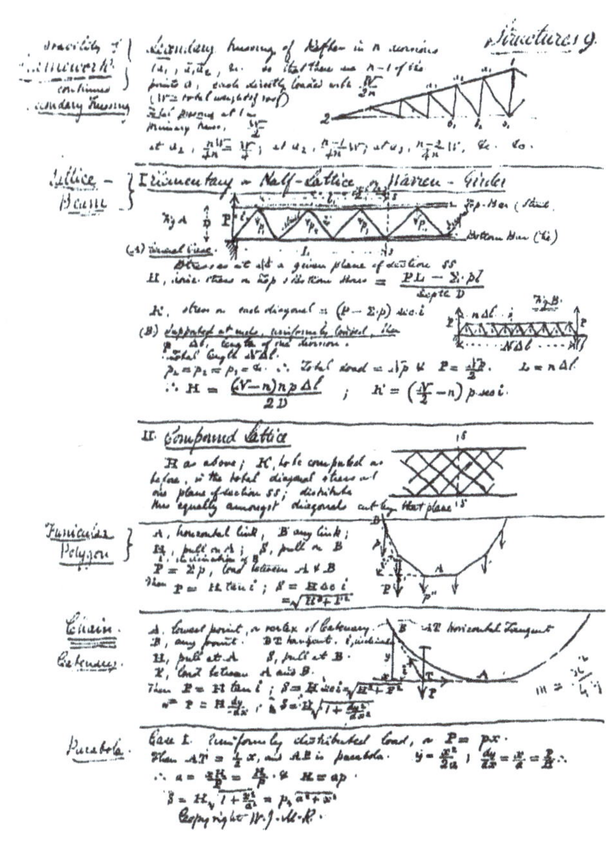

Fig. 2: A page from the lithographed notes prepared by W.J. McQuorne Rankine for the use of his Glasgow studens in the 1850s (Civil Engnering Library, Imperial College, London).

Structural Ironwork. 1850-1890

Most of the books in the first category, which were intended to provide some theoretical basis for study, were written by people who were teachers, and who were the main contributors to the discussions at the institutions that explored the mathematical analysis of structures. These included such people as Anderson, Pole, Tate, Unwin, Stoney and Reilly. Frequently their books were based on a series of lectures; in Anderson's and Unwin's case, those given at the Royal Engineering School at Chatham.[18] Of this group of publications perhaps the most famous was Rankine's *Manual of Civil Engineering*, published in 1862.[19] The *Manual*, and other books that Rankine wrote about statics and mechanical engineering, were based on the notes and lectures given to students at Glasgow University, and present the most comprehensive attempt to provide in one volume all the knowledge required by the civil engineer (Fig. 2).[20]

In the reviews of these books in the technical press there were frequent complaints that they were too theoretical in content, did not display the approach to design required by the practical man, and neglected to discuss those non-quantifiable aspects of construction detail that could have a profound effect on the solution selected.[21] Two volumes that escaped such criticisms, and indeed were highly praised at the time, were those by F.W. Sheilds on *The Strains on Structures of Ironwork* and J.H. Latham's *The Construction of Wrought Iron Bridges including the practical application of the principles of mechanisms to wrought iron girder work.*[22] These two books were both clearly set out, easy to follow, and contained a wealth of practical detail about connection methods, the qualities of material, and problems of fabrication. In the preface to his book, Sheild commented on one of the designer's problems:

> *'I have found great difficulty however in gaining a complete knowledge of the strains of several parts of iron frames as the works of previous authors, though displaying great talent and research have left much indetermined and obscure, which is necessary to the design of such structures.'*[23]

This is a view taken up by Latham in the Preface to his book, where he notes the almost total division of the practical from the theoretical in mid-century engineering, but pleads for both to be considered together saying, 'Yet it is only by their union that perfection can be obtained.'[24]

Criticisms of the theoretical content of textbooks suggest that the editors of journals saw their readership as consisting essentially of 'practical men' who would not be willing to attempt to assimilate new theories about design. Even an author such as Sheilds recognised that new research findings were not easily accessible to the practising engineer.

One writer whose works were of a practical nature, and who published extensively in mid-century, was William Fairbairn. His books appeared between 1850 and 1870, a period that included the publication of Rankine's books, but his style and content were in distinct contrast to Rankine.[25] Although Fairbairn contributed papers to the learned societies as well as writing books, it was the latter that reached the widest audience. His earliest book, *Useful Information for Engineers*, was in the mould of the engineering pocketbook. Perhaps his most famous was *On The Application of Cast Iron and Wrought Iron to Building Purposes*, first published in 1854 and reprinted and revised several times. The fourth edition in 1870 received very adverse reviews in *Engineering* which commented. 'If the 1870 edition was to be as useful as the 1854 edition it needed greater revising.'[26] This is in strong contrast to the reviews of Fairbairn's books published twenty years earlier. The main criticism appeared to be based on the use of what were considered to be out of date examples, such as the factory at Saltaire, the Spey Bridge and other tubular bridges, which were no longer accepted as appropriate structural forms. This volume, like others written by Fairbairn, contained much material reprinted from other sources. Because of their straightforward, non-mathematical approach his works were always easily understood by practical engineers. But although the style of writing was acceptable, the use of twenty year old examples, non-representative of the latest practice, meant that this works gradually lost approval.[27]

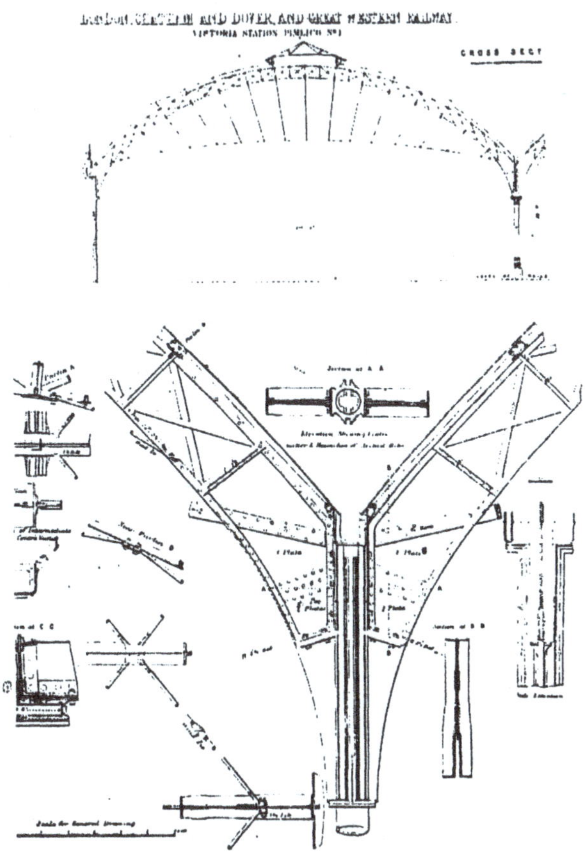

Fig. 3: John Fowler's roof for Victoria Station, a typical illustration from William Humber's *Record of the Progress of Modern Engineering* (1863).

Structural Ironwork. 1850-1890

The second group of books, those intended as direct aids to finding complete solutions, are exemplified by the various volumes written by William Humber, which came out between 1858 and 1870. All were large format volumes containing beautifully engraved and lithographed plates, accepted by contemporaries as an accurate record of the structure as built (Fig. 3).[28] Other books of this type by Dempsey and Grover contained, in addition to illustrations, tables of information about volumes of soil, and weights of materials.[29] Later in the century, when graphic statics had become an accepted technique, several books appeared, such as Olander's *A New Method of Graphic Statics*, where detailed drawings of a structure were accompanied by a worked example of the appropriate graphic analysis.[30] Volumes by Timmins and Tarn contained similar information, often presented so that one diagram could be used for varying loads and spans.[31]

The prevalence of books of this type and texts on draughtsmanship by people such as Binn, Bently and Charnock, with details of structural connections and simple calculations to determine the size of members, suggest that they were all intended for a readership, in the classroom or the office, with mathematical abilities that were limited.[32] It would appear that in the craftsman tradition these readers were prepared to copy from the examples of successfully completed structures, and not attempt to derive solutions from first principles.

It is difficult to judge the extent of the use of the third group of books, the manufacturer's handbooks, as so few have survived. One of the best known, Matheson's *Works in Iron* (from Andrew Handyside and Co.), seems to have been intended for the 'practical engineer' judging by its non-mathematical content.[33] The *Viaduct Handbook* from the Crumlin Iron Works, and Hutchinson's *Girder Making and Bridge Building*, were written in a similar style.[34] The first handbook of standard rollings was that published by Dorman Long in 1887 and was basically a catalogue. It was not until 1892 that a design manual, in the modern sense, was published by Redparth Brown and Co.[35] This contained safe load tables, provided on the assumption that there were very few engineers who would be able to produce calculations to justify the size of the beam to be used.

The final type of printed material considered here is the compendium or engineer's pocket book. Volumes of this type have been available for many years and went through a large number of editions. *The Practical Mechanics Workshop Compendium* by Templeton had fifteen editions in twenty-five years, and Molesworth's *Pocket Book of Useful Formulae and Memorandum* had its thirty-third edition in 1938.[36] The intended reader can be deduced from the comment in the Preface to Molesworth, which said:

> *Complex and difficult formulae have been as far as possible avoided, and in many cases, to the formulae have been added an easy approximate rule. Throughout the book there is scarcely a formula which cannot be mastered by anyone possessing little more than a mere knowledge of arithmetic'.*

Much of the comment in the journals and books talked of the need to copy what had been done before, and much of the information available, as described above, was presented in a form that made this possible. With design dependant on the use of factors to control the size of members, coupled with the use of small scale tests,

Stanley Smith

design education had only to develop the skills of draughtsmanship and an understanding of the disposition of structural members to succeed.

The approach adopted by many of these books suggests that until the end of the century the intended readership was not expected to be very familiar with complicated mathematical processes. The fact that some, such as Baker's *Long Span Railway Bridges*, were published in weekly parts in the technical press implies that they were intended for the aspiring artisan, hoping to improve his level of expertise.[37] As William Hutton wrote in the Preface to the fifteenth edition of Templeton's:

> *'... as the recognised text book and well worn and thumb marked Vade Mecum, of several generations of intelligent and aspiring workmen, it has the reputation of being the means of raising many of them in their position in life'.*[38]

One further source of information for the designer was the technical journal. By the mid-century the weekly technical paper was an established part of the engineering scene. The earliest magazines, *The Artizan*, *The Builder* and the *Mechanic's Weekly*, all catered for the general technician and constructor. *The Civil Engineer and Architect's Journal*, the *Engineer*, and *Engineering* were intended more specifically for the engineering design professions (Fig. 4).[39]

Fig. 4: The construction of Blackfriars Bridge (from the *Engineer*, February 18, 1876).

Education and Training

The ability of the practitioner of 1885 to make use of the research that was by then available can to some extent be measured by considering the content and quality of the education and training that was provided for the student twenty-five years earlier. By the mid-sixties the quality of the education for the engineer was the concern of many in the professions and was discussed by several of the Presidents of the Institution of Civil Engineers.

It was partly this concern that led the Institute to undertake a survey of training during 1868 and 1869. The results were published in a report in 1870, which expressed the Institution view that the education and training provided in the United Kingdom was the best in the world. The preface to the detailed information about courses gives a valuable statement of how the engineer of the time saw his profession and the preparation required for it. It endorsed the view that the education of an engineer should follow that adopted by other trades, namely a four year apprenticeship, and that this would provide the opportunity for the student 'to acquire competency'. It was considered that theoretical knowledge was not 'absolutely necessary', but the good pupil was expected to undertake private study to remedy any lack of theoretical knowledge. The report made very clear that it was considered that practical experience would compensate for any deficiencies in theoretical attainments.[40]

That this was not the universal view is apparent from the contribution made by Fleeming Jenkins in an appendix to the Report. Jenkins was at the time Professor of Engineering at Edinburgh, so was no doubt expressing the academic view of the situation. He described how at the age of eighteen the pupil was accepted, although a nuisance in the office, because of the fee; that the majority were ignorant of algebra, could not apply a simple formula from the pocket books, and that 'their arithematic is very shaky and a knowledge of physics, geology or the higher mathematics is wonderfully rare'. A further paragraph describes how little they were taught and how they picked up the minimum of information, only enough to enable them to do a limited class of work. As a contributor to *Engineering* pointed out, 'all the pupil in the office learnt was how to smoke cigars'.[41]

There was no mention in the report of what was considered to be the proper course of study, or type of education. This was perhaps because John Fowler had attempted to do this in his Presidential Address in 1866. He had described the training of the engineer under four headings.

1. General instruction or a liberal education.
2. Special education as preparation for technical knowledge.
3. Technical knowledge.
4. Preparation for conducting practical matters.

This course of study was based on the assumption that part one would be completed by the age of fourteen, the second after two years in an office, to be followed by a four or five year apprenticeship with an engineer to acquire the technical knowledge and practical experience outlined in the last two items.[42]

For the student in 1870 who did decide to undertake a formal course of study, the Institution report described ten courses that were available. Three, at the Royal School of Mines, the Royal School of Naval Architecture and Owens College did not award degrees. Of the seven university courses, two were in London, one in Edinburgh, one in Glasgow and three in Ireland. All covered to differing degrees subjects germaine to structural engineering, and all were of different lengths. The course in Glasgow required one year of four to be spent in an office, and the three year course at Cork provided, it was said, all the training required before setting up in practice.[43]

Stanley Smith

Throughout the latter half of the nineteenth century a series of investigations were made into the provision of technical education, as well as into the general availability of primary, secondary and higher education. What all these reports make clear, from the Yolland Report on *Engineering Education in the Army* to the seven reports of the *Royal Commission on Scientific Instruction* 1872-77, was that there was very little science or technical study conducted at any level, and that entrants to the professions and the armed forces were particularly ignorant in those areas involving science and mathematics that were considered important.[44]

By 1878, when the Livery Companies' Committee reported on technical education, in addition to the London based courses listed above they were able to describe classes at thirteen other institutions, ranging from Crystal Palace School of Practical Engineering to that at the South London Working Men's College.[45] These all seem to have been colleges that provided education for the artisan or the technician grade office assistant. Paradoxically, if the articled pupil received no formal tuition, the office draughtsman, if attending one of these courses, may well have been better informed about some of the technical matters discussed in the office.

Much of the evidence in the Livery Companies' Committee's Report notes how unsatisfactory apprenticeship could be with respect to any breadth of experience.

> *'... and the greater his skill in this one article the more chance there will be that he will be kept in it, both because he can earn more and also because he becomes more and more expert at this one speciality'.*

The Report goes on to note, in relation to 'abstract theory', that this should be obtained during the apprenticeship but:

> *'The practical fact however is, that such instruction is not given to apprentices except in rare instances.'*[46]

By the late 1880's, notwithstanding the defects in the system, the venerable William Pole could state:

> *'... this practical course furnished the most direct and expeditious entry to the professions. It was not the custom to consider any scientific knowledge as particularly necessary'.*[47]

Even those such as Fleeming Jenkins, who as Professor of Engineering was certain to be in favour of more formal tuition, were equivocal on this point. In his address to the B.A.A.S. in 1871, Jenkins pointed out how in the past good sense and long practice had provided the answers to simple problems, but in the seventies problems were more complex and the help of science and theory were required for their solution. However he went on to say:

> *'The business of the school is to teach those things which practice as an art will not teach a man.... one kind of knowledge of the properties of materials can only be acquired, by actually handling them.... Colleges cannot give him this; he must serve an apprenticeship, in fact if not in form.'*[48]

Structural Ironwork. 1850-1890

Theory, Science and Practice

What becomes clear from the evidence presented in these various reports is how the theoretical, scientific and technical education available in Continental colleges exceeded the scope and content of that available in the United Kingdom. The European courses were far more comprehensive and usually longer; an academic engineering qualification led to government posts and bettered social status, and was it seen as vitally important to the engineer. Many commentators, such as Jenkins, Calcott Reilly and Latham appreciated that continental engineers were going to be better prepared to confront new problems and derive solutions for them from first principles.[49]

This lack of formal technical education was very much the outcome of the English attitude to the provision of education and its content.[50] It was the result of several different influences: on one hand primary and secondary education was left to private enterprise, on the other the teaching of technical subjects was considered undesirable for reasons ranging from the wish to not raise the worker's aspirations, to the possible revealing of the firm's secrets.[51] It was also a reflection of two other attitudes that existed at the time. First, the anti-intellectual view of science and technical education, that was reinforced in the building industry by a widespread respect for the craftsman.[52] Secondly, the widely held mid-century opinion that all progess up to that time was soley due to the efforts of the 'practical man', the man on the job, whose education had been acquired on the shop floor, on the site, or in the office.[53]

It is clear that the need for a theoretical scientific education was understood in the 1870s and 1880s, but papers such as William Anderson's, given to the Junior Society of Engineers in 1887 at South Kensington, (where there was science teaching), with its plea that engineers should study nine sciences, suggests that at the time these were not part of the normal curriculum. Although Anderson's list is perhaps over ambitious, it presents a distinct contrast to that of Fowler twenty years earlier, where the emphasis was on the practical skills, but it still illustrates that those in practice in the nineties had not had the opportunity to acquire any substantial scientific knowledge in their youth.[54]

If it is assumed that the universities were the most likely source of scientific knowledge for engineers, the limited number of people on university engineering courses suggests that such knowledge was not very widespread. It would appear that in 1870 about 35% of those who described themselves as scientists or engineers attended university, about 200 in all, and the majority of these were scientists not engineers. Even in 1914, out of a total university population of 33,000, only 1622 (roughly 5%) were taking engineering courses. Outside the traditional centres, Cambridge, Edinburgh and Glasgow, the numbers in the civic universities were still very small, 800 in 1880, 3,300 in 1893, with only 93 graduating in that year. Because of the way the totals are aggregated there is no way of establishing how many were studying engineering, not science, but it seems that the number of engineering students was very small.[55]

One of the first things that Rankine attempted to do when appointed professor at Glasgow in 1855, was to define engineering knowledge and its relation to practice. He described as follows what he saw as the three parts of knowledge:

'Mechanical knowledge may obviously be distinguished into three

> *kinds; purely scientific knowledge, purely practical knowledge and that intermediate knowledge which relates to the application of scientific principles to practical purposes, and which arises from understanding the harmony of theory and practice.'*[56]

This was the theme taken up by J.H. Latham in 1858, in the preface to his book, where he attempted a definition of the 'theoretical man' and the 'practical man', based on the use of analysis, the activity which most mathematical techniques are used for in structural design. He saw analysis for the practical man being used to classify events 'actually passing before the eyes', and for the theoretical man being used to draw out the facts so that 'the bearing they have on the object in view may be connectedly stated in sentences whose logic is easily seen...'. He went on to say, with regret, that the reasoning of mathematics and the experience of the workshop 'are rarely considered as in union, but rather as opposed to one another'.[57]

One result of Rankine's efforts, and those of others, to use a more scientific approach was the change in the 1850s and 1860s to the use of stress as a design parameter, in place of a breaking load. In 1869 the credit for this change in girder design was given by W.C. Unwin to J.H. Latham, when he said the latter was:

> *'...the first to abandon completely the formulae for breaking weight and to proportion girders simply to the actual stress due to the load, and his work is still probably the best treatise in our language on wrought iron bridges'.*[58]

Although a change of attitude was taking place, it seems that for most engineers at the time, the design of structures was based on the assumption that the best course of action was to copy an example that had successfully performed the function required. Much of the comment in the journals and books talked of the need to copy what had been done before, and much of the information available was presented in a form that made this possible. With design dependant on the use of factors to control the size of members, coupled with the use of small scale tests, design education had only to develop the skills of draughtsmanship and an understanding of the disposition of structural members to succeed.

One indication of the attraction of techniques based on craft skills, such as draughtsmanship, can be seen in the way the use of graphic statics was taken up in the design office in the last decades of the century, in preference to mathematically based design techniques. As steel became more widely used, it became important to be able to define the stress in individual structural members, to ensure the economical use of an expensive material, and the technique described by Karl Culmann in his book, *Die Graphische Static* (1866), enabled this to be done.[59] It was of course a design tool that required the minimum of theoretical knowledge and was dependant on draughting skills.[16]

Today the practice of design is seen as requiring a theoretical background before a solution to a large scale problem can be produced. However design is a skill, exercised in a climate of knowledge derived from a variety of sources, which requires practice for its development. It was this additional requirement that was described by Zerah Colborn in *Engineering* in 1869, in an editorial entitled 'Experience in Design', where he said:

> *'Engineers are often called upon to design structures, the proper pro-*

> *portions of which cannot possibly be determined by any process of mathematical investigation, no matter how refined or labourous'*[60]

The whole article highlights the difficulties of design, where the quality of the material and distribution of members can falsify decisions based on exact calculation a similar point was made by Rankine:

> *'In practical science the question is – 'what are we to do' – a question that involves the necessity of the immediate adoption of some working rule'.*

He went on to accept that technology could not wait for the result of scientific research, that designers had to make decisions on the best data available, and that prompt and sound judgement in such cases was the characteristic of the 'practical man'.[61]

Although as the extracts quoted above show, the combination of theory and practice was seen as desirable, change was slow. Benjamin Baker, in his Presidential Address to the British Association for the Advancement of Science, made a strong plea for the application of scientific principle, and of Rankine's 'Intermediate Knowledge'. This was intended to qualify the student '... to plan a structure or mechanism for a given purpose, without the necessity of copying some existing example, and to adapt his designs to situations to which no existing example offers a parallel'.[62]

To the foreign engineer the situation in Britain was hard to understand. As Gustav Eiffel pointed out in 1888:

> *'The English engineers have almost bypassed calculations, and they fix dimensions of their members by trial and error, and by experiment... and small scale models'.*[63]

This view was reiterated in 1892 when William Anderson gave the first Forrest Lecture to the Institution of Civil Engineers:

> *'...this country is not keeping pace with its neighbours, and we shall, in the future, have to pay more attention to abstract science and its application to practice, than we have been so far in the habit of doing...'.*[64]

Notwithstanding these attempts to get new ideas about structural design adopted, all the technical journals were very critical of books about theory and judging from the tone and content of the editorials about design, they considered that their readers were more interested in the practicalities of fabrication and erection. Several editorials commented on the way the man on the site could, by the distortion and overstressing of components during erection, completely nullify the assumptions and intentions of careful calculation.

Conclusion

It is apparent that people involved in the design, fabrication and erection of iron and steel structures in the nineteenth century, did not see their activities as the result of a combination of either science, theory, practice or technology. For them applied science, although a term used by people such as Rankine, did not have its present-day meaning. For the mid-century designer all these concepts were separated, and his interest in theory was only for its power to forecast future events, or future performance. This is clear from a quotation from Bender's text book of 1877:

> *'...the thinking bridge engineer, who in the school or practice has learned to sift rubbish, both analytical and graphical, from the few principles of natural philosophy which are really needed... will decide for himself when... the application of the theory of the continuous girders can lead to any economy in construction'.*[65]

It must be remembered that the best developed theories that involved the performance of structures were those concerned with the properties of materials, especially elasticity, and their use in structural design was limited. Even when analytical methods were available it is clear that few designers were prepared to use them. In engineering, a major part of practice is design, which is essentially action, and action requires a decision to do one thing rather than another; but action is based on intentions, on a design achieving the desired aim, which in much of construction involves what often appears to be subjective choices about form, geometry and style. But in artisan technology, which it is suggested was the basis for much of engineering and structural design, these choices were based on action which had a logical basis and an intuitive one. The latter would be better defined as communication by tacit knowledge, being understood without being expressed, and was to a large extent controlled by craft rules and respect for past examples, and did not depend on any scientific knowledge.[66]

The general level of understanding, and the concentration of science on those disciplines where observations could affect the classification and understanding of natural phenomena, meant that the 'practical man' had little use for science or theory. This was the view that Huxley put forward in 1850, when he wrote:

> *'... practical men believed that the idol whom they worship – rule of thumb – has been the source of the past prosperity, and will suffice for the future welfare of the arts and manufactures. They were of the opinion that science is speculative rubbish; that theory and practice have nothing to do with one another; and the scientific habit of mind is an impediment, rather than an aid, in the conduct of ordinary affairs'.*[67]

and that same view would seem to be just as prevelent in 1890 as in 1850.

In the mid nineteenth century it was often assumed that physical science had an explanation for the observed behaviour of a phenomena and an exact theory that permitted the application of the general to the particular, providing a correct explanation of what had taken place. Frequently however when designs were based on theory, the theory was wrong and in a scientific setting the result would have been unacceptable. Fortunately in technology generally, and in construction in particular, rough and ready theories supplied quick estimates of orders of magnitude, and the traditional methods of construction and connection adopted, successfully in the majority of instances, masked the incorrect results of theoretical assumptions.[68] In this type of situation the view of a writer such as Layton is correct, when he comments that in the mid-nineteenth century there was very little of what the twentieth century would describe as applied science.[69]

This suggests that the premise put forward by some historians of technology, that the last half of the nineteenth century, in Britain, was a time when there were extensive developments in structural theory and scientific engineering are unfounded, and that the need to design large span structures did not create a demand for theories to explain structural performance.[70] Although by 1890 there had been an extensive body of research carried out into structural performance and the theoretical design of beams and trusses by continental engineers working in university and technical high school laboratories, the comments of practising designers do not suggest that the general view of the use of theory in design had changed much between 1840 and 1890.

The relative importance in relation to each other of science and technology, theory and practice, is still the subject of an extensive literature, in which numerous writers in the 1960s and 1970s, have attempted to resolve the question and to produce an acceptable philosophy of technology.[71] More recent writers have attempted to place the problem, as far as structural design is concerned, within the context of the design process. Thus whilst emphasising the creative aspects of design they have related the work of the practical man to the theoretical developments of the time and the range of problems to be solved.[72] However, there are still writers who are prepared to argue, as does Michael Forres, that all eighteenth and nineteenth century technical developments were the result of the efforts of the practical man.[73]

The preceding discussion demonstrates that for most of the professions in the construction world, the terms science, technology and theory had limited significance and that in the United Kingdom, perhaps in contrast to the rest of Europe, the 'practical man' remained highly respected. That this was true in education and design, as well as on site and in the workshop, is confirmed by a brief perusal of the biographies of late nineteenth century engineers, which show that in all aspects of their education and career, the practical was considered important.

The majority in the design professions had received an education that had been dominated by a craft approach to design, where respect for practical experience and tradition were paramount. This coupled with the scepticism for any contribution that academic study or science could make, ensured that many considered that practical experience and knowledge learnt in the workplace were more useful than anything else.

Nevertheless, by the end of the century, the general climate of ideas was slowly changing, and engineers and designers were aware that in many activities, even if not their own, science and theory had a contribution to make in providing explanations and developing analysis. This is apparent in the more theoretical content of the papers that were read to the institutions and that appeared in the technical press.

Although by 1890 there was a proliferation of institutions intended to serve the interests of a variety of engineering specialisms, it is apparent that the evidence does not substantiate the claim that there was an established discipline of scientific engineering. However changes were occurring and the dominance of the practical man, with his craft approach to education, design, and practice were being challenged. In design, as in many professions, the small size of the majority of practices has ensured that this preference for the practical or the craftsman approach has endured into the present century and can still be heard in discussions about professional education.

Correspondence:
Dr Stanley Smith, University of Greenwich, Dartford Campus, Oakfield Lane, Dartford, Kent DA1 2SZ, United Kingdom.

Notes and References

1. R.A. Buchanan, *The Engineers: A History of the Engineering Professions in Britain, 1750-1914* (1989).
2. Ove Arup, 'The World of the Structural Engineer', *The Structural Engineer*, 47, (1969) pp3-12. The Institution of Structural Engineers was founded as the Concrete Institute in 1908, and took the present name in 1922.
3. The whole of the 'Description of an Engineer' by Thomas Tredgoldwasprinted as part of C.H. Gregory's Presential Address in the *Minutes of the Proceedings of the Institution of Civil Engineers* 27 (1867-68), p181.
4. John Fowler, Presidential Address in *Min. Proc. Inst. Civ. Eng.*, 25 (1865-66), pp.180-228.
5. W. Fairbairn, 'An Experimental Enquiry into the Strength of Wrought Iron Plates and their Riveted Joints...', *Philosophical Transactions of the Royal Society*, (1850), pp.677-725, and W.H. Barlow, 'On the Existence of an Element of Strength in Bars Subjected to Transverse Strain...', *Phil. Trans. Royal Soc.*, 145 (1855), pp.225-242.
6. R. Turner, 'Description of the Iron Roof over the Railway Station, Lime Street, Liverpool', *Min. Proc. Inst. Civ. Eng.*, 9 (1849-50), pp.204-215.
7. B.H. Blyth, 'The Caledonian Railway Viaduct over the River Clyde at Glasgow', *Min. Proc. Inst. Civ. Eng.*, 61 (1880), pp.190-199.
8. J. Barton, 'On the Economical Distribution of Material in the sides or Vertical Portion of Wrought Iron Beams', *Min. Proc. Inst. Civ. Eng.*, 14 (1854-55), pp.443-490; W. Fairbairn, 'On Tubular Girder Bridges', *Min. Proc. Inst. Civ. Eng.*, 9 (1849-50), pp.223-287.
9. C. Reilly, 'On Uniform Stress in Girder Work', *Min. Proc. Inst. Civ. Eng.*, 24 (1864-65), pp.391-457; J. Gaudard, 'On Theory and Details of Construction of Metal and Timber Arches', *Min. Proc. Inst. Civ. Eng.*, 31 (1870-71), pp73-174.
10. H. Hayter, 'The Charing Cross Bridge', *Min. Proc. Inst. Civ. Eng.*, 22 (1862-63), pp.512-539; J. Wolfe Barry, 'The City Terminus Extension of the Charing Cross Railway', *Min. Proc. Inst. Civ. Eng.*, 27 (1866-67), pp.410-442; W.H. Barlow, 'Description of the St. Pancras Station and Roof' *Min. Proc. Inst. Civ. Eng.*, 30 (1869-70), pp78-93.
11. B. Baker, 'Practical Strength of Beams', *Min. Proc. Inst. Civ. Eng.*, 62 (1879-80), pp.251-271.
12. J. Gaudard, 'The resistance of viaducts to sudden gusts of wind', *Min. Proc. Inst. Civ. Eng.*, 69 (1882), pp.120-37; C. Bender, 'The Design of Structures to Resist Wind Pressure', ibid., pp.80-119.
13. *Spon's Engineers Price Book* (1888).
14. C. Graham Smith, 'Wrought Iron Girder Work', *Transactions of the Liverpool Engineering Society*, 33 (1881), pp.32-47.

15. H.W. Pendred, 'Designs, Specifications and Inspection of Ironwork', *Transactions of the Society of Engineers*, (1883), pp.87-112.
16. B.A. McFarlane, 'The Chartered Engineer: A Study of the Recruitment, Qualification and Conditions of Employment and Professional Assessment of Chartered Civil and Mechanical Engineers in Great Britain', (Ph.D thesis, London, 1961); B.P. Cuddy, 'The Royal Indian Engineering College, Coopers Hill, 1871-1906', (Ph.D thesis London, 1980).
17. I. Todhunter and K. Pearson, *A History of the Theory of Elasticity and of the Strength of Materials*, (2 Vols, Cambridge, 1886, 1893).
18. J. Anderson, *The Strength of Materials and Structures* (1872); B.B. Stoney, *Theories of Stresses in Girders and similar Structures* (2 Vols., 1866); W.C. Unwin, *Wrought Iron Bridges and Roofs* (1869).
19. W.J. McQuorne Rankine, *Manual of Civil Engineering* (1862).
20. W.J. McQuorne Rankine, a set of lithographed notes handed to students at Glasgow, 1856-57, now in the School of Civil Engineering Library, Imperial College, London.
21. Critical reviews are to be found in *Engineer* 1 (1856), p.57; 23 (1866), p.203; 28 (1869), p.56; 29 (1870), p.199.
22. J.H. Latham, *The Construction of Wrought Iron Bridges* (1861) and F.W. Sheilds, *The Strains on Structures of Ironwork* (1861).
23. See Sheilds, Sheilds was a friend of C.H. Wild, shared his mathematical interests, and had been involved in the re-erection of the Crystal Palace at Sydenham.
24. Latham, *Construction of Wrought Iron Bridges*.
25. W. Fairburn, *The Life of William Fairbairn*, edited and completed by W.Pole, (1877). There is no biography of Rankine, but in K.R. Hutchinson, 'W.J.M. Rankine and the Rise of Thermo-Dynamics', (D.Phil. thesis, Oxford, 1976) there are a brief summary and references to contemporary sources.
26. *Engineering*, 10 (1870), p.6.
27. W. Fairbairn, *Useful Information for Engineers* (First Series, 1856); *On the Application of Cast Iron and Wrought Iron to Building Purposes* (1854).
28. W. Humber, *A Practical Treatise on Cast Iron and Wrought Iron Bridges and Girders as Applied to Railway Structures* (1857); *A Record of the Progress of Modern Engineering* (1863). Other volumes were published in 1865 and 1868.
29. J.W. Grover, *Estimates and Diagrams of Railway Bridges for Turnpike, Public and Occupation Roads* (1866); G. Drysdale Dempsey, *Iron Roofs* (1862)
30. E. Olander, *A New Method of Graphic Statics Applied to the Construction of Wrought Iron Girders (1880)*.
31. T. Timmins, *A Simple and Practical Method of Obtaining by Designers the Strains in, and Strength of, Riveted Girders and Curved Roofs* (1882); E.W. Tarn, *An Elementary Treatise on the Construction of Roofs of Wood and Iron* (1882)
32. W. Binns, *The Second Course of Orthographic Projection* (1876); W. Bently, *Sketches of Engine and Machine Details* (1898); G.F. Charnock, *Exercises in Graphic Statics* (1900).
33. E. Matheson, *Works in Iron, Bridge and Roof Structures* (1869).
34. H.N. Maynard, *Handbook to the Crumlin Viaduct* (1862); E. Hutchinson, *Girder Making and the Practice of Bridge Building in Wrought Iron* (1879).

35. Part one of *The Dorman Long Company Handbook*, (1887) has been republished in W. Bates, *Historical Structural Steelwork Handbook* (1984).
36. W. Templeton, *The Practical Mechanic's Worshop Companion* (15th ed 1887); G.L. Molesworth, *Pocket Book of Useful Formula and Memoranda* (1st ed. 1862, 33rd. 1948).
37. B. Baker, 'Long Span Railway Bridges', *Engineering* 3 (1867); 'Elementary Papers on Construction in 19 Parts', *Engineer* 26-29, (1869-70).
38. Templeton, *Workshop Companion*.
39. Mechanic's Magazine (1823-1873), subsequently *Iron* (1873-1893); *The Artizan* (1843-1873); *The Builder*, (1842-); *The Civil Engineer and Architect's Journal*, (1837-1868); *The Engineer* (1856-) and *Engineering*, (1866-).
40. Institution of Civil Engineers, *The Education and Status of Civil Enginers in the United Kingdom and Foreign Countries* (1870), p.vii.
41. *Engineering*, 1 (1866), p.14.
42. Fowler, Presidential Address, as note 4.
43. *The Education and State of Civil Engineers.*
44. *Royal Commission on Scientific Instruction and the Advancement of Science*, (1872-77), First to Seventh Reports, Minutes of Evidence. *Report of the Commission Appointed to Consider the Best Mode of Reorganising the System for Training officers for the Scientific Corps, Together with an Account of Foreign and other Military Education* (1857). Lt. Col. Yolland was the Chairman.
45. Livery Companies' Committee, *Report on Technical Education* (1878).
46. G.C.T. Bently, 'Statistics and Suggestions on Technical Education', in *Report on Technical Education*.
47. William Pole, 'On the Status and Training of Civil Engineers' (1889), a paper prepared for the Institution of Civil Engineers and never published. The proof with handwritten corrections is in the Institution Library.
48. Prof. Fleeming Jenkin, 'Presidential Address, Mechanical Section' *British Association for the Advancement of Science* (Edinburgh, 1871), pp.225-229.
49. F. Jenkin and C. Reilly in *The Education and Status of Civil Engineers;* Latham, *Construction of Wrought Iron Bridges.*
50. Margaret Gowing, 'Science, Technology and Education in England 1870' in *Notes and Records of the Royal Society*, 32 (1977), pp.71-90; M. Argles, *South Kensington to Robins* (1964).
51. G.W. Roderick and M.W. Stephens, 'Education and Training of Engineers in 19th Century Liverpool', *Annals of Science* (1971), pp.143-163; for Europe see F.K. Ringer, *Education and Modern Society* (1979), and C.R. Day, 'The Making of Mechanical Engineers in France: The Ecoles d'Art et Metiers, 1803-1914', *French Historical Studies* 10 (1978), pp.439-460.
52. M.J. Weiner, *English Culture and the Decline of the Industrial Spirit* (1981); J. Wellens, 'The Anti-Intellectual Tradition in the West', and R.H. Wilkinson, 'The Gentleman Ideal and the Maintenance of a Political Elite', both in P. Musgrave ed. *Sociology, History and Education* (1970).
53. Weiner, *English Culture*.
54. W. Anderson, 'An Engineer's Education', *Journal of Junior Engineering Society*, 1 (1887), pp.6-18.
55. The Senate, University of London, *List of Doctorial Theses and Examination Statistics* (1904); R.M. Birse, *Engineering at Edinburgh University: A short history, 1673-1983* (Edinburgh, 1983); Argles, *South Kensington to Robbins;*

McFarlane, 'The Chartered Engineer'.
56. W.J. McQuorne Rankine, Preface 'A Preliminary Dissertation on the Harmony of Theory and Practice', *Applied Mechanics* (1864), p.5.
57. Latham, *Construction of Wrought Iron Bridges, pp.v-vi.*
58. W.C. Unwin, Preface to *Wrought Iron Bridges and Roofs* (1869).
59. K. Culman, *Die Graphische Statik* (Zurich 1866).
60. 'Experience in Design', *Engineering*, 7 (1869), p.173.
61. Rankine, *Applied Mechanics.*
62. B. Baker, 'Presidential Address to Section G Mechanical Science', *British Association for the Advancement of Science* (Aberdeen, 1885), pp.1182-1192.
63. G. Eiffel, *Les Grandes Constructions Metalique* (Paris, 1888), pp.10-11.
64. W. Anderson, 'The Interdependance of Abstract Science and Engineering', The James Forrest Lecture, *Min. Proc. Inst. Civ. Eng.*, 114 (1892-93), pp.254-283.
65. C. Bender, 'Application of Theory of Continous Girders to Economy in Bridge Building', *Proceeding of the American Society of Engineers*, quoted in *Engineering*, 24 (1877), pp.403-404.
66. P. Caws, 'Praxis and Techne', in Bugliarello and Donner Eds., *The History and Philosophy of Technology* (1980).
67. T.H. Huxley, 'Science and Culture' (1880), in *Science and Education Essays* (New York, 1898).
68. This subject has been covered by numerous writers in *Technology and Culture* in Volumes 2 (1960-61) and 7 (1966).
69. E. Layton, 'Conditions of Technological Development', in I. Spiegal-Rosing and D. de Solla Price, eds., *Science, Technology and Society (1977).*
70. Buchanan, *The Engineers.*
71. J.M. Staudenmair, *Technology's Storytellers. Reweaving the Human Fabric* (Cambridge, Mass. 1985).
72. W. Addis, *Structural Engineering, The Nature of Theory and Design* (1990); R. Mark, *Light, Wind and Structure. The Mystery of the Master Builders* (Cambridge, Mass. 1990).
73. M. Forres, 'The Myth of a British Industrial Revolution', *History*, 66 (1981), pp.181-198; A.E. Musson, 'The British Industrial Revolution', History, 67 (1982), pp.252-258.

17
Structural design by observation of failures: how the Gothic master masons determined the dimensions of their structures

Henry J. Cowan

Structural design by observation of failures prior to the development of the theory of structures was entirely feasible, although by comparison with modern methods it is wasteful in material. However, it depended on a relatively slow change in design principles, and on adherence to certain standard dimensions.

The principles of Gothic structure are examined, and the methods by which structural sizes could have been determined are discussed. It would have been necessary to observe failures over a long period of time to arrive at the high span / thickness ratios which were used successfully in some of the cathedrals.

Design by Observation of Failures

I first became interested in Gothic structure in 1942 when I was on an army surveying course in Ripon in Yorkshire, which left me with plenty of spare time for sightseeing. Nearby was Fountains Abbey, destroyed when Henry VIII dissolved the Monasteries. There is a very tall and thin column in the ruined Chapel of the Nine altars, whose slenderness ratio is well above the upper limit that was permitted by the British Concrete Code valid in 1942, or for that matter by the then current Australian Code. In other words one could not at the time have built a reinforced concrete column with these dimensions, however much reinforcement one put into it. But here was a column consisting of stones piled vertically on top of one another, joined only by a lime mortar so weak that one could scratch it out with a match stick, and with very little support at the top, because the roof had collapsed:

A column with an extremely small factor of safety, designed by a masterbuilder who had no knowledge whatever of structural theory, but which is nevertheless still standing after more than 500 years.

There is, I think, only one way in which the Gothic masterbuilders, and indeed the masterbuilders before them, could have obtained structural design information, and that is by the observation of cracking and failures. A failure, in particular, is incontrovertible evidence that the structure is inadequate.

This was already used by the Ancient Egyptians, the first people to erect durable buildings about four and a half thousand years ago, by carving in stone what had obviously been conceived as timber structures. In particular they carved the timber planks for the roof structure in stone.

It is very difficult to understand today why the Egyptians used this highly inefficient form of construction consistently for more than two thousand years. It is a structure eminently suitable for timber, which has tensile strength, but they used it for stone, which is weak in tension. We know the Ancient Egyptians knew the arch, because some mudbrick arches have survived. I presume the Egyptians did not use it for masonry structures because they were copying a timber structure in stone, and mudbrick structures were something else again.

The columns and the beams in Egyptian temples, are from the structural point of view, grossly oversized. Evidently structure was not considered. But the stone planks are not oversized. One can check this where some of the stone planks have collapsed. If all have collapsed, or if all are still in position, it is difficult to determine sizes; but where there are gaps between planks, it is easy to measure them, and to determine the maximum tensile stress in the stone. In most cases the stone planks are only just thick enough to carry their own weight. Evidently the Egyptians, being used to much thinner timber planks, and considering the amount of material that had to be transported, kept them as thin as they could. They presumably had an empirical rule for the thickness/span ratio, derived from failures.

Difference between Gothic and Ancient Structures

The same explanation applies to Gothic construction. But there are five major differences between Egyptian and Roman construction on the one hand, and Gothic construction on the other.

First of all Gothic structure is much lighter; its lightness is unique among pre-engineered construction. Secondly the Gothic structures were far more complex than any used before. Indeed, in 1942, when I first got interested in them, there were some whose accurate design was beyond the competence of any structural engineer; it was only through the defence research on shell theory and limit design during the Second World War that we were able to solve these problems after the War.

The third factor is a general change in attitude to failure between the Ancient World and the Middle Ages. The Law Code of Hammurabi, a Babylonian king who lived 3800 years ago, is typical of the Ancient attitude. It is engraved on a stone column that is now in the Louvre in Paris, and it lays down, inter alia, that if a master builder builds a house, and it collapses and kills the owner, then the masterbuilder shall be put to death. Naturally any sensible builder would make quite sure that every part of the building is oversized, and charge his client for the extra cost.

In Medieval times, at lease in the initial stages of Gothic construction, there was a firm belief in divine providence, so that the builder was not necessarily to blame.

Records of constructional activity in the Middle Ages are poor, and generally a failure was recorded only if it had a religious significance. However, we know from the evidence of the buildings that collapses were frequent.

The fourth difference between the construction of Gothic cathedrals and the temples of Ancient Egypt or Rome, is the small size of the constituency. The cost of the construction was no longer backed by a great empire which could command large sums of money and a large workforce, but generally only a small town and the surrounding countryside. We are considering a cathedral built by the citizens of Bathurst, rather than Sydney.

Finally, there was an appreciable decline in technological skill after the fall of Rome, and the efficiency of the simple Gothic hoisting gear (Fig. 1) was greatly inferior to that of the Romans, who used multiple pulleys. The masterbuilders therefore had to use relatively small pieces of stone, and they had to be economical in the use of material.

Elements Of Gothic Structure

It is generally considered that the first Gothic Cathedral, which showed a significant departure from the heavy Romanesque style, was the church of the Abbey of St. Denis, which is in suburban Paris, at the terminus of one of the Metro lines. It was probably designed by the Abbot Suger in the 12th century.

In its day it was simply contemporary architecture. The name Gothic was introduced in the 16th century by Giorgio Vasari, the first historian of the Renaissance, as a term of abuse for the modern barbarism perpetrated by the masterbuilders north and west of the Alps, which was infecting Northern Italy. To him the Goths, who sacked Rome in the 5th century, were the worst barbarians of all time, and this modern architecture was as barbarian. He might equally well have called it Hun or Vandal architecture. It was only in the 19th century that Gothic architecture became a term of approval.

The characteristic structural features of Gothic architecture were these:-

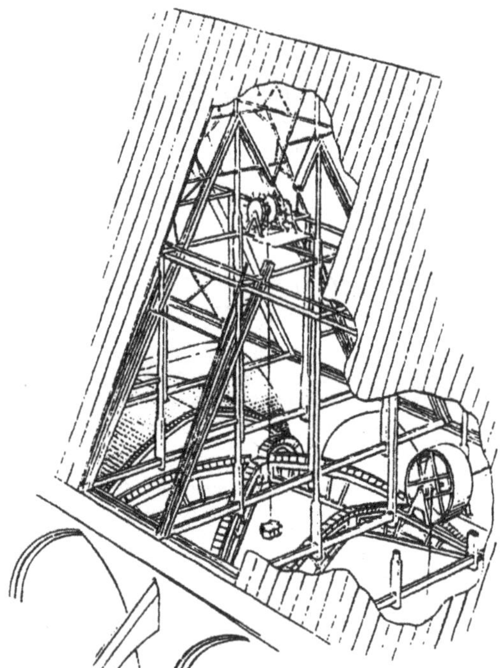

Figure 1. Method of handling blocks of stone for Gothic cathedrals suggested by Fitchen (Ref. 1, p. 286). The roof shown is that over Rheims Cathedral. the hoisting wheels are shown supported on the transverse tie beams of the timber roofs. The mechanical advantage of this hoising gear is greatly inferior to that available previously during the Roman Empire, or subsequently during the High Renaissance.

First, the pointed arch, which was developed around the 8th century by the Persians, and then taken over by the Arabs when they conquered Iraq. It is today generally called a Gothic arch, but Christopher Wren, who spent much time as the King's Surveyor repairing Gothic cathedrals, always called it the Saracen arch. It was brought to Europe by the returning Crusaders in the 11th century.

The strength of the pointed arch is shown by the fact that when Gothic cathedrals collapsed, the arches often remained standing.

The pointed arch consists of two circular arcs which are easy to set out, and its overall shape resembles that of a catenary, that is the shape assumed by a hanging chain, which is in pure tension when it hangs down, and must evidently be in pure compression, that is free from bending, when it is pointing up. It is most unlikely that the Persian, Arab, or Gothic masterbuilders appreciated that aspect, which was discovered only in the 17th century. However, there was a traditional form of construction in Iraq, which was still in use in the 1930's for building arches from reeds and I think the Gothic arch is simply a durable stone version of this combustible reed construction. Because it happens to be close to the catenary shape, it has remarkably good structural properties, and because it is made from circular arches, it is easy to set out.

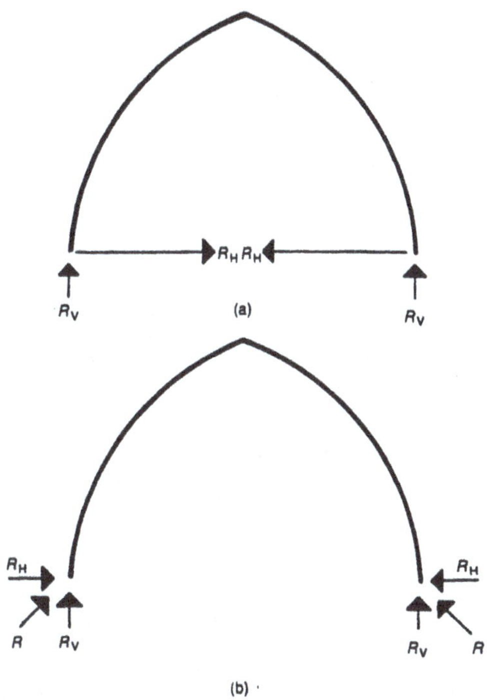

Figure 2. Every arch produces a horizontal thrust, which must be restrained by ties (a) or by buttresses (b) (Ref. 3, p. 183).

The second structural element is the cross vault, which has a more versatile geometry than the barrel vault, and replaced it in the final period of the Roman Empire.

There were still plenty of these vaults around in the 12th century, and indeed the Abbot Suger would not have needed to go very far to find one. The Cluny Monastery in Paris includes the remains of a Roman Imperial Bath, and its Tepidarium, the warm room, had a Roman concrete cross vault, which is still standing today. Abbot Suger could have got there in half an hour, if the metro had been running in his time, but even on a horse it would have taken less than a day.

The third ingredient of Gothic structure was the flying buttress. Every arch and vault produces a horizontal thrust (Fig. 2) which must be transmitted to the ground. Early Roman structures and most Romanesque structures are so massive, that the thrust can be transmitted to the ground within the existing structure. However, in late Roman and Byzantine construction flying buttresses were introduced, although these were relatively short and rather thick. In Gothic construction the flying buttresses became a conspicuous feature of the building, because of their slenderness; but they are basically just stone versions of the wooden props commonly used during erection.

The typical Gothic cathedral, like the Sydney Opera House, had two roofs on top of one another. The cross-vaulted stone roof was needed to protect the inside of the cathedral from fire, and particularly to protect the sacred relic which was usually the *Raison d' être* of a medieval cathedral. Because the rain water would not drain from the complex stone vaults, it needed a simple sloping timber roof on top of it for weather protection. Each of these roofs exerted an outward thrust, and thus there were generally two parallel sets of flying buttresses.

These three structural devices of Gothic made it possible to create a much lighter fabric for the building. The cross vault concentrated the forces from the roof at a few discrete points, where it was transmitted to the ground by the columns, by ordinary buttresses and by flying buttresses.

In between was a curtain wall supported by the very strong Gothic arches, and within these arches were large stained glass windows. The best of the cathedrals therefore have an appearance of structural lightness which had never been achieved before.

This was, however, only one aspect of the design. The other was a great reduction in the amount of material useed, compared to Romanesque construction, which was important because of the small workforce available to the small economic units that built most of the cathedrals, and because of the primitive handling equipment available after the dark ages.

Evidently, given the lightness of the construction, the sizing of the structural members presented a problem.

The sizing of the Gothic arches was relatively straightforward. The types commonly used in Gothic architecture, at least in the early period when the walls of the great cathedrals were built, were generally reasonably close to the catenary shape, and were therefore largely in compression. Since as a matter of aesthetic proportioning the arches could not be very thin, there was unlikely to be a structural problem.

The structural behaviour of the buttresses could not be analysed until the era of Neo-Gothic construction in the late 19th century. There is first of all the possibility that a buttress

Architectural Science Review Volume 35

might fail in shear (Fig. 3), and to suppress the tensile component of the shear stresses it may be necessary to add weight in the form of pinnacles (Fig. 4).

The forces are then transmitted to the ground and gradually turned towards the vertical by the addition of more and more dead weight.

The Mechanics of Gothic Buttresses

According to structural theory as it was conceived in the 19th century the resultant force had to remain always within the middle third. During the Neo-Gothic era the buttresses frequently looked much heavier than the lightest buttresses in the original Gothic construction, and conversely there were some Gothic structures that were, according to this analysis, unsafe, but which had not fallen down.

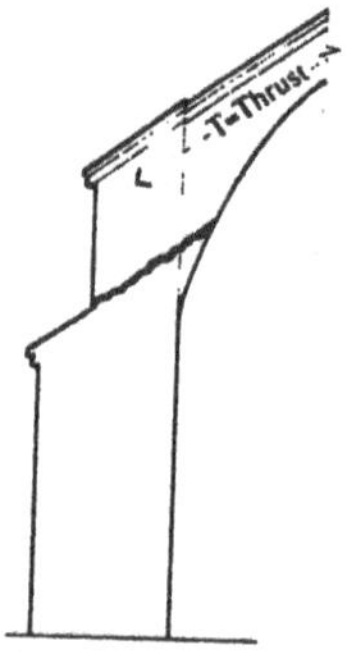

Figure 3. A buttress can fail in shear. This can be prevented by adding a pinnacle of sufficient weight to eliminate the diagonal tension due to shear (Ref. 4, p. 149).

As a simple model shows (Fig 5), the joints open up, but the structure does not actually overturn if the force lies outside the middle third, but within the section. The opening up of the joints does not present a problem in a structure which takes at the very least ten years to build, as the joints can easily be repointed. So the critical condition is that the resultant should lie, not within the middle third, but merely inside the structure to prevent overturning. There must, of course, be a factor of safety but, 1¼ is quite sufficient for a static dead load. The middle-third rule, it can be shown, gives a factor of safety against overturning of 3, which is much too high.

Some problems were caused by the simplistic geometry used to lay out many of the buttresses, which was based on four standard triangles (Fig. 6) that could easily be set out. Figure 7 shows on the right the buttresses of Ely Cathedral, built in the mid-13th century, which had been set out at angles of 30 degrees and of 39 degrees (the latter the angle of the 3-4-5 Pythagorean triangle) and still survive on the south side. In 1322 the central tower collapsed, and brought down the buttresses on the north side. They were then rebuilt as shown on the left at an angle which is obviously much more efficient. One does not need a knowledge of structural mechanics to see that.

The Mechanics of Gothic Vaults

The mechanics of the vaults is much more complex; however, there is one simplifying circumstance. The span of the vault and its geometry varied very little from one cathedral to the next.

The style of Gothic vaults changed gradually from the simple

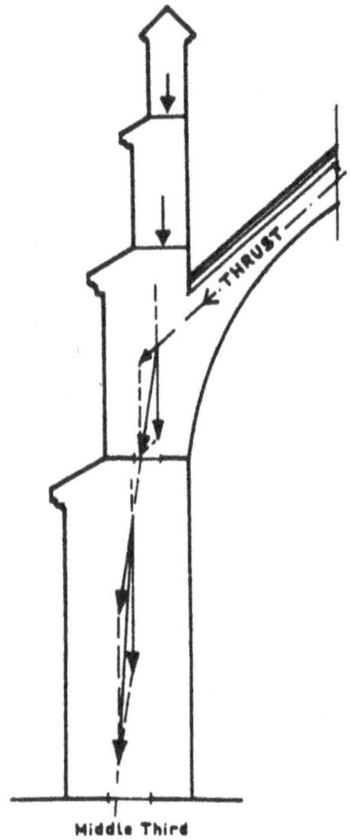

Figure 4. Analysis of a buttress by the middle-third rule. At any section the resultant of the thrust and the weight of masonry above it must lie within the middle third, if opening of the joints if to be avoided (Ref. 4, p. 149).

groined vault, to ever more complex geometries. But this change was gradual, and it occurred over a period of 4½ centuries, a period equal to that from the time of King Henry VIII of England to the present. A new fashion would not differ greatly from the then current orthodoxy, and there was therefore ample time to collect data of failures relating to it.

Furthermore, the span of Gothic cathedrals was remarkably uniform, and it was, by Roman or Renaissance standards, rather small. Some cathedrals were very tall, but they did not have great spans. Table 1 gives the spans of a number of major cathedrals.

There were few exceptions: Milan in Italy had a span of 16.8m. This presented plenty of problems, and 7 international consultants were called in succession to advise on its design.

Excluding Milan, these spans range from 9.4 to 13.4m, which means 11.4m + or - 17%. So one may reasonably assume that if a vault of a particular thickness gave no trouble in one cathedral, then a vault of the same thickness would give no trouble in another cathedral, if it was the same type of roof and had a similar span.

We can today with the help of finite elements solve most shell problems. Having determined the membrane forces and

Figure 5. Model illustrating the Middle-Third Rule. The joint opens up if the resultant falls outside the middle third, but collapse does not occur until the resultant falls entirely outside the section (Ref. 5, p. 21).

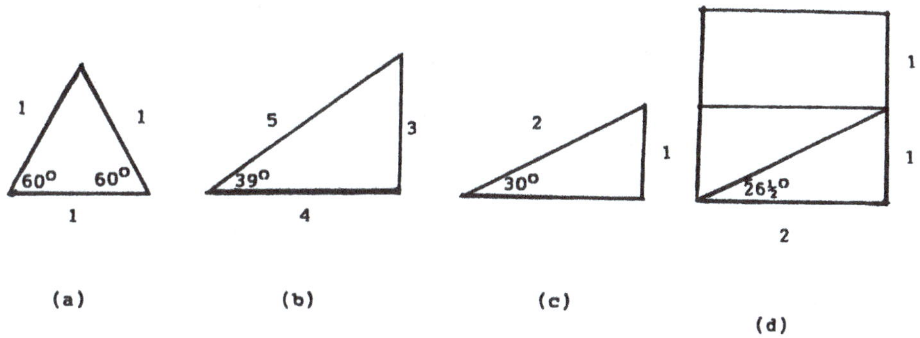

Figure 6. The four triangles used for setting out slopes in Gothic construction :-
(a) The equilateral triangle, set out by making all three sides equal. The angle to the horizontal is 60°.
(b) The Pythagorean triangle, set out by making the three sides 3, 4, and 5 units respectively. The horizontal angle is 39° or 51°.
(c)The right-angled 1-2 triangle. The horizontal angle is 30° or 60°.
(d) The 'halved-square'. The square is divided into two equal rectangles, and each is then halved by a diagonal. The horizontal angle is 26½° or 63½°.

Figure 7. Section through the choir of Ely Cathedral. The original buttresses of the mid-thirteenth century are shown on the right-hand side. The new buttresses built in the fourteenth century are shown on the left. The earlier design is geometrically simpler, but the greater structural efficiency of the later design is evident (Ref. 7).

bending moments, we then have to provide sufficient depth to ensure that the line of thrust at any particular section remains within the cross sections of the shell (Fig. 8).

The strength of the stone is not critical, because compressive stresses are quite small; it is the tension failure that is the problem, and we can calculate for any given geometry the ratio of span to thickness which is sufficient to prevent a collapse, because the resultant thrust remains entirely within the cross section of the shell. For each type of shell there is therefore a definite maximum ratio of span to thickness. The roof is structurally safe, provided this ratio is not exceeded. One could also determine experimentally what that ratio is by observing failures. There were plenty of those in the Middle Ages, particularly if one includes the occasions when a vault sagged as its formwork was lowered, and the props were hastily put back. If the roof failed, it was evidently too thin.

My theory is that information on these empirical span/thickness ratios was an important part of the mystery into which the builder was initiated when he was promoted from Journeyman to become a Master.

Table 1

Spans of the Naves of Some Major Gothic Cathedrals

France		England		European Continent	
Rheims	12.2m	Westminster Abbey	9.4m	S. Gudule, Brussels	10.7m
Notre Dame, Paris	12.8m	Winchester	10.0m	St. Stephen, Vienna	11.6m
Bourges	12.8m	Peterborough	10.4m	Toledo, Spain	11.9m
Amiens	13.1m	Salisbury	11.0m		
Chartres	13.4m				

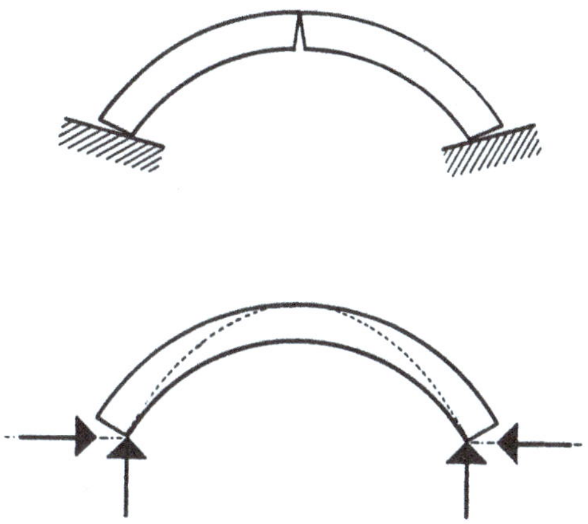

Figure 8. The arch or vault is structurally safe as long as the resultant thrust remains within the cross section (Ref. 8, p. 173).

We know that at least one masterbuilder was killed because he divulged a guild secret, and we know that masterbuilders travelled great distances as consultants at a time when the roads were bad and very dangerous, because armed robbery was a normal means of making a living. Master Builders from major centres, such as Strasbourg, received more invitations to travel during the Gothic era than during the Renaissance, which was far less secretive about the sizing of members.

Why did the Bishop or Duke or Cathedral Chapter go to the trouble of bringing a consultant from a great distance? I have no doubt that he advised on a variety of problems, but the really difficult one was the sizing of the stone vaults. I suppose the consultant Masterbuilder could on arrival have said the span/thickness ratio for this vault must be no more than 40, collected his fee, and gone home; but that would, of course, have given away the mystery. So he had to stay for some months, and gradually convey the information without disclosing what the secret was.

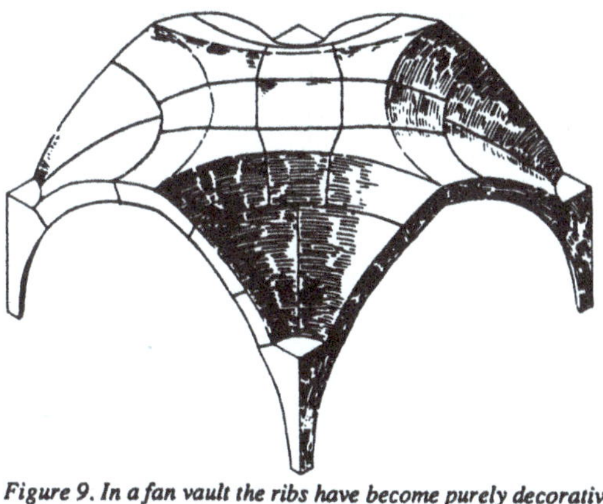

Figure 9. In a fan vault the ribs have become purely decorative, and the curvature changes gradually, so that the shell is almost flat at the top (Ref. 9).

In conclusion I would like to draw attention to the use of stone bosses in fan vaults, a brilliant structural solution to an extremely complex problem, which could only have been obtained by observation of failures.

Fan vaults were built only in England, during the perpendicular period of Gothic, mostly in the early 16th century. Kings College Chapel in Cambridge is one of the best known examples. In a fan vault the curvature changes smoothly and very gradually, and the ribs are purely decorative, and no longer coincide with an abrupt change in curvature (Fig. 9).

Because the shell is very flat on top, bending stresses, and consequently tension, develop there. The simplest answer with unreinforced masonry construction is to suppress the tension by adding some weight which superimposes compression.

This could be done by putting weight on top of the roof, but the Gothic master masons decided instead to hang a stone pendant underneath (Fig. 10). It seems at first sight totally irrational to hang a stone pendant from this thin shell roof, but there is in fact a good structural reason for doing so.

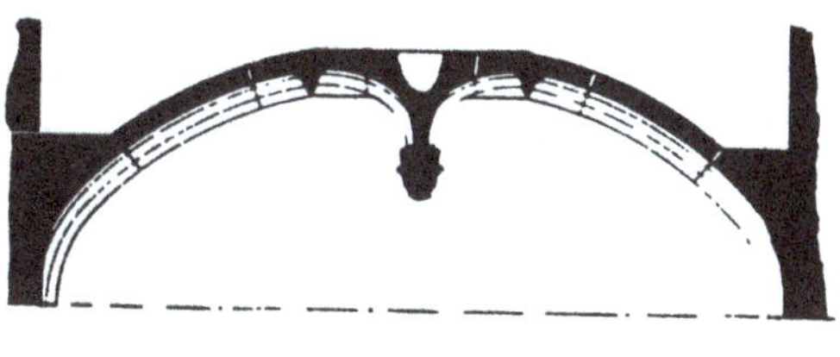

Figure 10. Stone pendant which supresses the tensile stresses in a fan vault. The stone is held in position by sloping the joints outwards, so that the pendant acts like a wedge (Ref. 9).

Conclusion

The method I described gives a possible explanation how Gothic structure could have been sized before the age of structural mechanics, and I do not know of any other. It is an essentially empirical approach, which was practicable only because the Master's time was not then a valuable commodity, and because the rate of progress and of change in architectural fashion was, by present standards, extremely show.

Acknowledgement

This paper is the substance of a lecture delivered to the Institution of Engineers, Australia, in Sydney on the 22nd of July, 1991.

References

1. J. F. Fitchen: The erection of French Gothic nave vaults. *Gazette des beaux arts*, Vol. 55 (1960), pp. 281-300.
2. Wilfred Thesiger: *The Marsh Arabs*. Longmans, London 1964. Figures 86-100.
3. H. J. Cowan and F. Wilson: *Structural Systems*. Van Nostrand Reinhold, New York 1981, 256pp.
4. H. J. Cowan: *The Masterbuilders*. Robert Krieger, Malabar Fie 1985, 299pp.
5. H. J. Cowan and J. Dixon: *Building Science Laboratory Manual*. Applied Science Publishers, London 1978, 156pp.
6. P. Frankl: *The Gothic - Literary Sources and Interpretations through Eight Centuries*. Princeton University Press, Princeton NJ 1960, 916pp.
7. D. H. S. Cranage: *Cathedrals and How they Were Built*. Cambridge University Press, London 1951, 42 pp.
8. J. Heyman: *Coulomb's Memoir on Statics*. Cambridge University Press, London 1970, 212pp.
9. F. E. Howard: Fan vaults. *Archeol. J.*, Vol. 68 No. 1 (1911), *quoted by* J. Hayman: Spires and fan vaults, *Int. J. solids Struct.*, Vol. 3 (1967), pp. 243-257.

Additional References

10. Theodore Bowie: *The Sketchbook of Villard de Honnecourt*. Indiana University Press, Bloomington 1968. 139pp.
11. Lon R. Shelby: *Gothic Design Techniques - The Fifteenth Century Design Booklets of Mathes Roriczer and Hanns Schmuttermayer*. Southern Illinois University Press, Carbondale 1977. 207pp.
12. John James: *Chartres - The Masons who Built a Legend*. Routledge and Kegan Paul, London 1982, 200pp.
13. Louis Groedecki: *Gothic Architecture*. Faber and Faber, London 1986, 221 pp.

18
The Britannia Tubular Bridge: a paradigm of failure-driven design

Henry Petroski

INTRODUCTION

Not all engineering failures take place suddenly and dramatically, accompanied by the crack and crash of steel and the tragic loss of life. Indeed, some classic errors in design have been all but ignored and forgotten in the context of what are generally hailed as tremendously successful projects. Yet these oversights can be no less endemic in the design process, and hence case studies elucidating them can be extremely instructive and valuable for the teaching and practice of design. One class of such errors may be described by the rubric of tunnel vision in design, and, like other types of errors, it is best introduced through a paradigmatic case study.

One of the most fruitful case studies of the role of failure in the history of engineering was hailed as a tremendous structural success and stood for 130 years as a monument to its engineer. Robert Stephenson's Britannia Bridge was generally regarded as the most significant construction project underway in the British Isles, if not the world, in the late 1840s, and engineers and designers made a point of visiting the site to see its progress. One such visit was the subject of a famous painting by John Lucas showing Stephenson in a site meeting with the partially completed Britannia Bridge in the background (Fig. 1).

What makes the Britannia Bridge so valuable as a case study is the twofold way in which it points out the important role that failure plays in design. Conscious and explicit considerations of structural failure were present (and well documented) throughout the entire design process, and this fact ensuréd the unquestioned structural success of the revolutionary bridge design. However, the tunnel vision created by so narrow a focus on the dominant structural problem of bridging the Menai Strait resulted in a monument to engineering shortsightedness. While the structural principles involved in the Britannia Bridge had a signal influence on the evolution of structures in the latter part of the 19th century, the bridge proved in retrospect to be such an economic and environmental failure that its own revolutionary form was obsolete almost immediately upon completion, as evidenced by that fact that only a few bridges like it were ever built.

How and why the Britannia Bridge could at the same time be both a tremendous success and a colossal failure are best understood in the context of circumstances surrounding the bridge's design and construction. As with all case studies meant to serve as paradigms, its story takes on greater significance as more and more details are known. While the following explication gives only some bare essentials of the case history, emphasizing in particular the role of failure in the design process, considerably more detail can be found in the invaluable contemporary account by the bridge's resident engineer Edwin Clark, "published with the sanction, and under the supervision of Robert Stephenson," in 1850 in two volumes plus a folio of plates.[1] Some introductory observations on the history of the design by Stephenson himself are included. Another contemporary account, by William Fairbairn, who conducted the crucial model experiments that determined the final configuration of the bridge's great wrought-iron tubes, was published in 1849 and reproduces many of the letters that Fairbairn and Stephenson exchanged as the design evolved.[2] A modern interpretation of the bridge's significance for the history of technology is contained in the 1978 monograph jointly authored by the economist Nathan Rosenberg and the engineer Walter Vincenti.[3]

Fig. 1. A commemorative painting by John Lucas shows Robert Stephenson (center) and Isambard Kingdom Brunel (seated, far right) at a meeting on the site of the Britannia Bridge, under construction in the background.

BACKGROUND ON THE INTRODUCTION OF IRON INTO BRIDGE BUILDING

The story of the Britannia Bridge is inseparable from the story of the development of the structural use of iron. What is generally regarded as the first iron bridge still spans the Severn River near Coalbrookdale in Shropshire, England. The 100-ft cast-iron arch was completed in 1779 after its proponents spent years convincing Parliament to authorize an iron bridge and further years outmaneuvering financial backers who were inclined toward a more conservative concept employing then-traditional bridge-building materials like timber and stone. The importance of the Severn River for the local iron trade was so great, however, that there was a reluctance among all concerned to obstruct river traffic with any permanent mid-stream piers that might be needed to support a wooden bridge or even the temporary timber falsework needed to erect a stone arch bridge. In the end, an iron bridge, whose large cast parts could be fitted together quickly and with the least disruption of commerce, was agreed upon.

In the 1770s there were few analytical tools for shaping or sizing the parts of an arch bridge, let alone one made of iron. Thus, it was natural to proportion bridge parts of an untried material as if they were parts of a bridge made of a proven material, and among the architectural sketches of proposed elevations for the bridge, there are clear attempts to make the bridge look as if it were made of stone or timber (Fig. 2). The profile of the bridge as built (Fig. 3) actually does have a strong resemblance to a Roman semicircular stone arch bridge, while the details of the cast iron ribs and how the parts fit together are strongly suggestive of the details of timber construction. The responsibility for supplying the parts for Iron Bridge fell to Abraham Darby III, whose inherited and direct experience with cast iron was unsurpassed.

Cast iron, like stone, was well suited to taking compressive loads, but it was known to be unreliable in tension. Wrought iron, on the other hand, could be forged into links and chains of great strength. Thus, in the early 19th century it was natural to advocate the use of increasingly available wrought iron to support the roadways of bridges from above rather than from below. In this way, water traffic would be virtually unimpeded over a wide channel not only during construction but throughout the life of the bridge. (Traditional arch bridges did, after all, have maximum headroom only at the center.) While the concept of a suspension bridge is said to have ancient Asian roots, with vines and other forms of vegetation first being employed in the principal structural members, a

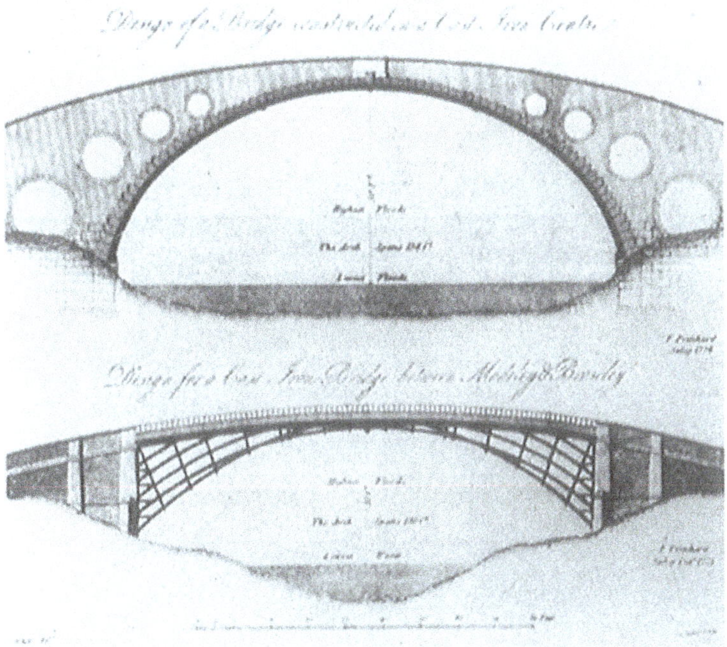

Fig. 2. Two designs for an iron bridge across the Severn River, unmistakably mimicking construction in the traditional materials of stone (top) and timber (bottom).

Fig. 3. Iron Bridge as built, completed 1779.

Fig. 4. Thomas Telford's Menai Strait Suspension Bridge, opened in 1826.

revitalization and extension of the form took place in the West in the early 19th century.

Among the most celebrated of wrought iron suspension bridges was the one completed across the Menai Strait in the mid 1820s (Fig. 4). Its designer, Thomas Telford, began his career as a stone mason, and the stonework of its approaches and towers, which still support and grace the structure, are among the features of the bridge that make it one of the most attractive in the world. But it was Telford the engineer and not the stone mason who had to wrestle with the hard question of how heavy a chain of flat wrought iron eye-bars would be needed to support the then-enormous span of approximately 580 ft between the towers. Typical of a new engineering endeavor, it was the failure characteristics of the chain that the engineer sought to understand in order to know how to avoid failure of the structure itself.

Telford suspended a specimen chain between test towers as much as 900 ft apart and hung weights from it. He employed an equation for the catenary developed by the British mathematician, physicist and optician Peter Barlow, and observed the stretch of the chain as a function of the load it bore. He noted that the iron chain began to stretch more rapidly after a certain level of strain was reached, and this occurred at about half the load that the fully stretched cable could support. He therefore incorporated a sufficient amount of iron in the bridge chains so that under the worst load that he could imagine on the roadway the strain in the cables would not exceed two-thirds of the critical strain. Thus, Telford employed a factor of safety (an explicit acknowledgement that failure was the limiting criterion) of 1.5 against yielding and 3 against ultimate failure of the chains. The most novel aspect of the design, the iron chains, naturally caused Telford to look with what proved to be tunnel vision at problems the chains themselves held for a successful structure. His concern with the failure of the wrought iron chains was so great, in fact, that he neglected to pay sufficient attention as to how the relatively light and flexible roadway might behave in the wind, and that in fact was to prove to be the bridge's weakest link.

In the meantime, the price of iron had dropped dramatically since its first use in Iron Bridge and its sparing use in the Menai Suspension Bridge. According to a contemporary observer looking at the development of bridge building in perspective:

- It is a distinguishing element of the engineer's art to adopt the material best suited, economically speaking, to the work he has to accomplish.
- In 1806 the price of bar iron, larger size, was 20

pounds per ton; in 1816 it was 10 pounds per ton; in 1828 it was eight pounds per ton; and in 1831 it was five–six pounds per ton.

- Thus, this material has gradually come into the domain of applications in construction, from which its high price had long excluded the consideration of its qualifications.[4]

In such a climate, it is no wonder that iron bridges came to be the structures of choice for the expanding railroads in Britain.

A period of economic depression hit Britain in the 1830s, however, and few bridges, innovative or otherwise, were built during that period. When the roadway of the Menai Suspension Bridge was blown down in the wind, it was essentially rebuilt as a matter of forced maintenance. The state of the art of iron-chain suspension bridge building in Britain was for some time to remain that of Telford's example across the Menai Strait. Iron arch bridge concepts had evolved considerably in the half century since Iron Bridge, and cast-iron girder designs had even been developed for more modest spans.

In the mid-1840s, in the resurgence of railroad and bridge building known as "railway mania," Robert Stephenson was called upon to design a number of bridges to meet a variety of water-crossing problems on the route of the Chester and Holyhead railway, a critical rail link between London and the mail boats to Dublin. Among the bridges needed was one to cross the River Dee at Chester. This bridge, whose design and failure have been discussed elsewhere,[5,6] presented structural problems that may have been considered almost trivial by Stephenson, especially in comparison to the monumental problems that had presented themselves in the design and construction of another bridge with which he was then engaged, a new crossing of the Menai Strait. The sudden collapse of the Dee Bridge in 1847 no doubt caused considerable concern that something was being overlooked in the much more ambitious design, and this may have caused the engineer to be more conservative than in his prior works.

THE BRITANNIA BRIDGE

In the early 1840s, amid the frantic activity associated with railway mania, the Chester and Holyhead railway wished to have a permanent rail link across the Menai Strait so that ferries would not have to be employed on the line until the terminus at Holyhead, on the western side of the Isle of Anglesey, where the mail from London could be transferred to a boat for Dublin. The existing Menai Suspension Bridge had been carrying carriage traffic for about two decades, with only infrequent interruptions to rebuild its roadway after severe storms. While this was a minor inconvenience that closed the carriageway only until the light roadway could be rebuilt, it would have been an embarrassment for a railroad whose reputation depended upon regularity and reliability.

Furthermore, the great flexibility of the suspension bridge's roadway made it problematic for heavy locomotives to pull railroad trains across it. The weight of the locomotive itself would deflect the bridge so much that a virtual valley would form under the load and a steep hill would have to be climbed out of to complete the passage. This was contrary to the objective of presenting a flat roadway, or at most a gradual grade, to the steam locomotive. Stiffening the bridge apparently was not considered a viable solution; that would have added a great deal of structural dead weight. George Stephenson, the great railway engineer and father of Robert, proposed unhitching the heavy locomotive when it reached the bridge, hitching up a team of horses to pull the cars across, and then connecting up another locomotive on the other side of the Strait. While this would have presented a resolution of the dilemma, it would hardly have been in keeping with the image the railways were trying to project. What was desired was a bridge that was strong and stiff enough to carry fully loaded trains without interruption across the strategic body of water.

Because the Menai Strait was so important to the British Admiralty, it had considerable control over what kinds of obstructions could be placed in the path of tall-masted ships. For almost half a century engineers had been confident that iron arch bridges with single spans of the order of 1000 ft (and thus capable of spanning the strait without intermediate supports) were perfectly feasible, but the Admiralty did not wish to have the shipping channels obstructed by any falsework for what might be years of construction. As early as 1810, Thomas Telford devised a scheme whereby an iron arch could be constructed without anchoring falsework in the water by suspending the incomplete arch from the top until it was self-supporting (Fig. 5), a scheme employed later in the century in America in the erection of the three-arch Eads Bridge across the Mississippi River at St Louis, and in Scotland in the great cantilever across the Firth of Forth. However, in Wales at mid-century, the Admiralty finally objected to an arch bridge regardless of its mode of construction, for the completed span would reduce the headroom toward the shores where the ships were sometimes driven by the tricky currents in the strait.

Stephenson was led to reconsider a suspension bridge, and in wondering if it were possible to stiffen the platform sufficiently for railroad trains he reflected upon some then-recent examples that employed trussing or trellising to carry turnpike roads and canals. Such schemes seemed to Stephenson to be difficult to extend to carry railroad traffic, however, for he believed that the

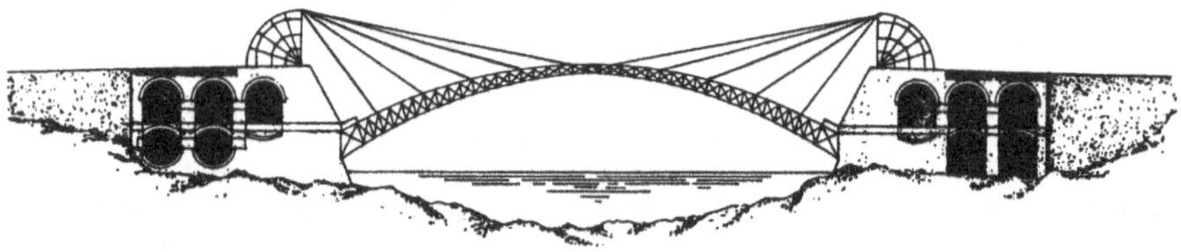

Fig. 5. Telford's 1810 scheme to construct an iron-arch bridge without falsework obstructing water traffic.

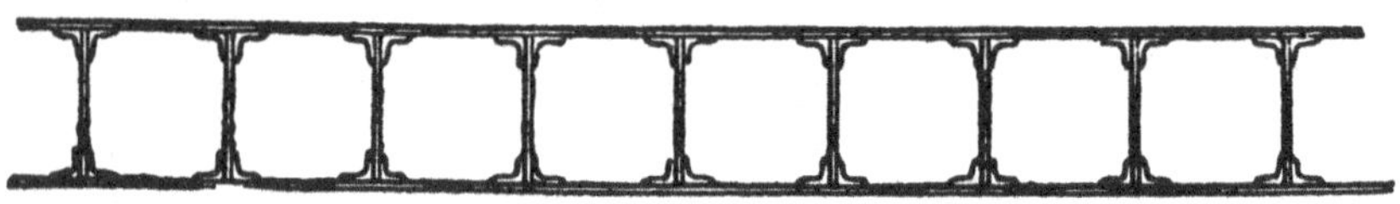

Fig. 6. Cross-section of Robert Stephenson's bridge over the River Lea.

extreme load on the suspension chain would alter its curvature so dramatically that "the direction and amount of the complicated strains throughout the trussing [would] become incalculable as far as all practical purposes are concerned".[1] In other words, in the absence of analytical or computational tools to predict how close to failure the design would be, Stephenson did not feel comfortable proceeding with a suspension bridge.

The functional and political objections to existing bridge designs forced Stephenson to look to novel ways to span the Menai Strait. As designers are wont to do in the conceptual stages of design, Stephenson thought about the ways in which similar problems had been solved in the past, and he recalled a structural scheme he had employed in 1841 for a small bridge that spanned the River Lea at Ware. In order to meet the dual design constraints of maintaining sufficient headway above the towing path along the river bank while not introducing a very great rise above street level, Stephenson had to come up with a flat yet shallow bridge deck to span the 50 ft distance. The scheme that he devised employed a wrought iron platform (Fig. 6) which he described as "a series of cells, . . . the whole being of boiler plate, riveted together with angle-iron, as in ordinary boiler building". While the bridge actually built at Ware employed parallel girders in a more conventional way, the cellular design remained in Stephenson's mind. At first, he thought to adopt the cellular structure to provide a stiff platform for a suspension bridge, "and the first form of its application was simply to carry out the principle described in the wooden suspended structure . . ., substituting for the vertical wooden trellis trussing, and the top and bottom cross braces, wrought iron plates riveted together with angle-iron".

Like most conceptual designs, once envisioned, that of the Britannia Bridge could be sketched and articulated and criticized. Indeed, the design process then became one of successive identification of failure modes and modifications to obviate them.[7] However, in the evolution of even the most radical of designs there is often an underlying constant, and Robert Stephenson's unchanging idea soon became to span the distances (of the order of 500 ft) between the shores and the Britannia Rock, a pre-existing natural obstruction upon which a tower could be built, with tubes of wrought iron that would be strong enough to support not only themselves but also the weight of railroad trains. These tubes would essentially be iron girders of such unprecedented size that the trains could pass through rather than go over the structure. Not only their form but also the manner of erection of the tubes would involve untried and thus unproven technology according to Clark's contemporary account.[1] Even as the bridge was being completed it was "difficult to retrace chronologically the numberless modifications which the general outline of this curious structure had undergone; but whether an elliptical or rectangular section, or a combination of the two, was the subject of discussion, the tube was always treated as a beam, of which the wrought iron top and bottom were analogous to the top and bottom flanges of a cast-iron T-girder. The calculations of its strength were made on this assumption." Even the accounts of Clark and Fairbairn, voluminous as they are, only touch on the variations considered.

Whatever the chronology, as with virtually all innovative designs the form of the Britannia Bridge evolved by successive considerations of failure. With each conceptual iteration, new modes of failure could be imagined possible and thus had to

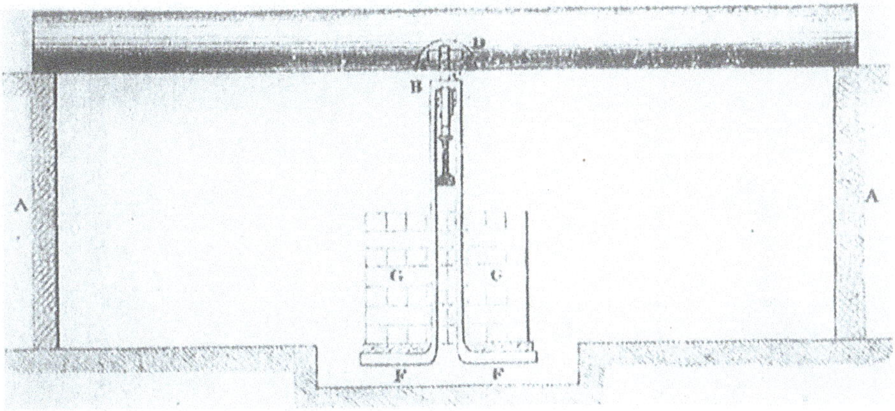

Fig. 7. One of William Fairbairn's tests of model wrought iron tubes.

be checked against or obviated. Even after their form was fixed, the manner of getting the tubes in place remained an obvious problem. Furthermore, in the interests of getting the railroad link finished as early as possible, the project proceeded on the principle of what today is called "fast track construction", and thus the stone towers were begun well before final details of the superstructure were known. An early concept was to employ suspension bridge-like chains to support the tubes during erection and provide supplementary support during use. Thus, when such chains would later prove to be unnecessary, the towers would remain confusingly tall and be found aesthetically objectionable by some structural critics.[8]

Because the use of tubular girders was an untried concept, Stephenson involved both Fairbairn and, later, Eaton Hodgkinson, whose experiments and analytical interpretation of the experimental data, respectively, provided guidelines for design and extrapolation from scale models. At one stage Stephenson thought a rectangular form to be undesirable because it would need diagonal bracing that would not be consistent with leaving a clear passage for trains; furthermore, the large flat sides of the bridge would present a great exposed surface to the wind. Such concerns over failure modes led Stephenson to favor a circular or oval form, as suggested by nature and adopted in such artifacts as bamboo fishing poles. Early experiments were in fact carried out on circular tubes (Fig. 7). What each experiment showed was that, under a central transverse load applied to the bottom of the tube (simulating a heavy locomotive on a full-scale bridge), the tube failed by buckling at the top or by being torn asunder at the bottom. Such failures were exactly what the experiments were designed to reveal, for without knowing how and under what load the model tubes would fail, Stephenson and his fellow engineers did not know how to design the details of the bridge itself. As each tube failed, the location of the failure was reinforced with additional thicknesses of wrought iron, or the design modified in some other way and the experiments repeated. It soon became evident that what was required best to obviate failure was a flange-like top and bottom, as used on cast iron beams, and so the original rectangular form evolved back into consideration as more logical than the round.

With an increasing accumulation of experimental data, Stephenson and Fairbairn sought the assistance of the theoretician Eaton Hodgkinson, who produced empirical formulae for predicting the strength of circular cylindrical, elliptical and rectangular wrought iron tubes employed as beams. Hodgkinson's summary of results, as reproduced in Fairbairn's account,[2] contains formulae for determining the empirical factor f representing the stress (which Hodgkinson called "strain") on the top and bottom of a tube when it gave way. His formula relating f to the "breaking weight" w and geometric parameters characterizing a rectangular tube of equal top and bottom thicknesses and arbitrary side thickness was as follows:

$$f = \frac{3wld}{2(bd^3 - b_i d_i^3)}, \qquad (1)$$

where l is the distance between supports, b and b_i are the external and internal breadths, and d and d_i are the external and internal depths, respectively.

Fairbairn's experimental results gave a much lower value of f than Hodgkinson's own experimental investigations, which were designed to explore effects of thickness of metal and scale. He found "a great falling off in the strength" of relatively thinner tubes and found the determination of a reliable value of f to be elusive, thus

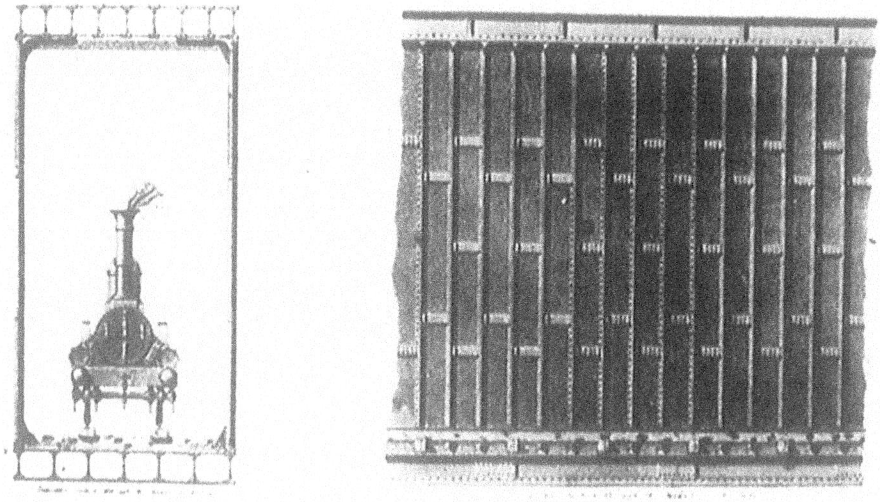

Fig. 8. A tube of the Britannia Bridge: left, in cross-section; right, a longitudinal view showing the assembly of wrought iron plates.

presenting "the chief obstacle to obtaining a formula for the strength of tubes of every form". It was such a negative finding, and his lack of a sophisticated understanding of the buckling phenomenon, that led Hodgkinson to "beg to recommend that suspension chains be employed as an auxiliary, otherwise great thickness of metal would be required to produce adequate stiffness and strength" in full-scale bridge tubes.

Eventually, the bridge tubes themselves were designed and assembled in the form shown in Fig. 8, with the towers having been made tall enough to receive chains, would they be necessary. To design an experiment to test the tubes fully was essentially to build the bridge itself, of course, and so, as with civil engineering structures generally, the final test would be in the erection and proof testing of the unique structure. Because Stephenson and his colleagues focused so much on the limit loads of the bridge tubes, which created failure at the critical section at midspan, little thought appears to have been given to the strength of the beam elsewhere along its length, although there was to be some debate between Stephenson and Fairbairn about how the tubes would behave over the piers. They were made continuous there, and this feature in part prompted Gustav Lindenthal to declare the Britannia Bridge "a milestone in bridge construction, not only because it was the first important iron bridge of the beam type, but also because it was the first representative of the continuous girder type".[9]

Center-loading was always considered to be a worst-case situation, however, and it dominated the sizing of the Britannia tubes, resulting in a conservative design. Such conservatism naturally came at the expense of economy, and in the rush to design and build the Britannia Bridge questions of optimization of resources seem not to have been given as much thought as they might. This occurred in spite of the fact that some iron rails had long been designed lighter at their ends than at midspan, thus economizing on material. In light of the fact that lifting the immense weight of the completed tube into place between the towers was to prove an enormous task, the tunnel vision that dominated the design of the tubes to enclose the trains and kept the opening almost constant in height throughout the length of the tubes proved to be a very costly error in design judgement. Furthermore, had the tube concept given way to one in which the train would ride atop a (box) girder, Stephenson's railroad bridge across the Menai Strait might have been more directly copied than it proved to be in the history of structural design.

THE SUCCESS AND FAILURE OF THE BRITANNIA BRIDGE

The great tubes of the Britannia Bridge showed themselves to be so robust during their floating and erection that supplementary suspension chains were never hung from the towers (Fig. 9). During proof tests, the bridge deflection was minimal and the structural success of the tubular concept was total. Indeed, the bridge might still be carrying trains across the Menai Strait had not the heat of an accidental fire in 1970 (started in a wooden roof erected to protect the iron from water leading to rust) so deformed the wrought iron tubes that it was deemed wiser to rebuild the bridge as an arch.

Fig. 9. The Britannia Bridge, completed in 1850.

While only a section of the original Britannia tubes was saved to be put on display beside the rebuilt structure, the shorter but conceptually similar Conway tubular bridge remains in service on the same rail line.

What made a failure of the bridge that stood for 120 years was its designer's (and his collaborators') apparent consuming distractions, if not obsession, with the singular design problem of carrying trains across the Menai Strait in a wrought iron tube. The enormity of this structural engineering problem in the climate of railway mania that existed when the project was conceived and approved seems to have so preoccupied everyone concerned that there was little time or inclination to ask questions relating to efficiency and use.

The total cost of the Britannia Bridge was approximately 600,000 pounds sterling. The experiments that played such a critical role in making the bridge a structural success cost a mere two-thirds of 1% of the total. The material alone for the tubes consumed about 25% of the cost; their construction, which included the driving of the order of 2,000,000 rivets, consumed another 37%. When the cost of floating the tubes and raising them into place was added, the wrought iron tubes of the Britannia Bridge accounted for almost 75% of the cost of the entire structure.

Even as the Britannia Bridge was being designed and constructed, alternative means of carrying railroad trains across 500 ft spans were under consideration by other engineers. Isambard Kingdom Brunel, Stephenson's British contemporary (Fig. 1), when faced with the problem of crossing the Tamar River at Saltash with a railroad bridge—a problem not unlike that of crossing the Menai Strait—came up with the solution of using large tubes as compression arches and wrought iron chains as suspension chains in a self-equilibrating arrangement that overall looks somewhat like a lenticular truss. Brunel's solution may also be imagined as one large open, as opposed to Stephenson's closed, iron tube. By the time Brunel's design was given the go-ahead, the free-spending days of railway mania had passed, and economy was at least as important as structure. Originally to be a two-track bridge, like Stephenson's across the Menai, Brunel's bridge across the Tamar was finally opened in 1859 as a one-track span (Fig. 10). Measured per foot of track length, the Saltash Bridge contained only 4700 pounds of iron compared to the Britannia's 7000. In terms of pounds sterling per ft of single track, the cost of the Saltash was only about half that of the Britannia crossing.[10]

In America, the engineer who had used the suspension bridge principle to carry canals over other bodies of water adopted the same principle to carry railroad trains across the Niagara Gorge. John Roebling's more than 800 ft span (Fig. 11) was opened in 1855 and provided a dramatic counter-example to the belief that a suspension bridge could not carry a railroad locomotive and heavy train effectively.[11] Roebling went on to

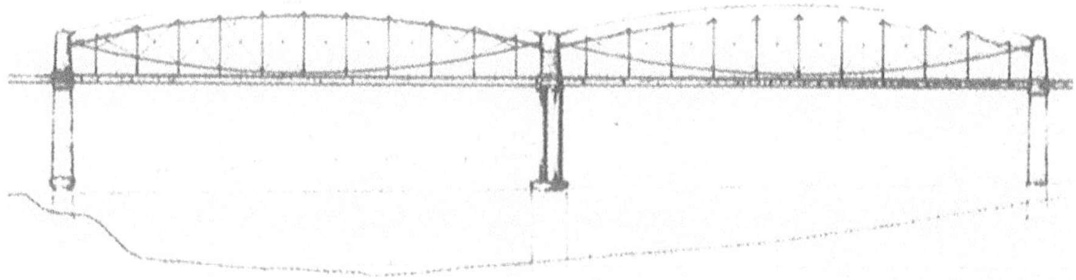

Fig. 10. Isambard Kingdom Brunel's Saltash Bridge, also known as the Royal Albert Bridge, opened in 1859.

Fig. 11. John Roebling's Niagara Gorge Suspension Bridge, opened in 1855.

design and build a suspension bridge across the Ohio River at Cincinnati and in his annual report for 1867 he pointed out that whereas each 460-ft-long tube of the Britannia Bridge weighed over 1500 tons, the 1057 ft main span of his Ohio Bridge weighed less than 1500 tons. Furthermore, the lighter suspension bridge was able to carry over 3000 tons which Roebling claimed was twice the live load that would "overtax the strength of a Britannia tube".[12]

American and British bridge building preferences were to diverge for the next century, with the British evolving variations on the truss and girder principle that were to lead both to disasters like the Tay Bridge and triumphs like the Forth Bridge. The Americans, on the other hand, were to favor the suspension bridge, which evolved from the Niagara Gorge into both the triumph of the George Washington Bridge and the embarrassment of the Tacoma Narrows. In all cases, economy was a principal determinant of form—until a colossal failure forced a reconsideration of the state of the art.

The concept of the Britannia Bridge proved to be a failure in more than economic terms. Because Stephenson's concept called for a tubular girder without penetrations, except at the extremities where railroad trains would enter and exit, each of the complete bridge's assembled tubes formed what was effectively a 1500-ft-long wrought iron tunnel supported 100 ft above the Menai Strait. This meant that the sun could heat the bridge's insides to unbearable temperatures, thus making the passage through it an uncomfortably warm experience at best. Furthermore, the exhaust of the coal-burning steam engines filled the air with smoke and soot. Passengers did not find the passage through the Britannia tubes very pleasant at all, and the bridge design was quickly identified as an environmental failure. One observer, writing in 1902, described what he saw as he watched for a train to emerge from one of the tubes:

> "Suddenly as one gazes, a hollow rumbling is heard, gradually increasing until with a hellish clang and the reverberation of a million echoes, a train dashes out, bringing with it a taste of the sooty air that lingers in the tubes, the product of 50 years, and abominably like that of an unswept chimney."[13]

This shortcoming of Stephenson's design was incontrovertible, and there was a clear attempt to rectify it in one of the few other tubular bridges built after the Britannia. The Victoria Bridge was to cross the St Lawrence River at Montreal, Canada, and it was to have about 6000 ft of wrought iron tubes as its major superstructure. Construction of the bridge was begun in 1854, and it opened five years later. One of the ways in which it differed from the Britannia Bridge was in having a slot formed in the top of the tube as a means of dealing with the foul air. This did not prove to be very effective, however, and smoke was a "major nuisance", especially when combined with the temperatures of 125°F recorded inside the tubes during the summer.[13]

CONCLUSION

While all innovative designs can be expected to be somewhat uneconomical in the sense that they require a degree of research, development, demonstration and conservatism that their technological descendants can take for granted, the case history of the Britannia Bridge provides a perspective on how the tunnel vision that can accompany a design project prevents designers from considering failure as carefully outside the narrow confines of the principal design challenge as they do inside it. This is not to say that there should be any less careful consideration of the major constraints of the problem; rather, it is to caution designers that they must make a special effort to step back from the design and consider what might appear to be some of the more mundane and less challenging aspects of the problem. It would not appear to be inevitable that revolutionary designs need be doomed to economic or human factors failure, for a circumspect design environment should properly consider those factors along with what appear to be the principal ones.

Of course, many design climates are subject to the "railway mania" atmosphere, in which getting the job done within tight time constraints precludes the exploration of optimizing a design so as to obviate its obsolescence upon completion. However, while the Chester and Holyhead railway may have been willing to accept an overly heavy (and hence overly expensive) bridge in the interests of getting a jump on the competition or merely in the interests of getting the bridge in place so that it could contribute to the revenue base, it seems unlikely that the railway would have chosen a hot and smoky tube over an open-air crossing like the bridge at Saltash proved to be. Indeed, in all the correspondence between Stephenson and Fairbairn documenting the testing of tube form after tube form and considering the structural principles of the Britannia Bridge, there appears to have been no consideration given to how it would function with people riding in the trains.

The value of a case study like this is in its potential for contributing to a common heritage of design paradigms whereby the designers of today and tomorrow can gain a circumspection that might otherwise be learned only by repeating the mistakes of the past. A design problem like bridging the Menai Strait will never again occur in exactly the same technological, economic and political climate as Robert Stephenson and his colleagues found themselves a century and a half ago. However, the process of design, from the conceptualization to the realization of artifacts, is essentially a timeless, placeless and thingless human activity that succeeds only insofar as it anticipates failure. Distilling the essence of that process (and especially its pitfalls) from classic case studies can provide an appreciation for its limitations that is unlikely to come in any other way, save repeating the mistakes of earlier designers.[14] By understanding the kinds of errors, lapses in judgement and tunnel vision that have limited the achievement of some of the great designers of the past, we are less likely to repeat those same kinds of errors in the future.

Acknowledgement—This work was supported in part by National Science Foundation Grant No. DDM-8922117.

REFERENCES

1. E. Clark, *The Britannia and Conway Tubular Bridges, with General Inquiries on Beams and on the Properties of Materials used in Construction*, Day and Son, London, 1850.
2. W. Fairbairn, *An Account of the Construction of the Britannia and Conway Tubular Bridges . . .*, John Weal, London, 1849.
3. N. Rosenberg and W. G. Vincenti, *The Britannia Bridge: the Generation and Diffusion of Technological Knowledge*, MIT Press, Cambridge, Massachusetts, 1978.
4. R. Hunt, ed., *Hunt's Hand-Book to the Official Catalogues: an Explanatory Guide to the Natural Productions and Manufactures of the Great Exhibition of the Industry of all Nations*, 1851, London, 1950.
5. P. G. Sibly, The Prediction of Structural Failures, Ph.D. Thesis, University of London, 1977.
6. H. Petroski, "The design and collapse of the Dee Bridge: a paradigm of success masking error," *Civil Engineering* (to appear), 1994.
7. H. Petroski, *To Engineer is Human: the Role of*

Failure in Successful Design, St Martin's Press, New York, 1985.
8. D. P. Billington, *The Tower and the Bridge: the New Art of Structural Engineering*, Princeton University Press, Princeton, New Jersey, 1983.
9. G. Lindenthal, "The continuous truss bridge over the Ohio River at Sciotoville, Ohio, of the Chesapeake and Ohio northern railway, with discussion," *Transactions of the American Society of Civil Engineers* **85**, 910–975 (1922).
10. Sir A. Pugsley, *The Works of Isambard Kingdom Brunel: an Engineering Appreciation*, Institution of Civil Engineers, London, 1976.
11. H. Petroski, "Failure as a source of engineering judgement: case of John Roebling," *Journal of Performance of Constructed Facilities* **7**, 46–58 (1993).
12. J. A. Roebling, *Report . . . to the President and Board of Directors of the Covington and Cincinnati Bridge Company*, 1876.
13. D. Beckett, *Stephensons' Britain*, David and Charles, Newton Abbot, Devon, 1984.
14. H. Petroski, *Design Paradigms: Case Histories of Error and Judgement in Engineering*, Cambridge University Press, 1994.

19

Structural failures and the growth of engineering knowledge

D.I. Blockley and J.R. Henderson

One aspect of the study of engineering failures, which has perhaps been somewhat neglected, is the relationship between them and the growth of engineering knowledge. A modern view of the nature of scientific and mathematical knowledge as being based on models or representations of our experience of the world, and its relationship to engineering knowledge, is presented. Following the ideas of Popper and Kuhn, science and engineering are seen as problem-solving activities within current paradigms. The essential difference between these activities is based on the consequences of incorrect conjectures about the solutions. The primary aim of the scientist is to falsify his conjectures as ingeniously and in as well controlled a manner as he is able; whereas the engineer is interested in safe, cautious conjectures which will not be falsified. Engineering failures are the occasions when the conjectures are in fact falsified and they are therefore central to the growth of engineering knowledge. Brief examples of failure which have affected the current engineering paradigm are given.

Introduction

It is important that the civil engineering profession, like any other profession, extracts as many lessons as possible from its failures. Enquiries into individual accidents often highlight the immediate technical causes of breakdown and periodic attempts are made at an overall review of the reasons for failure.[1-4] There is another aspect to this problem which in the past has been neglected: the relationship between failure and the growth of engineering knowledge. Philosophers have discussed at length the growth of scientific knowledge, almost certainly assuming that engineering is simply an applied science and that its problems are therefore subsumed under those of science; clearly, this is unsatisfactory. It is the purpose of this Paper to discuss a possible mechanism of the growth of engineering knowledge with failure as an essential component, which is analogous to the explanations of the growth of scientific knowledge as given by Popper[5] and Kuhn.[6]

2. An understanding of this process will also have other benefits. Effective communication between researcher and practising engineer[7] may be improved if the respective roles of these two groups in the development of engineering knowledge are understood; for instance, many disagreements stem from a lack of consensus view of the role of mathematics and science in engineering. Much disagreement in debates about what should be taught in undergraduate engineering courses flows from a lack of identification of the intellectual status of engineering

knowledge. A big research effort may be put into a topic which if seen in a wider context may only have a marginal relevance to the present difficulties in the continuing development of engineering knowledge.

Scientific knowledge

3. Science and mathematics are two key areas which form a major part (but not the whole) of engineering knowledge and so it is necessary to begin by briefly discussing them.

4. Since the turn of the century, there has been a revolution in the way philosophers view science. Before Einstein's relativity theory was accepted, and prior to the development of quantum mechanics, scientists and philosophers thought that scientific knowledge, and in particular Newtonian mechanics, was absolute—the truth. Kant, for example, was strongly under the influence of Newtonian mechanics and thought he had identified all the components of the categories which underpin all knowledge. This view is now untenable. Heisenberg's uncertainty principle, which effectively asserts that there are limits to what we can measure, is an example of the new attitude of uncertainty. In quantum mechanics a unique correspondence between precise positions and momenta of some postulated element, like an electron, at two different times cannot be established. It is possible only to discuss the problem in terms of probabilities. Modern research in physics has destroyed the hopes of the last century that science could offer us a picture and true image of reality. Science today has a much more restricted objective; it can now only hope to organize experience to enable some sort of prediction. Our scientific hypotheses are man-made devices in which symbols are used to represent features of our immediate experience, as well as defined concepts, which are used by us in a way decided by us. Scientists are free to construct their own systems and to use theoretical terms in them in any way they think most profitable. Thus science can no longer claim to be the 'truth'; we can treat it *as if* it is the truth but cannot assert that it is true. This leads us naturally to the idea that scientific hypotheses are models; they are models or representations of our experience and thoughts about the world.

5. In accepting this changed view of scientific knowledge. Popper has rejected the idea that induction is the hallmark of the scientific method. The traditional inductivist view of science is that the process of development of knowledge consists of

(*a*) observation and experiment
(*b*) inductive generalization
(*c*) hypothesis
(*d*) attempted verification
(*e*) proof or disproof
(*f*) knowledge

In Popper's opinion, this view of the power of generalizing from the particular to the general in the scientific method is untenable. He maintains that the growth of scientific knowledge is an evolutionary problem-solving process. The scheme expressed simply is

(*a*) problem
(*b*) proposed solution—a conjecture
(*c*) deduction of a testable proposition

(*d*) tests—attempted refutation
(*e*) preference established between competing theories

This scheme is a significantly different view of the growth of science and makes science very much more like other human activities. In science a problem arises from a breakdown of a previous theory and the proposed solution is a conjecture. This admits the value of any human facility to produce ideas whether by intuition, inspiration or any other creative faculty. The real key to the hallmark of science is whether the conjecture enables the deduction of a proposition which may be tested, either experimentally or by observation of the world. These tests will then either refute the conjecture, in which case it is rejected, or show that the conjecture can be tentatively accepted. The conjecture is not *verified*, it is merely *not falsified*. Clearly, some conjectures, the most powerful ones, will pass many of the tests despite the ingenuity of scientists. Newtonian mechanics was such a conjecture until Einstein suggested relativity theory. Thus highly tested theories or highly corroborated theories are dependable because they have been shown to be useful in many different circumstances, but they are not necessarily true theories. Popper's view of science is that it is simply a process of trial and error, of conjecture and refutation.

6. If a scientific hypothesis is to be used for prediction, certain assumptions have to be made; in particular, it has to be assumed that the world is regular. This assumption cannot be *logically* justified as Hume pointed out. The problem was neatly summarized by Magee.[8] Just because past futures have resembled past pasts, it does not follow that all future futures will resemble future pasts. The assumption, however, may be psychologically justified; for example, we all believe that the sun will rise tomorrow morning.

7. The influence of this difficulty concerning the necessary assumption of regularity is highlighted in the problems of the social sciences. Popper summarized this problem thus:[5]

> . . . long term prophecies can be derived from scientific conditional predictions only if they apply to systems which can be described as well-isolated, stationary and recurrent. These systems are rare in nature; and modern society is surely not one of them. . . . The fact that we can predict eclipses does not, therefore, provide a valid reason for expecting that we can predict revolutions.

8. Thus we can summarize the modern view of scientific knowledge. It is a set of models of the world which are not the 'truth'. This does not mean that they are not useful; the most dependable hypotheses are those which have been highly tested. Popper has also shown that any measure of this dependability is not a probability.[7,9]

9. When preference is established between competing hypotheses the chosen ones constitute part of what Kuhn called the current paradigm.[6] Within scientific knowledge there are many hypotheses of different levels of scope and generality, arranged in a hierarchical structure. For example, in structural mechanics Newton's three laws at the highest level are the most general and wide-ranging. Other hypotheses at a lower level vary from the principle of virtual displacements to formulae such as that for the Euler value for the stability of struts. As scientific knowledge develops, clearly the highest and most wide-ranging theories become more and more firmly established as part of the current paradigm. Thus when a more powerful theory is proposed to replace a high level hypothesis the

consequences are widespread. Einstein instigated a scientific revolution when he proposed relativity as a replacement for Newtonian mechanics. Smaller changes in the current paradigm have a less extensive effect and therefore constitute only minor revolutions. The growth of scientific knowledge depends on the investigation of the current paradigm through controlled testing. There will be periods of 'normal science' when the broad outline of the current paradigm is accepted. As the testing of hypotheses continues and evolves, ideas will change and anomalies will be found but they will be explained with only minor adjustments of the current paradigm. The volume of anomalies may grow and alterations to the paradigm become necessary at higher and higher levels, until only a major revolution will achieve a satisfactory resolution of the problems. This is exactly what has happened in modern physics.

10. Associated with this process is the human problem of the general acceptance of the new ideas. Inherent in all human systems is an inertia which will manifest itself in a reluctance to accept fundamental alterations to the current paradigm. Clearly there will be people who readily accept such an alteration but the majority will only be convinced after a considerable period of time.

Mathematical knowledge

11. A formula or a piece of mathematics which appears in a textbook or a technical paper may represent a piece of science, engineering or mathematics. The principles of deduction may appear to be so similar that it may be impossible at first sight to tell the difference. In fact, the difference lies in our grounds for believing the formula. In mathematics, the axioms stand at the head and everything else is deducible from them in a tight formal deductive system; in effect a theorem of mathematics is a useful restatement of the axioms. By contrast, in a scientific system the high level hypotheses are justified only to the extent that the propositions deduced from them agree with observations made in the world. Braithwaite used a zip fastener analogy:[10] 'The truth value of truth for mathematical propositions is assigned first at the top and then by working downwards, in a scientific system the truth value of truth (i.e. conformity with experience) is assigned at the bottom first and then by working upwards.'

12. A modern philosophical view of mathematics is that it is a strict formal language consisting of theorems which express no more than its axioms. It is only when the mathematics is *interpreted* by being used to discuss a scientific model that uncertainty (other than that contained in the validity of the axioms) is introduced.

13. Just as physics has undergone revolutionary change this century, so has mathematics. In 1930 Gödel wrecked the then existing notions of mathematical proof. He showed that if axiomatic set theory is consistent, there exist theorems which can neither be proved nor disproved, and that there is no constructive procedure which will prove axiomatic set theory to be consistent. In fact, later developments have shown that any axiomatic system, sufficiently extensive to allow the formulation of arithmetic, will suffer the same defect. In fact it is not the axioms which are at fault but arithmetic itself. Stewart concludes his book:[11]

> . . . so the foundations of mathematics remain wobbly despite all efforts to consolidate them. . . . For the truth is that intuition will always prevail over mere logic . . . . There is always the feeling that logic can be changed; we would prefer not to change the theorems.

14. Thus even mathematics can be conceived of as a model of the way we think about the world and is therefore open to change just as any other human activity.

Engineering knowledge

15. If scientific and mathematical knowledge is perceived as sets of models, then engineering knowledge cannot claim greater status and must also be so characterized. Engineering is also quite clearly a problem-solving process just as Popper describes science, but there are differences which must be explained.

16. The knowledge of the engineer has been characterized as 'know-how', whereas that of the scientist is said to be 'knowing-that'.[12] This, in the view of many, implies that scientific knowledge is superior. Superficially the difference may seem substantial, but at a deeper level it is not.

17. Firstly, consider methods of solving problems in the context of the needs and objectives of the problem-solver: de Bono[13] relates a particularly appropriate analogy.

> Consider a steep valley that has to be crossed. If you are on foot and in a hurry you could run across the flimsy bridge that spans the top of the valley. If you have a car you would use the shorter and stronger bridge that is set lower down the valley wall. If you had a truck you would want to use an even shorter and stronger bridge set nearer to the valley floor. If you want absolute safety and reliability you would descend to the valley floor, cross it and climb up the other side. These bridges of different strengths set at different levels correspond to levels of understanding. You use the bridge or level that is strong enough for your purpose. You do not need to descend to the valley floor every time you want to cross any more than you need to know the molecular structure of albumen in order to boil an egg. If you are in a hurry the long flimsy bridge across the top of the valley might be more practical.

18. Thus it is the usefulness of any particular method which is the main interest of the problem-solver. It is quite wrong to suggest that detailed explanations of phenomena are better or worse than those which are more vague. Often detailed explanation adds no more usefulness but does add a false appearance of validity. Vague answers are more likely to include the 'true' answer, but of course the answers need to have enough precision to be useful.

19. The engineer has quite a different set of problems from those of the scientist. In designing and building a structure, for example, the engineer comes across problems about which decisions must be made. If sufficient data or well developed models are not available then he must use his judgement and experience to overcome the difficulty. The scientist can organize his problems so that he works on those about which he has conjectures and about which he has deduced propositions which are testable in a fairly precise manner, usually in the well controlled confines of a laboratory. Thus, in brief, for the engineer, experience and judgement take over when scientific knowledge fails.

20. Now the way engineers characterize their collective experience (as distinct from each individual's experience which will not concern us here) is by using 'rules of thumb'. These derive from the craft origins of engineering and are really dependable, common sense hypotheses, but of very restricted scope and application. They have developed by a process of trial and error which is very similar to that described in the previous section, but with important differences

which are discussed later. 'Rules of thumb' suffer from the major feature of all common sense knowledge which is that while it claims to be correct, it is not often aware of the limits within which it is valid or successful. It is most effective when the underlying factors affecting it remain virtually constant, but since these factors are often not identified or recognized then it is incomplete. Scientific knowledge provides us with models which are much more general and therefore have much greater scope and application.

21. Historically the first rules to be developed were rules of proportion, based on the geometry developed by the ancient Greeks. Vitruvius and Palladio quote many examples of them.[7] More modern examples of rules are quoted in the engineer's handbook of 1859.[14] For example, for the deflexion of rectangular beams: 'Multiply the square of the length in feet by 0·02 and the product divided by the depth in inches equals the deflection.' For the strength of cast iron girders the rule was: 'the area of the bottom flange multiplied by the depth both in inches, and the product divided by the length in feet, equals the permanent load distributed in tons, allowing the permanent load to be one fourth of the breaking weight.' Empirical formulae such as these fitted to test data obtained in the laboratory contrast sharply with the approach of the French elasticians of the period.[7]

22. There are many modern equivalents of such rules. At the most simple level, for example, are the rules for determining the spacing of bolt holes in a steel joint. Other rules, based on some use of mechanics and some laboratory test data, seem authoritative but if the underlying assumptions are examined they bear only a partial relationship to the actual behaviour of the structural element. For example, in order to determine the number and size of bolts required in a moment-carrying steel end-plate connection, a common assumption made is that the joint rotates about the bottom row of bolts and the forces in the rest of the bolts are proportional to their distances from the bottom row. In reality the problem of understanding the detailed mechanics of the joint behaviour is very difficult because of the many stress discontinuities and unknown load distributions. The method adopted in design works satisfactorily because it produces reasonably safe and economic solutions.

23. Clearly many approximating assumptions are found in all design calculations. Joints are assumed to be pinned or fixed, loads are assumed to be uniformly distributed, wind loads are assumed to be static pressures. The approximations are justified only to the extent that they have been used in the past to produce designs which have not failed.

24. Thus three components of engineering knowledge have been identified, all of which are models of our collective experience. The first is scientific knowledge, consisting for structural engineers of Newtonian mechanics and tested in the precise confines of the laboratory. The second is the application of these models, with necessary approximating assumptions, to the design of actual structures. These applications are tested by the success or failure of actual structures and by some laboratory work on idealized components. The third component, 'rules of thumb', is also tested by these methods. The nature of this testing is discussed in the next section.

25. Just as there are current paradigms in science, so there are current paradigms in engineering. In order to design a given type of structure a current set of models are used which make up the current calculation procedure model (c.p.m.). For example, the design of steel pitched roof portal frame buildings involves the use of plastic theory and a number of simplifying assumptions including pinned or fixed joints, ignoring the stiffening of cladding and its effects on the postulated failure

mechanism. Among many other assumptions are that the roof loads are taken to be uniformly distributed, and checks are made to account for such effects as finite deflexions before collapse, strain hardening and buckling. Clearly, further research will modify this model at many different levels. Some will involve only minor changes; others may involve extensive alterations. Failure of a structure can lead to a major revolution in the c.p.m.

Failure as a test of a calculation procedure model

26. The similarity between the growth of scientific knowledge and the growth of engineering knowledge can now be recognized if Popper's problem-solving view is adopted. The scheme for the development of engineering knowledge analogous to that given (§ 5) for scientific knowledge is

(*a*) problem
(*b*) conjectural solution (by recognition of some approximate similarity with previous problems or by speculation—usually within the current paradigm)
(*c*) appraisal of the consequences of the solution
(*d*) decision on a set of actions (i.e., design)
(*e*) carrying-through of actions (i.e., construction)
(*f*) testing of solution (by the performance of the structure: if it fails, it is falsified)
(*g*) feedback concerning the dependability of the current paradigm

The components of this scheme are obviously not clear-cut and independent but the essential development is contained within it. The important testing phase is item (*f*).

27. The engineer designs and builds in the hope that his structure will operate successfully and will not fail. In using and developing the c.p.m., engineers are interested primarily in safe, cautious conjectures (while acknowledging economic constraints) because the consequences of failure are so severe. Thus a c.p.m. is rarely falsified directly in service: instead, great reliance is placed on partial testing of isolated aspects of it using idealized components in the precise conditions of a laboratory. The procedures are only falsified in the laboratory under conditions which do not completely reflect the conditions under which the actual structure operates.

28. This aspect of the scheme contrasts sharply with that of the scientist because the latter tries to falsify his conjectures as ingeniously as he is able. It is precisely because he can do this that science progresses rapidly. The engineer has no wish to falsify his rules directly and so the boundary of the c.p.m. is only indicated by the occurrence of an accident. The boundary will be very difficult to define exactly because failure will be due to a combination of circumstances: the role of any particular rule in the c.p.m. will not be directly isolable. Design rules are therefore only weakly not falsified. In fact a particular rule in a c.p.m. may be false but its effect may be masked by the conservative assumptions made in the rest of the design.

29. The difference between the work of the engineer and the scientist is not therefore a fundamental one, or even a substantial one. It is not due to the different nature of the methods each uses or even the way each perceives the world, but it is due to the differing consequences of error in the predictions they make. Engineers

do not want to falsify their conjectures; by contrast, this is the scientist's primary aim. Engineers are interested in safe, cautious theories; scientists are interested in detailed, accurate theories. Both are interested in solving problems. Structural engineering scientists tend to be dominated by the scientific interest in accuracy and, as a result, often frown on 'rules' as being intellectually inferior. Designers rely on 'rules' when science lets them down. Many misunderstandings arise because of the failure to appreciate this distinction.

Failures and the current paradigm

30. The discussion so far presents us with a strange antithesis: it is the success of engineering which holds back the growth of engineering knowledge, and its failures which provide the seeds for its future development. Not only is it necessary therefore to identify the particular causes of an accident and the dominant causes of a group of accidents, but it is also important to identify the overall changes in the current pagadigm which have followed. This latter aspect is discussed briefly in this section. The changes occur at many different levels with widely varying scope as has been mentioned, and are often difficult to define in any precise sense. Many of the changes mentioned here are well known and have been discussed individually at length. The purpose of this section is briefly to put them in the perspective of the previous discussion on the growth of engineering and scientific knowledge. One of the important consequences of an heightened awareness of the changes is that new developments in research and in practice can be seen as part of the continuing process of the growth of engineering knowledge. This aids an understanding and appreciation of the roles of those who contribute to that process.

31. The changes are identified briefly in five groups: materials, structural form, theories of response analysis, construction methods, and engineering organization. A number of changes which are at present in progress are also identified.

32. Historically the development of a new material has led to new structural forms, usually after some time delay, although sometimes new ideas, like prestressed concrete, have suffered early set-backs because of deficiencies in material quality. Two separate series of structural failures have led to an increased awareness of the potential deficiencies in materials. Firstly, the failures of the school buildings at Camden and Stepney, and other buildings, led to the banning of HAC concrete. Secondly, failures such as Kings Bridge, Point Pleasant Bridge and Sea Gem showed that brittle fracture and fatigue can cause failure of steel structures if adequate precautions are not taken. These failures have stimulated engineering science research into the mechanisms of the processes, and designers have been made aware of the possibility of other material deficiencies.

33. The failures of the Tacoma Narrows Bridge and the Ronan Point flats are classics and both showed deficiencies in the current paradigm concerning the behaviour of a structural form. The former accident demonstrated the necessity of wind tunnel tests to determine the aerodynamic stability of slender bridges. Wind tunnel testing now has an established place in civil engineering design, where appropriate. The Ronan Point failure demonstrated the possibility of progressive collapse which, although it is not an unpredictable phenomenon, was not part of the c.p.m. of the time.

34. The failures of Listowel Arena, of Tay Bridge and at Ferrybridge showed deficiencies in the models of loading for snow and wind. The piling-up of snow on one side of the Listowel Arena roof was not considered in the calculations and the

figure used for the equivalent uniformly distributed load was too small. The wind loading model for the Ferrybridge cooling towers ignored the fact that there was a group of towers rather than a single one.

35. The Ardeer cooling tower failure and that of the first Quebec Bridge, as well as some of the box girder incidents, showed up deficiencies in the models used for structural response predictions. The relatively large-scale research programme put into operation after the box girder incidents was one reaction to failure which has generated much valuable information and a new c.p.m. for future designers.

36. The complex situations surrounding the failures of the Tay, Quebec and Westgate bridges made it almost impossible to isolate single causes of failure. These bridge failures, more perhaps than any others, have demonstrated that human problems of effective organization and communication must be seriously considered as part of the c.p.m.

37. This brief identification of some of the paradigm changes after past failures leads naturally to a consideration of the current changes which have not fully filtered through the whole of structural engineering. For example, the realization that many failures are due to human error leads to a greater concentration on checking and control procedures and the inclusion of human error in the c.p.m. This leads to a reappraisal of risk control through regulations. It has also stimulated further research into types of uncertainty and ways of measuring it. The use of probability theory in structural engineering was almost unheard of 20 years ago, but now with the renewed interest in risk analysis and the prevention of failure it is becoming more and more an established part of the higher levels of some c.p.m.s in the current paradigm. Finally, the arguments presented in this Paper demonstrate vividly the importance of the full-scale testing of structures and the potential of proof-testing (although the cost of the latter is a serious drawback and has to be recouped).

Conclusions

38. It has been asserted that all knowledge consists of models of our thoughts and experience of the world. The development of scientific knowledge is an evolutionary problem-solving process.

39. The development of that part of engineering which is not engineering science is also a problem-solving process which has generated rules which are weakly not falsified.

40. The essential differences between engineers and scientists is that the former are interested in cautious, safe theories and do not wish to falsify their conjectures, while the latter are interested in falsifying their conjectures as ingeniously as possible. The development of engineering rules is therefore much slower than the development of scientific knowledge.

41. Failure is an essential component of the growth of knowledge. Clearly this is not desirable in any accidental sense but it points directly to an increased research expenditure on full-scale loading tests.

Acknowledgement

42. The Authors would like to thank the Science Research Council and the Department of Transport for their interest and support of the second Author under a Science Research Council CASE studentship.

References

1. SMITH D. W. Bridge failures. *Proc. Instn Civ. Engrs*, Part 1, 1976, **60**, Aug., 367–382.
2. SIBLY P. G. and WALKER A. C. Structural accidents and their causes. *Proc. Instn. Civ. Engrs*, Part 1, 1977, **62**, May, 191–208.
3. BLOCKLEY D. I. Analysis of structural failures. *Proc. Instn Civ. Engrs*, Part 1, 1977, **62**, Feb., 51–74.
4. THE INSTITUTION OF STRUCTURAL ENGINEERS. Structural failures in buildings. Symposium, April 1980.
5. POPPER K. R. *Conjectures and refutation*. Routledge & Kegan Paul, London, 1976.
6. KUHN T. S. *The structure of scientific revolutions*. University of Chicago Press, 1962.
7. BLOCKLEY D. I. *The nature of structural design and safety*. Ellis Horwood, Chichester, 1980.
8. MAGEE B. *Popper*. Fontana Modern Masters, 1978.
9. POPPER K. R. *The logic of scientific discovery*. Hutchinson, London, 1977.
10. BRAITHWAITE R. B. *Scientific explanation*. Cambridge University Press, 1953.
11. STEWART I. *Concepts of modern mathematics*. Penguin Books, Harmondsworth, 1975.
12. JARVIE I. C. (MITCHAM C. and MACKEY R. (eds)). Technology and the structure of knowledge. In: *Philosophy and technology*. The Free Press, New York, 1972, Chapter 4.
13. DE BONO E. *Practical thinking*. Penguin Books, Harmondsworth, 1978.
14. *Engineer's and contractor's pocket book for 1859*. John Weale, London, 1859.

20

The evolution of structural engineering design procedures: a history for that skill called design

William Addis

The word 'design' has become something of a 'Humpty Dumpty word', meaning just about anything the user wants it to mean. Early uses of the word, both as a verb and a noun, related to the production of drawings of some engineering artefact to be manufactured. Nowadays its use as a verb relates to almost any part of the act of creation of almost any product, from a building to a T-shirt or hairstyle, while as a noun the word now usually relates to an abstract quality of an artefact—something about the way it looks, its image, its style.

Even after restricting the field of interest to that of designing buildings, there can still be confusion or ambiguity. If one were to ask who designed the Sydney Opera House, some people might reply Utzon (an architect) while others might reply Ove Arup & Partners (a firm of consulting engineers).

The present paper will concentrate on a particular sense of the word 'design', relating principally to its use, as a verb, to denote the creative activities associated with the modern civil and structural engineer, and, as a noun, to denote the immediate products of this creativity. When looking at the past this will still lead to some ambiguity, since the boundaries between the professions have changed, and even now are often blurred. This will not, however, be important since the paper does not concern itself with this aspect of the history.

It is not the aim of this paper to try to establish a precise meaning of the word 'design'; rather it is to discuss what designing is, and what it is not, and what it has been in the past. It is, in fact, and fortunately, far easier to use words than it is to define them. It is, therefore, quite likely that most people would understand what the following sentences mean.

Cathedrals were designed without the use of statics.

Early suspension bridges were designed empirically.

Sir Benjamin Baker designed the Forth Bridge.

BS 8110 relates to the designing of buidings made of reinforced concrete.

This paper will try to clarify the subject by setting out a way of looking at engineering design which, while not resolving every ambiguity, may help a little. In particular, it aims to help us to look upon the history of engineering design with new eyes. Following the author's interest, examples will be taken largely from building and structural engineering design. It is, however, possible, indeed likely, that at least some of the argument could apply to other branches of engineering design.

NATURE OF STRUCTURAL DESIGN

Before discussion of what design is, a few remarks should be addressed to what it is not.

Design, especially nowadays, is often associated only with an act of original creation, much as one thinks of the work of a composer, a sculptor or an inventor, when creation and originality are fundamental goals of the activity.

This is not the case with structural engineering design: each act of design is making use of many well tried ideas. In the extreme, it could be argued that the entire repertoire of possible structural forms was established many thousands of years ago when the first beam, arch and suspension structure were 'invented': all structural design since then has been merely copying and combining copies in different ways. However, this argument is about as useful as claiming that ancient man had discovered all the musical notes possible and that Mozart was merely a plagiarist.

A variant on this theme is used by some philosophers of technology who imply that engineering design is a branch of technology, since an engineering artefact is 'merely' the agglomoration of atomic units of technical knowledge into complex technological 'molecules'. Thus a steam engine or Eiffel tower are seen as the mechanical or structural equivalents of DNA.

The structural engineer's work does involve creativity and originality and, like Mozart's, is a matter of devising a suitable, even aesthetically pleasing, arrangement of the various available elements. In place of Mozart's notes, the engineer has beams, columns, arches, ties, slabs, shells, membranes and so on.

However, structural engineering involves much more than the act of creating original combinations of elements. Unlike musical notes, they cannot be combined in any way the person chooses or which aesthetic criteria favour. The laws of nature are less tolerant than the most vitriolic of music critics. After the creative act of 'inventing' a possible structural solution to a problem, there follows the considerable job of checking that it meets the many constraints imposed by the problem—the physical laws of nature, a client's functional requirements, limitations of both time and cost—all in addition to any aesthetic considerations. Finally, there is the not inconsiderable task of turning the proposals into physical reality.

Between these exteme attitudes lie several which view design as a process rather than an instantaneous act. Such models can be very useful and they can also be misleading and problematic.

A common model of this type portrays design as a problem-solving process. An input is provided in the form of a client's requirements or problem to be solved, and the designer produces an output, a solution to this particular problem. The means by which this process occurs is then likened to problem-solving in science. An analogy is drawn with the way in which scientists carry out their work employing the 'scientific method', as described by philosphers such as Karl Popper.

According to this view, a scientific problem is solved by proposing a hypothesis, and then devising an experiment and carrying it out in order to test the hypothesis. Data are collected and used to try to falsify the hypothesis. If falsified it is rejected: if not falsified it is accepted, at least for the time being.

In the activity of design, the process is also described, by analogy with science, as rational and objective.[1] Information about the problem is gathered, and solutions are proposed and 'tested', initially in thought experiments and finally in reality, by building the design.

The inference in this view of design is clear: design is being compared with science in an attempt to render it a 'scientific process' and, therefore, somehow better. The products of such 'scientific design' are seen as virtually infallible, and as if, given the circumstances of a particular design problem, there is only one scientifically inevitable solution.

Despite its currency and popularity, this is not a tenable conclusion. Although post-rationalization of a design is both common and, indeed, necessary, this does not mean that the designer uses a totally rational route to achieve it.

APPLICATION OF THEORY IN PRACTICE

Perhaps the most prevalent version of design being viewed as a process is the notion that designing is the process of 'putting theory into practice'. We have an educational system which is largely based on this precept: at school and university you learn the 'theory' and then you go out and 'apply it'. There is only one problem: where does one learn how to apply theory? The attitude of many companies to young graduates is that however much theory they may know, most graduates have very little idea of how to make use of it in designing anything.

More fundamentally, there is seldom any attempt to establish or explain precisely what is meant by phrases such as 'to put theory into practice' or 'to apply theory', yet they have great currency both at universities and among professional engineers. Without elaborating the arguments here, there are severe problems in trying to find any useful meaning in such phrases.

THE EVOLUTION OF STRUCTURAL ENGINEERING DESIGN PROCEDURES

It is also worth reminding the reader that designing was going on long before any of our current engineering 'theory' was invented. Also, 'theory' of many kinds has been around at least since Roman times, and for most of the intervening period people have been calling for bridges across the enormous gap between theory and practice. Yet, if no one has found such a bridge in 2000 years, it must be a very difficult problem indeed. However, perhaps the question of how to bridge the gap is one which has no answer because it has no meaning.

This section has sought briefly to highlight a few problems with some common views of design, especially those which focus too strongly on some contingent facts about how design is carried out nowadays. Such approaches virtually prohibit the comparison of the process of design now with design 100 or 1000 years ago. In order to facilitate such historical study, a different approach is needed.

'DESIGN' AS A SKILL

Each of the above views of design tends to ignore the human beings involved. It does not look at what an engineering designer actually does or how she or he does it. The design process is analysed as one which is independent of people, and which could, therefore, perhaps be carried out without them. These views overlook the fact that design is a human skill.

Let us consider what type of a skill or, rather, bundle of skills design is. Some of the skills are physical:

—producing engineering drawings
—using calculation devices
—building prototypes or models of an artefact.

Most, however, are intellectual and include

—the ability to understand the significance of data
—the ability to foresee the consequences of a decision
—the ability to analyse a problem logically
—the ablity to create one or more conceptual models (often geometric or mathematical) of a physical stucture, its behaviour, the loads acting on it, or the materials of which it is made
—the ability to 'feel' how strong or stiff a material or structural component is
—the ability to imagine how a structure will deflect under a series of imagined loadings
—the ability to imagine how a structure will collapse if overloaded
—the ability to imagine an unbuilt artefact
—the ability to imagine how an artefact could be manufactured or built
—the ability to communicate ideas and information to others
—the ability to 'design' at different levels of detail—initially perhaps only approximately; later, in great detail
—the ability to decide how an artefact should be designed—whether a problem can be solved with a standard solution or whether it requires the development of a new approach.

Looking at these, and other, skills in another way, one can also describe some of them simply as 'knowing'—knowing some facts or some theories of engineering science—while others one would describe as 'knowing how'—for instance, knowing how to argue logically or knowing how to evaluate two alternative engineering structures.

A person also has the ability to refer directly to his or her own experience and to draw, indirectly, upon the experience of millions of other people accumulated over thousands of years. A person can also learn, both from his own experience and that of others. At a deeper level, humans have the ability to assess many data and select only some which are considered somehow 'relevant' to a problem: to see the wood or the trees as necessary. And at an even more fundamental level, a human being is capable of intention, of having a notion of which way forward is (in tackling a problem); of taking carefully selected actions which will achieve identified goals, both intermediate and ultimate; and finally, of recognizing when goals have been reached.

If the ability to design is a series of skills, of general applicability, there remains, finally, the question as to how these skills are synthesized or brought to bear upon a particular problem. In a very real sense, the ability to do this is the all-embracing skill of a design engineer: the ability to decide which skills to use, and how and when to use them, in order to proceed from a skilfully assessed initial position to a skilfully assessed final goal.

It is this route, from an initial position to a final one, wich the author calls a 'design procedure'. Being able to devise such a route (not *the* route, for there are always many) is the ultimate, all-embracing skill of the engineering designer.

Described in such a way, the skill of the engineering designer can be seen as timeless. It does not depend upon the actual tools or materials or theoretical concepts which are available at a particular time in history, or place in the world.

THE DESIGN PROCEDURE

In the context of this paper, the route, which is the design procedure, represents the way chosen by the designer of using the knowledge at his disposal to achieve the desired goal. The idea of the design procedure is perhaps best explained by proceeding backwards from the end of the design process.

The endpoint of design is taken to be a set of instructions to the builder, including lists of component parts and drawings of the way they are to be assembled. This will be called the 'description'.

In addition to the description, the designer will need to provide evidence of the adequacy of the proposed design to any of a number of interested parties. This might be the client, the project financier, the regulating authorities, the designer himself or, in the event of a collapse, a court of law. This constitutes the 'justification' of the proposed design.

The design procedure itself can be represented as a 'black-box' process. Its 'content' is chosen by the designer and it has the description and the justification as its outputs.

There is also an input to the process, namely all the knowledge, information and experience which the design engineer feels to be necessary and sufficient to enable the design procedure to generate a final description and justification of the proposed design. The relevant knowledge must be carefully distinguished from much knowledge which is irrelevant. Some of the information may already be known and available in books or from experts. However, the designer may also come to recognize that certain vital information is not known and will need to be generated. This situation can arise in many ways and often unexpectedly; for instance

—in the use of a new material whose properties are poorly understood (cast iron in 1800, reinforced concrete in 1900)
—in the use of new types of structure (such as concrete shells in the early 1950s)
—in ever-increasing spans of bridges and roofs
—in the need for lighter or cheaper structures
—in the imposition of new types of loading (such as locomotives in the 1830s) or in the increasing significance of loads (such as wind on ever-slenderer bridges).

Finally, there are many influences which act upon the design procedure in a regulatory way. For instance, a client will not want just any building, but rather one specifically tailored to his needs; and money, time and legislation might also affect the type of solution which the designer will produce.

The idea of a design procedure is illustrated in Fig. 1. An important characteristic of this model of the design process and the work of the engineering designer is that it is extremely general and could equally be used to help portray the designing of a bridge over a small stream or one over a major estuary; a bridge designed by an amateur or one designed by a professional; a building designed in 1390 or 1930 or even 1930 B.C.

What will be different in each of these cases is not the fundamental nature of the process itself but the precise details of the type of knowledge used, the type of description and justification provided, the types of economic or legislative influence upon the project, and so on. All of these details will differ from place to place, time to time, project to project and individual to individual, but their fundamental relationship, the nature of designing, does not change. This point is illustrated every time there is, for instance, a major bridge design competition. Several design teams each working to the same brief always produce different designs—sometimes very different indeed. It can be seen, therefore, that there is no logical necessity about the relationship between a design procedure and the artefact which is designed: similar design procedures can result in different designs and vice versa.

THE EVOLUTION OF STRUCTURAL ENGINEERING DESIGN PROCEDURES

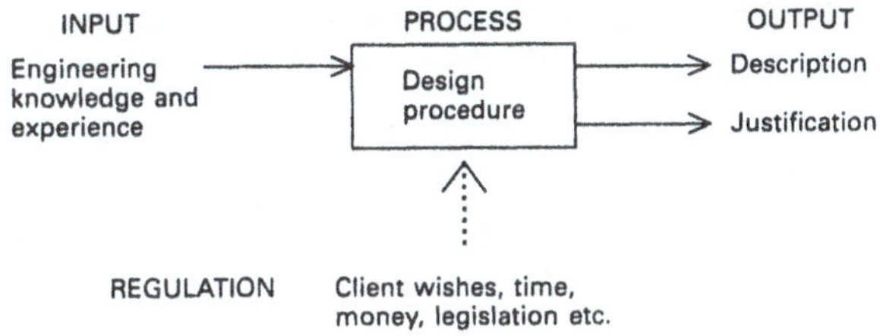

Fig. 1. The design procedure.

The preceding discussion enables us to define the 'ultimate all-embracing skill of the engineering designer' more precisely. The skill of a designer is not only to produce a design for an artefact, comprising a description of a proposed design and a justification of its viability. It is also to be able to devise an appropriate *design procedure* which will ensure the production of a viable design.

STUDY OF DESIGN

Having introduced some views on what is and is not meant by 'design', we can now see how these notions might help to guide the study of design and, therefore, its history also.

Isolating the notion of a 'design procedure', used during the design of a particular artefact, takes the focus of attention off the artefact itself and draws it towards the way in which the artefact was designed. It thus becomes of equal interest for someone studying 'design' to consider the physical and the intellectual side of design—the artefact itself and the design procedure its designer used.

It is particularly this intellectual side of the skill of a designer which has been largely ignored in the appreciation of both living engineering designers and the history of the skill. Yet our assessment of the skill of a designer is affected both by the product and how it was designed. It would considerably affect our opinions about the designers of Gothic cathedrals if a series of mediaeval calculations involving wind loads, overturning moments and foundation reactions were discovered. Similarly, our opinions of the designers of the Sydney Opera House would be altered if no evidence of such calculations could be found.

In the history of engineering it is only too seldom that the discussion of how and why Brunel or Brunelleschi were eminent designers goes beyond the fact that they designed famous buildings or bridges. This contrasts sharply with the state of affairs in painting or sculpture or music or science or (more closely related to building design) in architecture. All of these creative activities have stimulated considerable areas of study relating to the skills of the practitioners, criteria for assessing excellence, whole philosophies of the subjects and, most notably of all, critical histories of the subjects.

One line of investigation which needs to be developed is establishing why engineering design has not attracted the interest of critics, philosophers and historians: it is not, however, the theme of the present paper.

The present concern is to indicate an approach to our subject which will give due credit to the intellectual side of designing. It is proposed that this could be achieved by using the model of design and the concept of 'design procedure' which have been introduced above.

In order fully to study and evaluate the 'design' of an artefact such as a building one would need to investigate the following:

—the bodies of knowledge and information available to the designer
—the design procedure used by the designer
—the description of the building in the form of drawings, models and written documents
—the means used to justify the proposed design
—the various factors which may have had a crucial effect on the work of the designer, especially the client's precise brief, and contractual, legal, time and financial constraints

as well as, of course,

—the artefact itself.

THE EVOLUTION OF STRUCTURAL ENGINEERING DESIGN PROCEDURES

It should be mentioned at this point that, in order to discuss the entire design stage of a project, it has been implied that the entire design of a building or other structure is carried out by one person. This is, of course, seldom the case. Many people are involved throughout the design stage of a project and the whole design process may even transcend several professional demarcations. There are some architects who consider their responsibility stops when the overall look and shape of a building has been established, the rest of the design being undertaken by engineering designers and the construction engineers. Others, however, work closely with other professionals until every detail is specified. Similarly, some engineering designers deal with overall issues, leaving the details (perhaps, literally, the number of bolts) to specialist designers, who deal mainly with the production or construction stage.

Looking into the past, the precise boundaries between the various professions involved in building design and construction—craftsmen, architects and engineers—have changed markedly over the centuries. Cathedrals were not designed by professional architects. In the 19th century many engineering structures were designed within construction companies and there were few engineering design consultants.

Also, a designer has always been able to rely on the skill, knowledge and experience of the constructors. However, depending on the time and place in history, the type and degree of this expertise has varied considerably. Indeed, a designer might not even be aware of the value of this expertise since in many cases he never actually sees it: problems of construction only become manifest after the designer believes he has finished the design. The skill of a builder or construction engineer, for instance, who can make an impossible or difficult design buildable is seldom fully appreciated even in the present century.

Since this paper is concerned with the philosophy of engineering design, these professional practicalities and responsibilities associated with designing, which do not affect the overall argument, will not be addressed.

HISTORY OF DESIGNING

The previous section has provided an approach which might guide us in studying the history of the activity of engineering design. It has indicated how we can consider something more than just the physical appearance and content of engineering artefacts. It has indicated how we might look at the intellectual side of the subject, a history of ideas related to designing, a 'philosophy of engineering'. This can serve us in much the same way as the subject called 'philosophy of science' facilitates and guides the study of the subject called 'history of science'.

The study of the history of a subject is a combination of two views of that subject, one synchronic, the other diachronic; the one a snapshot in time, the other a picture over time; the one the identification and study of a parameter, the other a study of how the parameter varies in time. (A musical analogy might be the study of harmony and counterpoint in a piece of music.)

The 'design procedure' provides us with just such a parameter for studying the history of designing. In order to see how this parameter can be used, a useful parallel can be found in the study of the history and philosophy of science.

HISTORY OF SCIENCE

An important historical parameter in the history of science is the scientific theory or hypothesis. At a particular time in history the state of science is indicated by the theories and hypotheses which are believed and those which are being tested and investigated by means of scientific experiment. Scientific change over periods of history is indicated by which theories and hypotheses have been rejected and how and why they were superseded.

The means by which science progresses over time is clearly a key issue and is a question which has not been unequivocally resolved. There are, at present, two principal views, following the philosophies of Karl Popper and Thomas Kuhn.[2]

Briefly, Popper portrays science as a wholly objective process based upon a rigorous scientific method. Scientific data are collected, and alternative hypotheses are proposed to explain the data and are used to predict as yet unobserved or untested phenomena. Experiments are devised which try to falsify a hypothesis. If it is falsified it is discarded and an alternative hypothesis is tested, and so on. With time, certain hypotheses remain unfalsified and become established as scientific theories which are generally adopted. At any time, however, it remains possible that any established theory could be falsified and discarded.[3] For Popper, science is totally independent of subjective views and its progress is therefore held to be an entirely rational process.

Kuhn, on the other hand, portrays scientific activity as dependent upon the people who are undertaking it, both as individuals and as groups, and therefore open to subjectivity, or more generally, relativism—that is to say, not objective, but dependent upon the interpretation given by a particular person, in a particular scientific community, at a particular time.[4]

The particular strength of Kuhn's view is that it does not reject a discarded hypothesis as simply 'wrong'. The hypothesis that the sun orbits around the earth served science for many thosands of years until Kepler and others proposed the familiar alternative. Similarly, the modern thermodynamic analysis of the heat engine, developed by Sadi Carnot, was originally based entirely on the hypothesis that heat was a weightless fluid called caloric, a belief which is now usually dismissed as incorrect.

Kuhn's model of scientific progress is somewhat Darwinian. Scientific beliefs, indeed the very notion of what constitutes science, are particular to a time, place and society and have the function of explaining the world as perceived by that society and the individuals which make up its communities. As societies, their individuals and their beliefs change, so the perception of science changes, with some views dying out, others being born and others surviving by adapting to the new environment.

Kuhn distinguishes two types of scientific progress. His 'normal' science develops gradually within a scientific community and comes to be embodied in a range of beliefs about and attitudes towards scientific issues. These he calls the 'scientific paradigms' of the community and they represent what is, by and large, agreed within that community and about which there is not much dissent.

From time to time, however, periods of normal science enter a state of crisis when their very foundations come under threat. The period just before Kepler's hypothesis of the heliocentric universe was such a time. Progress following such a state is not smooth and gentle: it involves radically different types of explanation, new concepts, new paradigms. It is characterized as a time of turmoil and, for this reason, Kuhn calls it a period of 'scientific revolution'.

Eventually, but not in a wholly rational way, the new paradigms come to be accepted by more and more individuals, and finally a new period of normal science, with new beliefs and concepts (or paradigms) and scientific communities, is reached.

ENGINEERING DESIGN REVOLUTIONS

Kuhn's idea of scientific progress by scientific revolutions provides a suitable model for progress in engineering design as encapsulated in the historical parameter, the engineering design procedure.

We can, by analogy with the structure of scientific revolutions, propose periods of 'normal engineering design' which, from time to time, reach 'crisis' and are followed by an 'engineering design revolution', during which utterly new approaches to design are proposed, involving utterly new concepts and leading to utterly new types of design procedure.

Immediately following a revolution in design procedure, there need not, of course, be a dramatic change in the look of the artefact produced by the design procedure. It was illustrated earlier that there is no logical connection between a design procedure and the artefact to which it leads. Two identical objects can be designed in different ways.

What is likely to be different is the way the proposed design is justified: justification is the engineering design equivalent to the scientific explanation. Just as the relationship between the sun and earth can be explained differently by Ptolomy's and Kepler's models of the universe, so the proposed design of a timber roof truss can be justified differently by a skilled carpenter and a modern, mathematically literate design engineer.

SOME EXAMPLES OF STRUCTURAL ENGINEERING DESIGN REVOLUTIONS

Throughout the history of building design, precedent has been the most common basis of design procedures. In providing both an adequate description and justification of a proposed design, a successfully executed example is clearly most satisfactory.

However, both the changing needs of societies and their building requirements, and man's increasing tendency to want change for its own sake, render inadequate precedent alone. Inherent in most design procedures, then, is the need to be able to experiment with new designs with no precedent.

THE EVOLUTION OF STRUCTURAL ENGINEERING DESIGN PROCEDURES

Experimentation with large-scale structures has always been costly, and alternative means involving the use of representations or models of buildings have been long established. In addition to the use of small-scale physical models, buildings have also been represented by mathematical models, especially geometric ones, at least since Greek times. During the last few centuries other types of mathematical model, involving algebra, statics and matrices in conjunction with theories of engineering science, have become current. The mathematics used in representing a building has also been of use in providing the description and justification of proposed designs.

Description often involves the production of a geometric model of the proposed design—a drawing, for instance. This contains within it the relevant logical links between the constituents of the artefact. The rules of geometry validate the correspondence between the lines on a piece of paper and the position of lines in three-dimensional space in the finished artefact.

Any justification contains an implicit or explicit logical argument which leads from certain assumptions, by means of logical steps, to certain desired conclusions, hence proving these conclusions to be valid consequences of the initial assumptions. Mathematics is the tool by which such logical aruments can be represented and, being rational, this process increases our faith in the conclusions, even though they are still utterly dependent on the validity of the initial assumptions!

The history of structural engineering design can, then, be summarized in the development of different ways in which models of proposed designs have been created and used in the process of experimenting with new designs and in the design procedures themselves. A few brief examples will illustrate some major structural engineering design revolutions. These and other design revolutions are treated more fully elsewhere.[5]

Before the examples are presented, a word of explanation is called for. The author admits that a detailed and full analysis has been made of only one design revolution: the 'plastic' design revolution of the early 1950s. Data are plentiful since the revolution is very recent: much has been written down and many of the key persons are still alive. For earlier revolutions, fewer data are available and certain assumptions have had to be made. Historians who insist upon a higher density of hard facts may find such an approach distasteful. Such an approach can nevertheless be justified on the grounds that the study of history is not simply the objective collection of all relevant data. This view contains the highly subjective judgement as to what is relevant. The discovery of historical data presupposes a hypothesis that certain data may exist, just as the discovery of scientific data, such as the existence of oxygen, presupposes a hypothesis that something exists with properties which can be detected.

Future historians might unearth further data relating to the detailed structure of engineering design revolutions if they have been persuaded by the arguments in this paper and believe there may be evidence waiting to be discovered.

Greek Design Procedures. It can perhaps be claimed that the Greeks invented design procedures for buildings, in Europe at least. In around 600 B.C. the nascent Hellenistic culture led to a need for a new type of building: a temple, which had to be larger and more permanent than previous buildings. The traditional building crafts were unable to meet the needs; in particular, the ability to describe (specify) and justify a new building before it had been begun.

Design procedures based upon the use of simple numbers and the ratios between them were evolved.[6] Using a very few key dimensions such as the diameter of a column and the intercolumnar spacing, the entire plan and elevation of a temple could be described. If a modification to a previous design was required, the ratios between dimensions could be changed and the effect on the final design could be predicted.

In fact, two different major types of design procedure developed, the Doric and the Ionic. The first was based upon the use of a single base dimension and subsequent multiples of it, while the Ionic design procedures were sequential—a dimension being derived from one or more already established ones.

The Greek design procedures were able to provide more than just descriptions and specifications of a proposed design. The use of numbers, ratios and (from about 300 B.C.) simple geometric figures such as circles and squares also provided justification of the results, in conjunction, of course, with experience gained from precedent. The philosophical beliefs of the time held that such numbers, ratios and figures had been used by the gods in their creation of the world, and that what was seen in nature, for example in numbers, geometry and harmonics, was the physical manifestation of the very rules used by the gods. Such exemplars therefore provided excellent models and rules upon which to base the design of buildings. While this reliance on number and geometry may appear strange to us, it is not so very different from modern attitudes to our 'laws of nature'.

Doric and Ionic design procedures coexisted for several centuries, separately to begin with in different parts of the Greek empire, and later with Doric designs being increasingly influenced by Ionic as the sophistication of Ionic design procedures began to show. By the year 25 B.C. the Doric methods were being severely criticized, indeed almost ridiculed, for being too inflexible.[7] It seems very likely that, had the Greek empire continued, an Ionic design revolution would have continued to completion. As it was, developments fell into the hands of the Romans, who let the Doric pass into history and took up both Ionic and the neglected Corinthian design procedures, which they developed in their own ways in the context of their own traditions.

The 'Gothic' Revolution in Design Procedures. The area around Paris during the twelfth century was the setting for a major design revolution. There were already well established design procedures for churches which had been developed continuously during the previous centuries. The mood, however, was ripe for change. New philosophies were being developed, especially by the neo-platonists Hugh of St. Victor and Thierry at Chartres, for understanding and explaining the world. These were based mainly upon the Bible and its interpretation and continued the earlier work of St. Augustine and Boethius during the fifth and sixth centuries. There was, however, a new ingredient: geometry.

Geometry was already well established as a practical skill but it came to be more than merely a practical skill. It came to be seen as the wordly manifestation of the laws upon which God had based his creation (design) of the world. The designs of buildings reported in the Bible (such as the Ark, Solomon's Temple and the designs for the Celestial City) were fully analysed in order to understand the rules better and were used as models to be emulated.

Although the precise dates are not known, it was during this very period of philosophical work, sometime between 1126 and 1142, that Adelard of Bath translated Euclid from Arabic into Latin, making it available for the first time to the scholars at Chartres. It seems to have had a profound effect on their work: it has even been said that under Thierry's influence 'the school of Chartres attempted to transform theology into geometry'.[8]

Practical geometry was already used in the processes of setting out and in the construction of buildings. What Euclid added was the idea of being able to prove the correctness of a certain design, though not just any design: precedent was of course still a major influence. Geometry could be used as an important part of the process of justifying proposed designs and thereby giving the designer additional confidence in them. A geometry textbook of around 1140 states that the specific duty of theoretical geometry is 'to give reasons and dispel doubt'.[9]

While we do not know exactly how geometry was used to give reasons, the explosion of interest in geometry during the 1120s and 1130s did occur at the very beginning of what we now know as the Gothic era. The cathedrals of St. Denis, Sens and Chartres were all designed in the 1130s; another dozen followed during the next 25 years. It is interesting to wonder if it was simply the added confidence provided by the geometric justifications which encouraged the proposal of bolder and bolder designs for the cathedrals. Whatever the details, the writings we do have from the time talk of geometry in just the way we nowadays talk about the laws of physics,[9] and it is difficult not to be struck by the similarity between the increasing boldness of structures during the 'Gothic' design revolution and that which followed the 'statical revolution' of the mid-19th century.

Rankine's Design Revolution. Even in the late 1840s the majority of bridge and building structures were still being designed using design procedures based upon precedent. A great many reliable design rules had been developed from experience and careful experimenting using test structures and models. Yet there were also considerable bodies of engineering theory which were not being used. These included much of the elasticity, statics and strength of materials which is in regular use today. There was also the strong feeling that such theory should be used to help justify proposed designs. Many engineering scientists berated the practical engineers for not making use of their theories, while the designers complained that the scientists' theories were of no use for real structures, which were always far more complex than the simplified ideal cases the scientists considered.

At the same time there was also the feeling that design procedures needed to be improved and developed: structural failures were relatively frequent and it was realized that precedent was of limited use in dealing with structures of increasing span and having to withstand increasing loads.

What was lacking was an understanding of just how the theoretical knowledge could be used, or of how the gap between theory and practice could be bridged. Expressed in another way, a state of Kuhnian 'crisis' existed: there was pressure for change in ways which current methods (paradigms) seemed not to allow.

It was W. J. M. Rankine, Professor of Engineering at the University of Glasgow, who demonstrated how theory and practice could be 'harmonized'. Perhaps because he had had both a practical and a theoretical background, he proposed a new way of interpreting the 'theory' of the scientists.[10] He saw the theory as a way of creating a simplified and idealized model of a real structure. The creation of such a model of a real structure required many assumptions to be made which, for any given real case, would be inaccurate. He also proposed that these inaccuracies could be allowed for by using the well established concept of the 'factor of safety' in a new way. Rather than being literally a measure of the safety of a structure (a predicted strength divided by the largest load it was likely to encounter), he saw that it could be used to take into account the inevitable inaccuracies inherent in the process of creating a theoretical model of a real structure. Its magnitude would need to be established by experience and experiment.

This new approach allowed engineering designers, for the first time, to make use of the scientists' theory in their design procedures. The factor of safety (sometimes referred to as the factor of ignorance) provided the empirical bridge between the theories, which had previously been seen as only relevant to the controlled world of the laboratory, and the real structures the engineers were building, with all the imperfections and uncertainties of real materials, loads, members and joints. Rankine had shown how design procedures could incorporate the idealized mathematical theories to help to provide a rational justification of proposed designs for structures which lay outside the limits set by precedent.

The 'Plastic' Design Revolution. Theories of the elastic behaviour of materials were first developed in the late 17th century. These enabled engineering scientists to calculate the internal stresses in structures such as columns and beams from measurements of applied load and deflections of a structure under load. It was also accepted that the useful strength of a material was defined by a maxiumum stress above which the material, and hence the structure, would fail either by permanent deformation or fracture.

During the 19th century, especially due to the work of Navier, this approach was gradually adopted by designers as part of the procedure for justifying the adequacy of more and more types of structure under load. These procedures replaced previous methods which had relied entirely upon the accumulated experience of precedent and controlled experiment. By the turn of the present century this 'elastic design revolution' was fully complete. Even today, such elastic design procedures are still the norm in many circumstances.

Already in the late 1920s, however, there had arisen certain dissatisfactions with the elastic procedures. In particular, it was felt that many of them were irrational and often led to the inefficient and uneconomical use of steel. In 1929 the British Steelwork Association set up the Steel Structures Research Committee to review the situation and make recommendations for new design procedures.[11]

The main concerns of the design professions were that

—it was well known that steel had a considerable reserve of strength beyond its 'failure' at the limit of its elasticity

—it was unrealistic and pessimistic to design a steel-frame building as a series of independent columns and beams

—it was well known that minute differential movements of column foundations were inevitable in all buildings and that such movements invalidated the major assumption behind the design of any beam resting on more than two supports

—it was well known that parts of a steel structure, for instance around the rivet holes of joints, could and did deform permanently without any cause for concern

THE EVOLUTION OF STRUCTURAL ENGINEERING DESIGN PROCEDURES

—it was well known that most steel structures as built already contained significant internal stresses resulting from the manufacturing and assembly processes.

In Kuhnian terms, a large number of 'anomalies' were apparent.

The work of the Steel Structures Research Committee was initially directed at investigating just how much the usual assumptions upon which steel frame design was based deviated from the actual conditions found in simple frames. The deviations were found often to be considerable. The second stage was directed towards developing design procedures which were based upon more rational assumptions and which would lead to an improved correspondence between predicted stresses and those found in actual structures. These procedures were still based upon a consideration of the elastic behaviour of the steel and to some extent they were an improvement over previous procedures: they were generally agreed to be more rational and they did give a better correspondence between predicted and measured stresses.

The new design procedures, however, involved the use of more complex mathematics by the designer and the resulting steel structures were never significantly safer or cheaper. Nor did they deal satisfactorily with many of the 'anomalies' listed above. After a series of in-service trials of the new design procedures and simplified versions of them, they were finally rejected by the profession.

In other words, attempts at dealing with the 'anomalies' had failed. Some of the research workers came to perceive the situation as very serious indeed. The accepted elastic design procedures seemed to be incapable of further development which would satisfy the design profession. A state of Kuhnian 'crisis' had been reached.

The dramatic breakthrough came with the realization that the problems with the exisiting elastic design procedures (the 'anomalies') could only be answered by means of a radically different approach. The key to the solution came in the form of a new idea of what constituted failure. The idea that it was the point at which a certain limiting stress (the limit of elasticity) was reached was replaced by the notion that collapse could be the limit. A certain amount of plastic deformation could therefore be tolerated as long as sufficient parts of the structure remained below their yield point, thus preventing the structure from becoming a mechanism.

The consequence of this alternative view was to force an entire rethink of design procedures. No longer was the level of stress in every part of the structure of interest; the main concern was the location and relative magnitude of local bending moment maxima as the applied loads were all increased, in proportion, up to the point when sufficient plastic hinges had formed to transform the structure into a mechanism which could collapse. The safety of the structure under working loads could be achieved by ensuring that the applied loads were less than the collapse load by a margin which came to be known and the 'load factor'.

The new approach led to the need for new concepts, such as 'plastic hinge', 'full plastic moment', 'shape factor' and 'incremental collapse', as well as new uses of old ideas such as 'failure' and 'collapse'. In Kuhn's language, a new set of 'paradigms' was born and this led to the establishment of a new 'community' of structural designers.

The new 'plastic' design procedures (as they came to be known) required an utterly different approach to the behaviour and design of structures, and this was not easily accepted by many of the established 'community' of designers. Indeed, even 30 years after this 'design revolution', the procedures have still not been universally accepted.

The new design procedures did, however, resolve the problems which had been identified with the elastic procedures, the 'anomalies' mentioned above. For a certain class of design situations, the plastic design procedures are now widely accepted as more rational and they are frequently much simpler to use than elastic procedures, as well as often leading to substantial economies of material. Final recognition of the new approach has been accorded with the most recent design codes of practice, for both steel and reinforced concrete structures, which allow the use of plastic design procedures, nowadays included under the more general heading of 'limit state design procedures'.

CONCLUSION

The paper has sought to present an alternative view of the history of design methods for buildings and other engineering structures. This has been based on an understanding of design as a skill which

THE EVOLUTION OF STRUCTURAL ENGINEERING DESIGN PROCEDURES

has, in a sense, been carried out in the same fundamental way for some 2000 years. Differences have been in the type of information and intellectual tools available to designers at different times, and in the details of the 'design procedures' they have used; in particular, in the means of justifying the adequacy of a proposed design.

The progress of design through the ages has been described in a manner similar to the well established view of scientific progress developed by Thomas Kuhn. History has consisted of periods of 'normal' design by established 'design communities' punctuated by radical changes of view and approach associated with 'design revolutions'.

The argument has been illustrated by brief descriptions of four major revolutions in design procedure. A fuller history of structural engineering design would call for a similar treatment of other structural design revolutions, such as

—the statical revolution in masonry arch design procedures during the late 17th century
—the elastic revolution in beam design procedures during the early 19th century
—the revolution associated with the concept of statical determinacy during the mid-19th century
—the graphical statics revolution of the mid-19th century
—various revolutions associated with the use of physical models such as structural models in the 19th century and models for wind-tunnel testing following the collapse of the Tacoma Narrows bridge in 1940
—the safety factor revolution during the last 20 years, involving the use of partial safety factors and statistical methods in design procedures.

It is hoped that the present paper has made a contribution to the history and philosophy of engineering in a manner parallel to the considerable work which has already been undertaken in the history and philosophy of science.

As suggested at the beginning of the paper, the author also believes that a similar approach could be applied to the study of other branches of engineering design history. There have already been somewhat similar treatments of aero-engine[12] and locomotive[13] design, while other fields, such as soil mechanics and dam design, wind and water wheel design, naval architecture and chemical and mechanical engineering, still await the attentions of other historians and philosophers of engineering design.

NOTES AND REFERENCES

1. D. I. Blockley, *The Nature of Structural Design and Safety* (Ellis Horwood, Chichester, 1980).
2. Some people hold their different views to be rivals, though others, including the author, do not consider this, since they do not deal with the same problem.
3. K. R. Popper, *The Logic of Scientific Discovery* (Hutchinson, London, 1959).
4. T. S. Kuhn, *The Structure of Scientific Revolutions* (University of Chicago Press, Chicago and London, 1970, 2nd edn.).
5. W. Addis, *Structural Engineering: the Nature of Theory and Design* (Ellis Horwood, Chichester, 1990).
6. J. J. Coulton, *Greek Architects at Work* (Paul Elek, London, 1977).
7. Vitruvius, *Ten Books on Architecture* (c. 25 B.C.).
8. O. G. von Simson, *The Gothic Cathedral* (Pantheon, New York, 1956).
9. S. K. Victor, "Practical Geometry in the High Middle Ages", *Memoirs of the American Philosophical Society*, Vol. **134** (1979).
10. W. J. M. Rankine, *Introductory Lecture on the Harmony of Theory and Practice in Mechanics* (Griffin, London and Glasgow, 1856).
11. J. F. Baker, *The Steel Skeleton—1. Elastic Behaviour and Design* (Cambridge University Press, 1954); J. F. Baker, M. R. Horne and J. Heyman, *The Steel Skeleton—2. Plastic Behaviour and Design* (Cambridge University Press, 1956).
12. E. W. Constant, *The Origins of the Turbojet Revolution* (John Hopkins University Press, Baltimore and London, 1980).
13. M. C. Duffy, "The Evolution of Engineering Design Technique", *Engineering Designer*, Vol. **5** (1979), Jan.-Feb., pp. 19-22; Mar.-Apr., pp. 19-22; May-June, pp. 21-26; July-Aug., pp. 31-35; Sept.-Oct., pp. 19-23; Nov.-Dec., pp. 21-26.

Name and Place Index

Please note: references to illustrations and tables are indicated by italic page numbers, and notes by the use of square brackets (e.g. 82 [n. 94] means page 82, note 94).

Subject Index

Please note: references to illustrations and tables are indicated by italic page numbers, and notes by the use of square brackets (e.g. 82 [n. 94] means page 82, note 94).